Maureen Royce

Selected Titles in This Series

168 **Junjiro Noguchi,** Introduction to complex analysis, 1998

167 **Masaya Yamaguti, Masayoshi Hata, and Jun Kigami,** Mathematics of fractals, 1997

166 **Kenji Ueno,** An introduction to algebraic geometry, 1997

165 **V. V. Ishkhanov, B. B. Lur′e, and D. K. Faddeev,** The embedding problem in Galois theory, 1997

164 **E. I. Gordon,** Nonstandard methods in commutative harmonic analysis, 1997

163 **A. Ya. Dorogovtsev, D. S. Silvestrov, A. V. Skorokhod, and M. I. Yadrenko,** Probability theory: Collection of problems, 1997

162 **M. V. Boldin, G. I. Simonova, and Yu. N. Tyurin,** Sign-based methods in linear statistical models, 1997

161 **Michael Blank,** Discreteness and continuity in problems of chaotic dynamics, 1997

160 **V. G. Osmolovskiĭ,** Linear and nonlinear perturbations of the operator div, 1997

159 **S. Ya. Khavinson,** Best approximation by linear superpositions (approximate nomography), 1997

158 **Hideki Omori,** Infinite-dimensional Lie groups, 1997

157 **V. B. Kolmanovskiĭ and L. E. Shaĭkhet,** Control of systems with aftereffect, 1996

156 **V. N. Shevchenko,** Qualitative topics in integer linear programming, 1997

155 **Yu. Safarov and D. Vassiliev,** The asymptotic distribution of eigenvalues of partial differential operators, 1997

154 **V. V. Prasolov and A. B. Sossinsky,** Knots, links, braids and 3-manifolds. An introduction to the new invariants in low-dimensional topology, 1997

153 **S. Kh. Aranson, G. R. Belitsky, and E. V. Zhuzhoma,** Introduction to the qualitative theory of dynamical systems on surfaces, 1996

152 **R. S. Ismagilov,** Representations of infinite-dimensional groups, 1996

151 **S. Yu. Slavyanov,** Asymptotic solutions of the one-dimensional Schrödinger equation, 1996

150 **B. Ya. Levin,** Lectures on entire functions, 1996

149 **Takashi Sakai,** Riemannian geometry, 1996

148 **Vladimir I. Piterbarg,** Asymptotic methods in the theory of Gaussian processes and fields, 1996

147 **S. G. Gindikin and L. R. Volevich,** Mixed problem for partial differential equations with quasihomogeneous principal part, 1996

146 **L. Ya. Adrianova,** Introduction to linear systems of differential equations, 1995

145 **A. N. Andrianov and V. G. Zhuravlev,** Modular forms and Hecke operators, 1995

144 **O. V. Troshkin,** Nontraditional methods in mathematical hydrodynamics, 1995

143 **V. A. Malyshev and R. A. Minlos,** Linear infinite-particle operators, 1995

142 **N. V. Krylov,** Introduction to the theory of diffusion processes, 1995

141 **A. A. Davydov,** Qualitative theory of control systems, 1994

140 **Aizik I. Volpert, Vitaly A. Volpert, and Vladimir A. Volpert,** Traveling wave solutions of parabolic systems, 1994

139 **I. V. Skrypnik,** Methods for analysis of nonlinear elliptic boundary value problems, 1994

138 **Yu. P. Razmyslov,** Identities of algebras and their representations, 1994

137 **F. I. Karpelevich and A. Ya. Kreinin,** Heavy traffic limits for multiphase queues, 1994

136 **Masayoshi Miyanishi,** Algebraic geometry, 1994

135 **Masaru Takeuchi,** Modern spherical functions, 1994

134 **V. V. Prasolov,** Problems and theorems in linear algebra, 1994

133 **P. I. Naumkin and I. A. Shishmarev,** Nonlinear nonlocal equations in the theory of waves, 1994

132 **Hajime Urakawa,** Calculus of variations and harmonic maps, 1993

131 **V. V. Sharko,** Functions on manifolds: Algebraic and topological aspects, 1993

130 **V. V. Vershinin,** Cobordisms and spectral sequences, 1993

(*Continued in the back of this publication*)

Introduction to Complex Analysis

Translations of
MATHEMATICAL
MONOGRAPHS

Volume 168

Introduction to Complex Analysis

Junjiro Noguchi

Translated by
Junjiro Noguchi

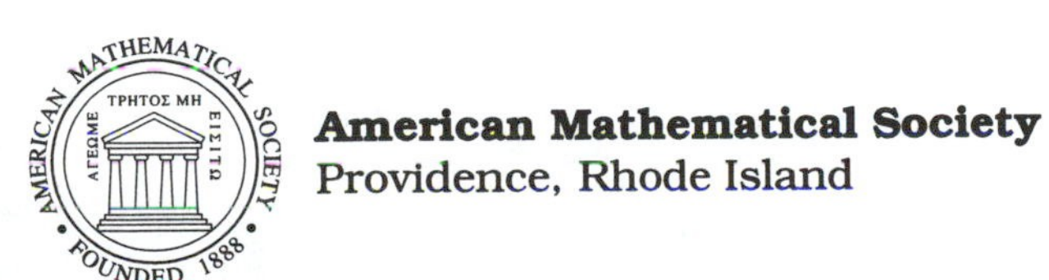

複素解析概論

FUKUSO KAISEKI GAIRON

(Introduction to complex analysis)

by Junjiro Noguchi

Originally published in Japanese by Shokabo Publishing Company, Ltd., Tokyo, 1993

Translated from the Japanese by Junjiro Noguchi

1991 *Mathematics Subject Classification*. Primary 30–01.

Library of Congress Cataloging-in-Publication Data

Noguchi, Junjirō, 1948–
[Fukuso kaiseki gairon. English]
Introduction to complex analysis / Junjiro Noguchi ; translated by Junjiro Noguchi.
p. cm. — (Translations of mathematical monographs ; v. 168)
Includes bibliographical references (p. –) and index.
ISBN 0-8218-0377-8 (alk. paper)
1. Functions of complex variables. 2. Mathematical analysis. I. Title. II. Series.
QA331.7.N6413 1997
515′.9—dc21 97-14392
CIP

∞ The paper used in this book is acid-free and falls within the guidelines
established to ensure permanence and durability.
Visit the AMS home page at URL: `http://www.ams.org/`

10 9 8 7 6 5 4 3 2 1 02 01 00 99 98 97

TO AKIKO

Contents

Preface

Complex analysis is an active research subject by itself, but, even more, it provides the foundations for broad areas of mathematics, and plays an important role in the applications of mathematics to engineering.

This book is intended to describe a classical introductory part of complex analysis for university students in the sciences or engineering, and in particular to serve as a text or a reference book for juniors, seniors, or first-year graduate students in the sciences. The prerequisites are elementary calculus, from the real numbers through differentiation and integration, and linear algebra; set theory and some general topology are not required, but would be helpful. Historically, the contents of this volume had been discovered by the beginning of the twentieth century, so they are not very new. Nevertheless, the author tried to arrange the presentation and choose terminology to agree with modern work in the field.

Many books have been written on this subject, in many languages. Roughly speaking, they can be divided into two types. One tries to give an understanding of the subject using intuitive arguments. The other puts more emphasis on rigorous proofs, presenting the subject as a fundamental theory of mathematics. This book falls in the latter group. It is well-known that there is some difficulty in dealing with curves, related to Cauchy's integral theorem, a point on which several published books are somewhat unsatisfactory. To deal with it rigorously, we give detailed descriptions of the homotopy of plane curves. This material on curves is something the reader may well have encountered in previous courses, so it is scattered through the text where needed. Readers whose mathematical background already includes this material may simply confirm the theorems and go on.

Since residue theorem is important both in pure mathematics and in applications, we give a fairly detailed explanation of how to apply it to numerical calculations; this should be sufficient for those who are studying complex analysis for applications.

After this book, the student should be ready to take up value distribution theory, the theory of Riemann surfaces, complex analysis in several variables, the theory of complex manifolds, and other subjects. Complex analysis will also provide fundamental methods for the theory of differential equations, algebraic geometry, number theory, and other fields. The author wrote Chapters 6 and 7 with these transitions in mind.

The author will be very pleased if this book helps students to understand the classical theory of complex analysis, to relish its beauty, and to master the rigorous treatment of mathematical demonstrations.

This volume is based on lectures for third-year students given by the author at the Tokyo Institute of Technology. The class contained not just mathematics majors, but also physics and applied physics students. The author is very grateful to all of them.

Finally, but not least, the author would like to express his deep thanks to Professor Mitsuru Ozawa, who led him to complex analysis, and to Professor Shingo Murakami, who recommended him to write this book. He is also obliged to Mr. Shuji Hosoki, of the Shokabo publishing company, for the proofreading.

April, 1993 at Ohokayama

Junjiro Noguchi

Added in English translation:

The publication of this English translation was made possible by the suggestion and the recommendation of Professor Katsumi Nomizu. The author expresses his sincere gratitude to him.

February, 1997 at Ohokayama

Junjiro Noguchi

CHAPTER 1

Complex Numbers

In this chapter we assume that the real numbers are known, and we explain the elementary facts of complex numbers.

1.1. Complex Numbers

The set $\mathbf{R}$ of all real numbers cannot be enlarged as long as its order " $\leqq$ " and the Archimedean axiom are kept (completeness of the system of real numbers). The order implies that $x^2 \geqq 0$ for all $x \in \mathbf{R}$. Therefore the following simple algebraic equation

$$x^2 + a = 0 \tag{1.1.1}$$

cannot be solved in the system of real numbers for any $a > 0$. To overcome this inconvenience, we drop the order of the number system, and extend $\mathbf{R}$ by taking a solution i of equation (1.1.1) normalized with $a = 1$, so that we obtain a *complex number* $z = x + iy$ $(= x + yi)$ $(x, y \in \mathbf{R})$. We denote by $\mathbf{C}$ the set of all complex numbers. This extension makes it possible not only to solve equation (1.1.1), but also to develop a grand and beautiful theory that cultivates the essence of mathematics. The number i is called the *imaginary unit* and satisfies

$$i^2 + 1 = 0. \tag{1.1.2}$$

Then $-i$ is also a solution of (1.1.2), but it does not matter which is the imaginary unit once we choose one solution of (1.1.2). Now we define $z = 0$ if and only if $x = y = 0$. For two complex numbers $z = x + iy, w = u + iv$ we define $z = w$ if and only if $x = u$ and $y = v$. The following operations are defined in $\mathbf{C}$:

$$\begin{aligned} z \pm w &= (x \pm u) + i(y \pm v), \\ zw &= (xu - yv) + i(xv + yu). \end{aligned}$$

If $z \neq 0$, we set $z^{-1} = x/(x^2 + y^2) + i(-y)/(x^2 + y^2)$. Then

$$zz^{-1} = 1.$$

We also write $1/z$ for z^{-1}, and call it the *inverse* of z. Thus $\mathbf{C}$ carries the four operations of addition, subtraction, multiplication, and division. This fact is referred to by saying $\mathbf{C}$ forms a *field.* We may identify $z = x + iy \in \mathbf{C}$ with the point $(x, y) \in \mathbf{R}^2$, so that we may also consider $\mathbf{C}$ as a 2-dimensional vector space over $\mathbf{R}$. This vector space is called the *complex plane*, or *Gaussian plane*, and z is called the *complex coordinate.* In this complex plane, the x (resp., y) coordinate axis is called the real (resp., imaginary) axis.

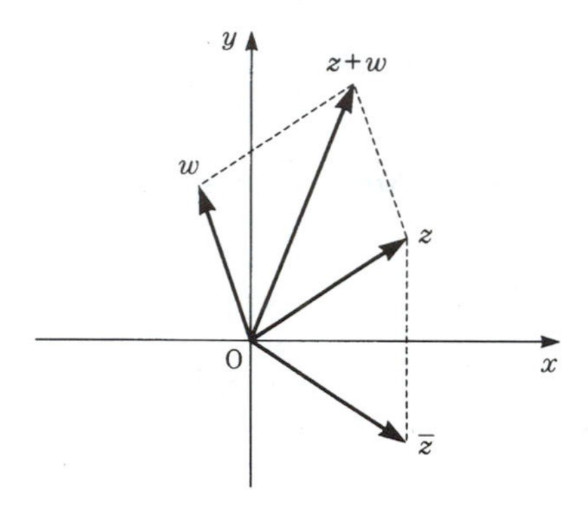

FIGURE 1

A point $z = x + iy \in \mathbf{C}$ of the complex plane is called a rational point (not rational number!) if $x \in \mathbf{Q}$ and $y \in \mathbf{Q}$. In general, for $z = x + iy$ we set

$$\overline{z} = x - iy,$$

which is called the *conjugate* of z. We have

$$\overline{z+w} = \overline{z} + \overline{w}, \quad \overline{zw} = \overline{z}\,\overline{w}, \quad \overline{z^{-1}} = \overline{z}^{-1}. \tag{1.1.3}$$

It also follows that

$$x = \frac{1}{2}(z + \overline{z}), \quad y = \frac{1}{2i}(z - \overline{z}),$$

which are respectively called the real and imaginary part of z, and denoted by $\operatorname{Re} z$ and $\operatorname{Im} z$. If $\operatorname{Im} z = 0$, z is called a real number, and if $\operatorname{Re} z = 0$, z is called a *purely imaginary* number:

z is real if and only if $z = \overline{z}$;
z is purely imaginary if and only if $z = -\overline{z}$.

Note that $z\overline{z} = x^2 + y^2 \geqq 0$. The non-negative square root of $z\overline{z}$ is called the *absolute value* of z, and is denoted by $|z|$:

$$|z| = \sqrt{x^2 + y^2} \geqq 0.$$

It follows that

$$\begin{aligned} &|-z| = |z|, \qquad |zw| = |z| \cdot |w|, \\ &|z \pm w| \leqq |z| + |w|, \\ &|z| = 0 \Longleftrightarrow z = 0. \end{aligned} \tag{1.1.4}$$

For $z, w \in \mathbf{C}$ we set $d(z,w) = |z-w|$; this is called the *distance* between z and w, and satisfies

$$
\begin{aligned}
&d(z,w) = 0 \Longleftrightarrow z = w,\\
&d(z,w) = d(w,z),\\
&d(z_1,z_2) + d(z_2,z_3) \geqq d(z_1.z_3).
\end{aligned} \tag{1.1.5}
$$

By making use of the trigonometric functions, $\cos\theta$ and $\sin\theta$, we may write a complex number $z = x + iy$ as follows:

$$
\begin{aligned}
&z = x + iy = r(\cos\theta + i\sin\theta),\\
&\theta \in \mathbf{R}, \qquad r = |z| \geqq 0.
\end{aligned} \tag{1.1.6}
$$

If $z \neq 0$, θ is uniquely determined up to integral multiples of 2π (that is, θ is uniquely determined "modulo 2π"). We call θ the *argument* of z, and denote it by $\arg z$. The pair (r,θ) is called the *polar coordinate* system, and (1.1.6) the polar coordinate representation of z.

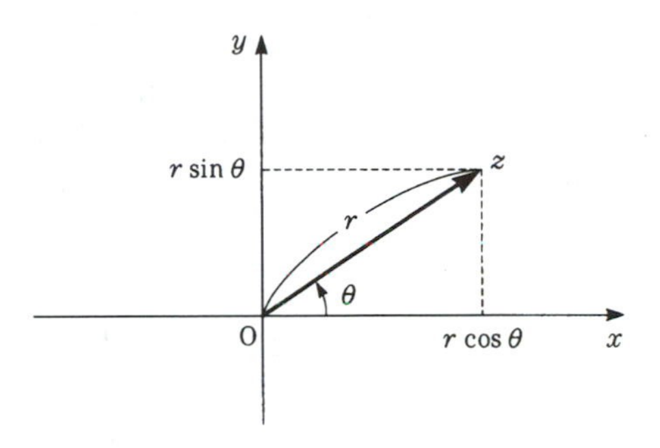

FIGURE 2

REMARK. Here we dealt with the trigonometric functions $\cos\theta$ and $\sin\theta$ by geometric intuition. This is not consistent with our aim of presenting the theory rigorously. We have to define the trigonometric functions without geometric intuition, and show that they are periodic and satisfy the above stated geometric intuition. This will be done in Chapter II, §5.

1.2. Plane Topology

For a point $a \in \mathbf{C}$ of the complex plane and a real number $r > 0$, we define the *disk* of radius r with center a by

$$\Delta(a;r) = \{z \in \mathbf{C}; d(z,a) < r\} \qquad (\{|z-a| < r\}).$$

In particular, when $a = 0$, we set

$$\Delta(r) = \Delta(0;r)$$

and $\Delta(1)$ is called the *unit disk*.

A subset $A \subset \mathbf{C}$ of $\mathbf{C}$ is said to be *bounded* if there is $r > 0$ with $A \subset \Delta(r)$. The set A is said to be *open* if for any $a \in A$ there is a $r > 0$ with $\Delta(a;r) \subset A$. An open set containing a point $z \in \mathbf{C}$ (resp., a subset $B \subset \mathbf{C}$) is called a *neighborhood* of z (resp., B). The disk $\Delta(a;r)$ itself is an open set, since $\Delta(z;r - |z-a|) \subset \Delta(a;r)$ for all $z \in \Delta(a;r)$. Hence, $\Delta(a;r)$ is called a *disk neighborhood* of a. We denote the set of all open subsets of $\mathbf{C}$ by $\mathcal{A}$. Then we have

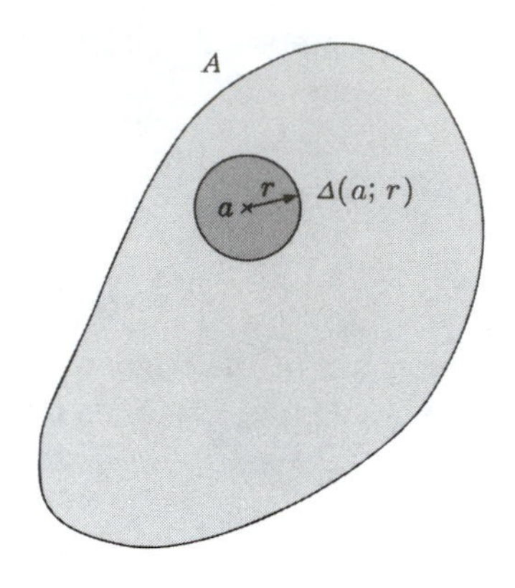

FIGURE 3

(1.2.1) i) $\emptyset \in \mathcal{A}$, $\mathbf{C} \in \mathcal{A}$;
ii) for finitely many $A_1, \dots, A_n \in \mathcal{A}$, $A_1 \cap \cdots \cap A_n \in \mathcal{A}$;
iii) for an arbitrary subfamily $\{A_\alpha\}_{\alpha \in \Gamma} \subset \mathcal{A}, \bigcup_{\alpha \in \Gamma} A_\alpha \in \mathcal{A}$.

In general, "defining a *topology*" means that a family of subsets which satisfies the properties (1.2.1), i)~iii) is given. A subset $A \subset \mathbf{C}$ is called a *closed* set if the complement A^c is open, and a point $z \in \mathbf{C}$ an *accumulation point* of A if for any $r > 0$

$$(\Delta(z;r) \setminus \{z\}) \cap A \neq \emptyset.$$

A point of A which is not an accumulation point of A is called an *isolated point* of A. The set consisting of all points of A and all accumulation points of A is called the *closure* of A and is denoted by $\overline{A}$. The naming comes from the following theorem.

(1.2.2) THEOREM. i) *A subset $A \subset \mathbf{C}$ is closed if and only if $A = \overline{A}$.*
ii) $\overline{\overline{A}} = \overline{A}$.
iii) *$\overline{A}$ is the smallest set among the closed subsets containing A.*

PROOF. i) Assume that A is closed. Let $a \in \overline{A}$ be an accumulation point of A. If $a \notin A$, then $a \in A^c$, so that $\Delta(a;r) \subset A^c$ for some $r > 0$; that is, $\Delta(a;r) \cap A = \emptyset$. This is absurd. Thus $A = \overline{A}$.

To show the converse, we take an arbitrary $z \notin A$. By assumption, z is not an accumulation point of A, either. Therefore, there is $r > 0$ such that $\Delta(z;r) \cap A = \emptyset$; that is, $\Delta(z;r) \subset A^c$. This shows that A^c is open.

ii) It suffices to show that $\overline{\overline{A}} \subset \overline{A}$. Take an arbitrary accumulation point z of $\overline{\overline{A}}$. Since for any $r > 0$

$$\Delta(z;r) \cap \overline{A} \neq \emptyset,$$

we may take $z_1 \in \Delta(z;r) \cap \overline{A}$. Set $r_1 = r - |z_1 - z|$. Since $z_1 \in \overline{A}$, we have

$$\Delta(z_1;r_1) \cap A \neq \emptyset.$$

Hence, $\Delta(z;r) \cap A \neq \emptyset$, and then $z \in \overline{A}$.

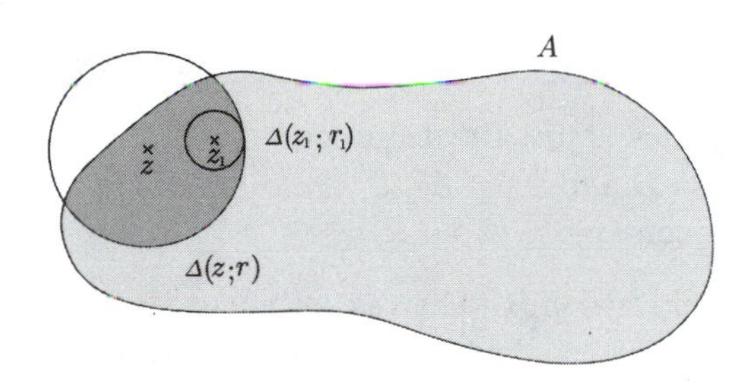

FIGURE 4

iii) Taking an arbitrary closed set E with $E \supset A$, we have to show that $E \supset \overline{A}$. This is easy and is left to the reader. □

For a subset $A \subset \mathbf{C}$, a point $z \in A$ is called an *interior point* of A if there is a neighborhood U of z with $U \subset A$. Therefore, the set of all interior points of A, denoted by $\overset{\circ}{A}$, is open. An interior point of A^c is called an *exterior point* of A. The set $\partial A = \overline{A} \setminus \overset{\circ}{A}$ is called the *boundary* of A, and a point of it a *boundary point.* In particular, ∂A is a closed set.

We set

$$C(a;r) = \partial\Delta(a;r) = \{|z-a| = r\},$$

which is called the *circle* of radius r with center a; in particular, $C(0;1)$ is called the *unit circle.*

Let B be a subset of A. Then B is called an *open set in* A if there is an open set $U \subset \mathbf{C}$ with $B = A \cap U$. Similarly, B is called a *closed set in* A if there is a closed set $E \subset \mathbf{C}$ with $B = A \cap E$. In this way we get the *relative topology* of A. If B is an open (resp., closed) subset of A, then $A \setminus B$ is a closed (resp., open) subset of A. If B has no accumulation point in A, B is said to be *discrete in* A.

In general, a family $\{U_\alpha\}_{\alpha\in\Gamma}$ of subsets $U_\alpha \subset \mathbf{C}$ is called a *covering* of A if $A \subset \bigcup_{\alpha\in\Gamma} U_\alpha$. In particular, if all U_α are open, the covering is called an *open covering* of A. In this case, $\{U_\alpha \cap A\}_{\alpha\in\Gamma}$ is called an open covering of A (with respect to the relative topology), too.

If there is an open covering $\{U_1, U_2\}$ of A satisfying

$$U_1 \cap A \neq \emptyset, \quad U_2 \cap A \neq \emptyset, \quad U_1 \cap U_2 \cap A = \emptyset, \tag{1.2.3}$$

we say that A is *not connected.* If there is no such covering, then A is said to be *connected.* A connected open subset of $\mathbf{C}$ is called a *domain.*

(1.2.4) THEOREM. *The following three conditions are mutually equivalent.*

i) *A is connected.*

ii) *Let* $\{U_1, U_2\}$ *be any open covering of* A *with respect to the relative topology, such that* $U_1 \cap U_2 = \emptyset$. *Then either* $U_1 = A$ $(U_2 = \emptyset)$, *or* $U_2 = A$ $(U_1 = \emptyset)$.

iii) *If any non-empty subset* B *of* A *is open and closed in* A, *then* $B = A$.

PROOF. i) $\Leftrightarrow$ ii) This is just the definition.
ii) $\Rightarrow$ iii) Put $U_1 = B$ and $U_2 = A \setminus B$.
iii) $\Rightarrow$ ii) Let the non-empty U_i be B. □

Let I be a closed interval of $\mathbf{R}$. Then a continuous mapping $\phi : I \ni t \to \phi(t) = x(t) + iy(t) \in \mathbf{C}$ (i.e., $x(t)$ and $y(t)$ are real-valued continuous functions on I) is called a *curve* or *arc*. In the case where I is a general interval (open interval, semi-closed interval, etc.), it is called a *general curve*. The image $C = \phi(I)$ is also called a *curve*, but, strictly speaking, ϕ should be called a curve. We say that "C is given by ϕ". In fact, the two curves

$$\phi_1 : [0, 2\pi] \ni t \to \cos t + i \sin t \in \mathbf{C},$$
$$\phi_2 : [0, 2\pi] \ni t \to \cos 2t + i \sin 2t \in \mathbf{C},$$

have the same image $\phi_1([0, 2\pi]) = \phi_2([0, 2\pi])$. But ϕ_1 goes around the unit circle once, and ϕ_2 twice, so they should be distinguished. The variable t is called the *parameter* of ϕ.

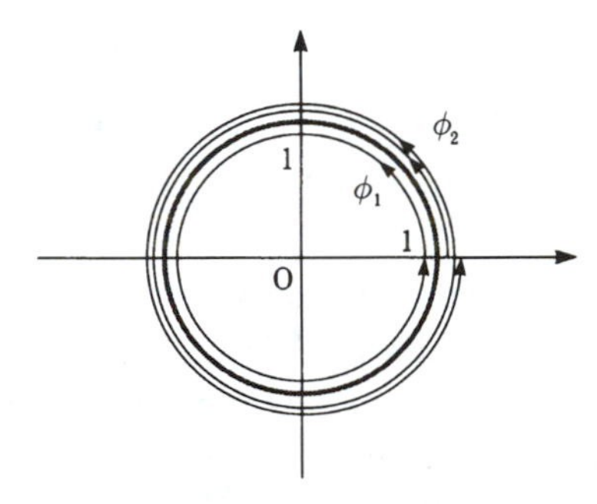

FIGURE 5

A curve $\psi : J \to \mathbf{C}$ is called a *parameter change* of ϕ if there is a strictly monotone continuous function $\tau : J \to I$ from J onto I satisfying $\psi(s) = \phi(\tau(s)), s \in J$.

If $I = [T_0, T_1]$ and $\phi(T_0) = \phi(T_1)$, then ϕ is called a *closed curve*. In this case, we consider also the following ψ as a parameter change. Let $J = [S_0, S_1]$ $(S_0 < S_1)$ be an arbitrary closed interval, and take $\tau_0 \in (T_0, T_1)$ and $\sigma_0 \in (S_0, S_1)$. We put

$$\psi(s) = \phi\left(\tau_0 + \frac{s - S_0}{\sigma_0 - S_0}(T_1 - \tau_0)\right), \qquad S_0 \leqq s \leqq \sigma_0, \tag{1.2.5}$$
$$\psi(s) = \phi\left(T_0 + \frac{s - \sigma_0}{S_1 - \sigma_0}(\tau_0 - T_0)\right), \qquad \sigma_0 \leqq s \leqq S_1.$$

Furthermore, a parameter change of this ψ is also considered as a parameter change of ϕ.

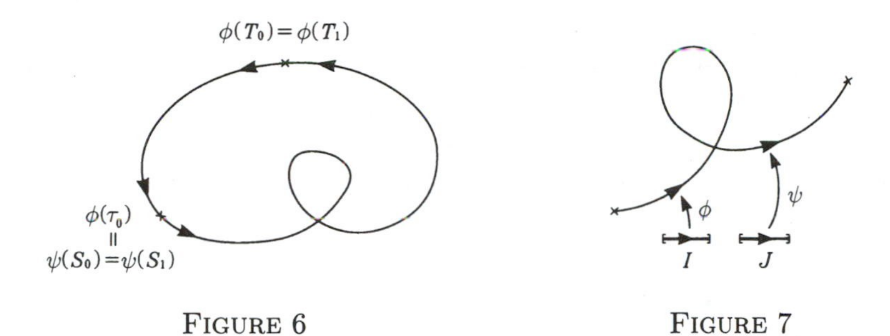

FIGURE 6 FIGURE 7

It is clear that parameter changes of curves give rise to an equivalence relation. In our arguments, it is inconvenient to consider a curve as a continuous mapping. Therefore, we call the equivalence class of ϕ a *curve*, and denote it by C, $C(\phi)$, or $C(\phi : I \to \mathbf{C})$. Here ϕ stands for the representative. Still, we may call ϕ and the image $\phi(I)$ a curve, but the meaning will be clear.

EXERCISE 1. Let $\phi : [T_0, T_1] \to \mathbf{C}$ be a curve, and $[S_0, S_1]$ $(S_0 < S_1)$ an arbitrary closed interval. Then there is a parameter change $\psi : J \to \mathbf{C}$ of ϕ with $J = [S_0, S_1]$.

Let $C(\phi : [T_0, T_1] \to \mathbf{C})$ be a curve. Then $\phi(T_0)$ is called the *initial point*, and $\phi(T_1)$ the *terminal point*. The initial and terminal points of C are called the *ends* of C, and we say that C *connects* $\phi(T_0)$ and $\phi(T_1)$. If ϕ is one to one, C is called a *simple curve* (or a *Jordan curve*). If C is closed and ϕ is one to one on $[T_0, T_1)$, C is called a simple (or Jordan) closed curve. When C is a closed curve, an arbitrary point of C can be the initial (terminal) point by a parameter change. Thus, in this case we consider that the initial (terminal) point of C is determined if a representative ϕ of C is fixed. If ϕ is constant $(\phi(t) \equiv \phi(T_0))$, then ϕ is called a *constant curve*.

We say that $\phi : [T_0, T_1] \to \mathbf{C}$ is continuously differentiable if the real part $x(t)$ of $\phi(t)$ and the imaginary part $y(t)$ of $\phi(t)$ are defined in an open interval containing $[T_0, T_1]$, are differentiable there, and their derivatives are continuous. We set

$$\frac{d\phi}{dt}(t) = \phi'(t) = \frac{dx}{dt}(t) + i\frac{dy}{dt}(t).$$

Here, in case $\phi(T_0) = \phi(T_1)$, we also assume that $d\phi(T_0)/dt = d\phi(T_1)/dt$. Moreover, ϕ is said to be *non-singular* if $d\phi(t)/dt \neq 0$ for all t. If there is a partition $T_0 = t_0 < t_1 < \cdots < t_n = T_1$ such that the restrictions $\phi|[t_{j-1}, t_j]$ to $[t_{j-1}, t_j]$, $1 \leqq j \leqq n$, are continuously differentiable, ϕ is said to be *piecewise continuously differentiable*. A curve C is (piecewise) continuously differentiable if there is such a representative ϕ of C. A non-singular curve and a piecewise non-singular curve are similarly defined.

Now, let $C(\phi : [T_0, T_1] \to \mathbf{C})$ be a curve. On the image of C, $\phi(t)$ moves from the initial point toward the terminal point as the parameter $t \in [T_0, T_1]$ increases. Assume that C is not a constant curve. By the above Exercise 1 we may assume $[T_0, T_1] = [-1, 1]$. Put $\psi(t) = \phi(-t)$. Then $\psi : [-1, 1] \to \mathbf{C}$ is not a parameter change of ϕ. If $\phi(-1) \neq \phi(1)$, then $\phi(t)$ moves from $\phi(-1)$ toward $\phi(1)$ as t moves from -1 toward 1, and $\psi(t)$ from $\phi(1)$ toward $\phi(-1)$. Therefore we may consider that the curve $C(\phi)$ already has an *orientation* in this sense. The curve given by ψ is called the curve with inverse orientation and is denoted by $-C$.

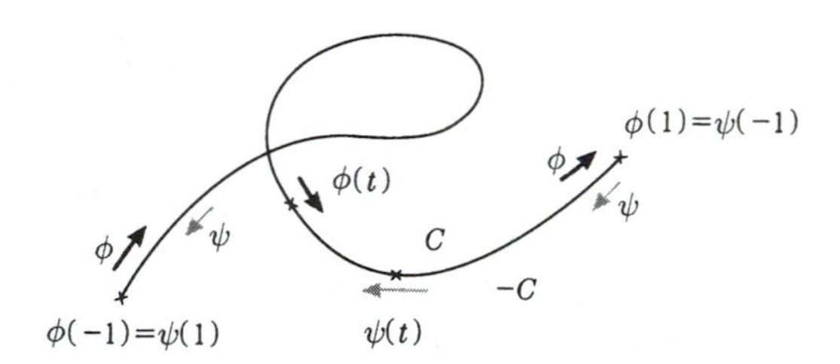

FIGURE 8

Now, suppose that the terminal point of a curve C_1 is the same as the initial point of another curve C_2. Suppose that C_1 is given by $\phi_1 : [0, 1] \to \mathbf{C}$ and C_2 by $\phi_2 : [1, 2] \to \mathbf{C}$. Then we define a curve $C_3(\phi_3 : [0, 2] \to \mathbf{C})$ so that it is equal to ϕ_1 over $[0, 1]$ and to ϕ_2 over $[1, 2]$. The curve C_3 is called the *sum* of C_1 and C_2, and we write

$$C_3 = C_1 + C_2.$$

For a subset $A \subset \mathbf{C}$, C is called a curve in A if its image is contained in A. If two arbitrary points of A can be connected by a curve in A, A is said to be *arcwise connected.* For instance, a disk $\Delta(a; r)$ and $\mathbf{C}$ itself are arcwise connected.

EXERCISE 2. A closed interval $[T_0, T_1]$ $(\subset \mathbf{R} \subset \mathbf{C})$ is connected in the sense of Theorem (1.2.4).

(1.2.6) THEOREM. *An open subset of* $\mathbf{C}$ *is connected if and only if it is arcwise connected.*

PROOF. The "if" part follows from Exercise 2. To show the "only if" part, we assume that A is a connected open set and $A \neq \emptyset$. Fix a point $z_0 \in A$. Let U_1 be the totality of all points of A connected to z_1 by curves in A, and let U_2 be its complement. For any point $a \in U_j$ $(j = 1, 2)$, we take $r > 0$ with $\Delta(a; r) \subset A$. Since $\Delta(a; r)$ is arcwise connected, $\Delta(a; r) \subset U_j$. Thus the U_j are both open sets. Since $A = U_1 \cup U_2$ and $U_1 \neq \emptyset$, we have $U_2 = \emptyset$ by definition, and so $A = U_1$. $\square$

A curve $\phi : [T_0, T_1] \to \mathbf{C}$ is called a *piecewise linear curve* (Streckenzug) if there is a partition $T_0 = t_0 < t_1 < \cdots < t_n = T_1$ of $[T_0, T_1]$ such that on each $[t_{j-1}, t_j]$ ϕ is written as

$$\phi(t) = \phi(t_{j-1}) + a_j(t - t_{j-1}), \qquad t_{j-1} \leqq t \leqq t_j,$$

where $a_j \in \mathbf{C}$. We immediately obtain from the proof of Theorem (1.2.6)

(1.2.7) COROLLARY. *An open subset A of $\mathbf{C}$ is connected if and only if two arbitrary points of A can be connected by a piecewise linear curve.*

In general, let A be a subset of $\mathbf{C}$. For two points $z_1, z_2 \in A$ we write $z_1 \sim z_2$ if there is a connected subset $A' \subset A$ with $z_1, z_2 \in A'$. We show that this is an equivalence relation. The reflexive law, "$z_1 \sim z_2 \Rightarrow z_2 \sim z_1$", is trivial. Assume that $z_1 \sim z_2$ and $z_2 \sim z_3$. Take connected subsets $A', A'' \subset A$ such that $z_1, z_2 \in A'$ and $z_2, z_3 \in A''$. It suffices to show that $A' \cup A''$ is connected. Let $\{U_1, U_2\}$ be an open covering of $A' \cup A''$ with respect to the relative topology such that $U_1 \cap U_2 = \emptyset$. Since $\{U_1, U_2\}$ is also an open covering of A', $A' \subset U_1$ or $A' \subset U_2$; we may assume $A' \subset U_1$. In the same way, we see that $A'' \subset U_1$ or $A'' \subset U_2$. Since $z_2 \in A' \cap A'' \subset U_1$, $A'' \subset U_1$. Therefore, $A' \cup A'' \subset U_1$, and so its connectedness follows.

For $z \in A$ we set $A_z = \{z' \in A; z' \sim z\}$; this is called the *connected component* of A containing z. If $A_{z_1} \neq A_{z_2}$, then $A_{z_1} \cap A_{z_2} = \emptyset$. Taking representative elements z_α from all connected components of A, we have

$$A = \bigcup_{z_\alpha} A_{z_\alpha}. \tag{1.2.8}$$

This is called the decomposition of A into connected components. In particular, if A is open, then so are the connected components of A, so that their representatives $z_\alpha = x_\alpha + iy_\alpha$ may be chosen as rational points. Since there are only countably many such points, we may enumerate them, and hence denote them by $z_i, i = 1, \ldots, N(\leqq \infty)$. Put $A_i = A_{z_i}, i = 1, \ldots, N$. Then they are disjoint domains and

$$A = \bigcup_{i=1}^{N} A_i. \tag{1.2.9}$$

1.3. Sequences and Limits

From now on, a complex number is simply called a number.

An infinite sequence $\{z_0, z_1, \ldots, z_n, \ldots\} = \{z_n\}_{n=0}^{\infty}$ of numbers (or points) is called a *sequence*. A sequence $\{z_n\}_{n=0}^{\infty}$ is said to *converge* to $a \in \mathbf{C}$ or to have *limit* a if for an arbitrary $\epsilon > 0$ there is a number $n_0 \in \mathbf{N}$ such that

$$|z_n - a| < \epsilon \quad \text{for all } n \geqq n_0.$$

In this case we write

$$a = \lim_{n\to\infty} z_n \quad \text{or} \quad z_n \to a \ (n \to \infty). \tag{1.3.1}$$

Since

$$\max\{|\mathrm{Re}\, z|, |\mathrm{Im}\, z|\} \leqq |z| \leqq |\mathrm{Re}\, z| + |\mathrm{Im}\, z| \tag{1.3.2}$$

for $z \in \mathbf{C}$, (1.3.1) is equivalent to

$$\alpha = \lim_{n\to\infty} x_n, \qquad \beta = \lim_{n\to\infty} y_n, \tag{1.3.3}$$

where $z_n = x_n + iy_n$ and $a = \alpha + i\beta$. A convergent sequence $\{z_n\}_{n=0}^{\infty}$ is always bounded (that is, for some M, $|z_n| < M$ for all n).

In particular, for a sequence $\{r_n\}_{n=0}^{\infty}$ of real numbers we write

$$\lim_{n\to\infty} r_n = +\infty$$

if for an arbitrary $K > 0$ there is a number $n_0 \in \mathbf{N}$ such that $r_n > K$ for $n \geqq n_0$.

If $\lim z_n$ and $\lim w_n$ exist, and $a, b \in \mathbf{C}$, then we have

$$\begin{aligned}
&\lim(az_n + bw_n) = a \lim z_n + b \lim w_n, \\
&\lim(z_n w_n) = (\lim z_n)(\lim w_n), \\
&\overline{\lim z_n} = \lim \overline{z}_n, \qquad |\lim z_n| = \lim |z_n|.
\end{aligned}$$

EXERCISE 1. Show the above four equalities.

A sequence $\{z_{n_\nu}\}_{\nu=0}^{\infty}$ formed by a part of $\{z_n\}_{n=0}^{\infty}$ (the order is not changed) is called a *subsequence* of $\{z_n\}_{n=0}^{\infty}$. If $\{z_n\}_{n=0}^{\infty}$ converges to a, then so does any of its subsequences.

(1.3.4) THEOREM. *Every bounded sequence has a convergent subsequence.*

This is clear by the corresponding Weierstrass' theorem for real numbers, (1.3.2) and (1.3.3).

A sequence $\{z_n\}_{n=0}^{\infty}$ is called a *Cauchy sequence* if for an arbitrary $\epsilon > 0$ there is an $n_0 \in \mathbf{C}$ such that

$$|z_n - z_m| < \epsilon, \qquad n, m \geqq n_0.$$

In this case, $\{\mathrm{Re}\, z_n\}_{n=0}^{\infty}$ and $\{\mathrm{Im}\, z_n\}_{n=0}^{\infty}$ are both Cauchy sequences by (1.3.2). Thus by Cauchy's theorem for sequences of real numbers we have

(1.3.5) THEOREM. *A sequence $\{z_n\}_{n=0}^{\infty}$ converges if and only if it is a Cauchy sequence.*

Let $A \subset C$ be an arbitrary subset. Note that A is closed if and only if any convergent sequence of points of A has a limit in A. We say that A is *compact* if any sequence of points of A has a convergent subsequence.

(1.3.6) THEOREM. *The following are equivalent for $A \subset \mathbf{C}$:*

i) *A is compact.*

ii) *A is bounded and closed.*

iii) *Let $A \subset \bigcup_{\alpha \in \Gamma} U_\alpha$ be an arbitrary open covering. Then there are finitely many $U_{\alpha_1}, \ldots, U_{\alpha_l}$ such that*

$$A \subset \bigcup_{i=1}^{l} U_{\alpha_i}.$$

PROOF. The equivalence of i) and ii) follows immediately from Theorem (1.3.4).

ii)$\Rightarrow$iii) Put $S = \{\operatorname{Re} z; z \in A\}$. Then S is a bounded closed subset of $\mathbf{R}$. Let σ be the minimum of S. Then we put

$$A[\sigma, \tau] = \{z \in A; \sigma \leqq \operatorname{Re} z \leqq \tau\}$$

for $\tau \geqq \sigma$. The corresponding statement for bounded closed subsets of $\mathbf{R}$ is known as the Heine-Borel theorem. Therefore $A[\sigma, \sigma]$ is covered by finitely many U_{α_j}, $1 \leqq j \leqq k$:

$$A[\sigma, \sigma] \subset \bigcup_{j=1}^{k} U_{\alpha_j}. \tag{1.3.7}$$

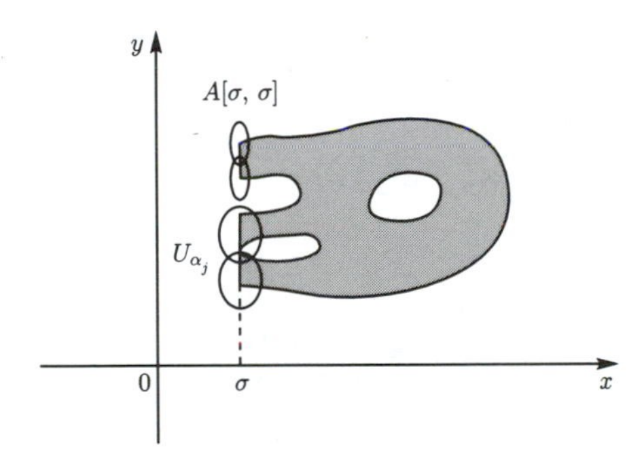

FIGURE 9

It follows that there is a $\tau_1 > \sigma$ such that

$$A[\sigma, \tau_1] \subset \bigcup_{j=1}^{k} U_{\alpha_j}.$$

(If there is no such τ_1, we can find $z_n \in A[\sigma, \sigma + 1/n] \setminus \bigcup_{j=1}^{k} U_{\alpha_j}$, $n = 1, 2, \ldots$. By "i) $\Leftrightarrow$ ii)" shown above we may assume that $\{z_n\}_{n=1}^{\infty}$ converges to a. Then $a \in A[\sigma, \sigma]$, so that some $U_{\alpha_j} \ni a$. Thus $z_n \in U_{\alpha_j}$ for large n. This is absurd.) Now, let τ_0 be the supremum of τ such that $A[\sigma, \tau]$ is covered by finitely many U_α. Suppose $\tau_0 < \max S$. It follows from the same reason as (1.3.7) that $A[\tau_0, \tau_0]$

is covered by finitely many U_α, and hence so is $A[\tau_0 - \delta, \tau_0 + \delta]$ for some $\delta > 0$. Since $A[\sigma, \tau_0 - \delta]$ is covered by finitely many U_α, so is $A[\sigma, \tau_0 + \delta]$. This contradicts the choice of τ_0. Thus $\tau_0 = \max S$. Again by the same reasoning we see that $A = A[\sigma, \max S]$ is covered by finitely many U_α.

iii)⇒ii) Consider the open covering $\{\Delta(a; 1); a \in A\}$ of A. By the assumption there are finitely many points $a_1, \ldots, a_l \in A$ such that

$$A \subset \Delta(a_1; 1) \cup \cdots \cup \Delta(a_l; 1).$$

Therefore A is bounded.

We next show that A is closed. Let a be an accumulation point of A. Suppose $a \notin A$. Put

$$U_0 = \left\{ z \in \mathbf{C}; |z - a| > \frac{1}{2} \right\},$$

$$U_n = \left\{ z \in \mathbf{C}; \frac{1}{n+2} < |z - a| < \frac{1}{n} \right\}, \quad n = 1, 2, \ldots.$$

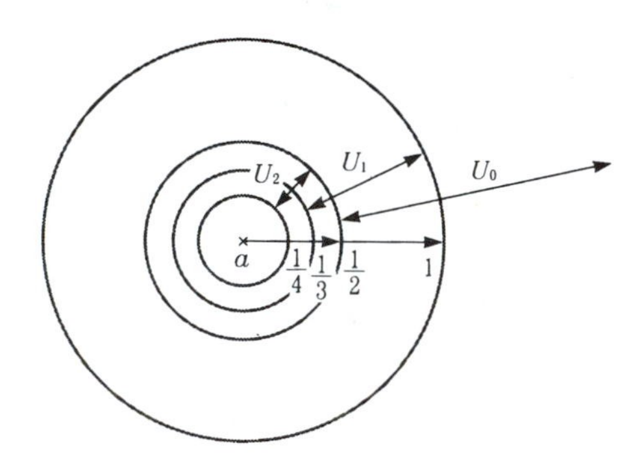

FIGURE 10

Then $A \subset \mathbf{C} \setminus \{a\} = \bigcup_{n=0}^{\infty} U_n$, and so $\{U_n\}_{n=0}^{\infty}$ is an open covering of A. Since a is an accumulation point of A, A cannot be covered by finitely many U_n. This is a contradiction, and hence A is closed. □

A subset B of A is said to be *relatively compact* if the closure $\overline{B}$ of B is compact and $\overline{B} \subset A$, and in this case we write $B \Subset A$.

EXERCISE 2. In Theorem (1.3.6), iii) we may take relatively compact open subsets $V_i \Subset U_{\alpha_i}$ with $A \subset \bigcup_{i=1}^{\infty} V_i$.

Let $\{z_n\}_{n=0}^{\infty}$ be a sequence. We call the formal sum $\sum_{n=0}^{\infty} z_n$ a *series*, and each z_n a term of $\sum_{n=0}^{\infty} z_n$. For $N = 0, 1, 2, \ldots$, the sum $s_N = \sum_{n=0}^{N} z_n$ is called the Nth *partial sum*. If $\{s_N\}_{N=0}^{\infty}$ converges, $\sum_{n=0}^{\infty} z_n$ is said to be convergent, and the limit $\lim_{N\to\infty} s_N$ is denoted by $\sum_{n=0}^{\infty} z_n$:

$$\sum_{n=0}^{\infty} z_n = \lim_{N\to\infty} \sum_{n=0}^{N} z_n.$$

If $\sum_{n=0}^{\infty} z_n$ and $\sum_{n=0}^{\infty} w_n$ are convergent, and $a, b \in \mathbf{C}$, then

$$\sum_{n=0}^{\infty}(az_n + bw_n) = a\sum_{n=0}^{\infty} z_n + b\sum_{n=0}^{\infty} w_n,$$

$$\overline{\left(\sum_{n=0}^{\infty} z_n\right)} = \sum_{n=0}^{\infty} \overline{z}_n.$$

We say that $\sum_{n=0}^{\infty} z_n$ satisfies the *Cauchy condition* if the sequence $\{s_N\}_{N=0}^{\infty}$ of the Nth partial sums is a Cauchy sequence. If $\sum_{n=0}^{\infty} |z_n|$ converges, we say that $\sum_{n=0}^{\infty} z_n$ *converges absolutely.*

Let $\lambda : \mathbf{Z}^+ \to \mathbf{Z}^+$ be an *injective* and *surjective* mapping; that is, λ is one to one, and $\lambda(\mathbf{Z}^+) = \mathbf{Z}^+$. Then $\sum_{n=0}^{\infty} z_{\lambda(n)}$ is called a *series of order change* of $\sum_{n=0}^{\infty} z_n$.

(1.3.8) THEOREM. i) *A series $\sum_{n=0}^{\infty} z_n$ converges if and only if it satisfies the Cauchy condition.*

ii) *If $\sum_{n=0}^{\infty} z_n$ converges, then $\lim_{n\to\infty} z_n = 0$.*

iii) *If $\sum_{n=0}^{\infty} z_n$ converges absolutely, it converges.*

iv) *Any series of order change of an absolutely convergent series converges absolutely, and has the same limit as the original one.*

EXERCISE 3. Prove the above theorem.

Let $\sum_{n=0}^{\infty} z_n$ and $\sum_{n=0}^{\infty} z'_n$ be two series. Put

$$w_n = \sum_{i=0}^{n} z_i z'_{n-i}.$$

Then the series $\sum_{n=0}^{\infty} w_n$ is called the *Cauchy product* of $\sum_{n=0}^{\infty} z_n$ and $\sum_{n=0}^{\infty} z'_n$. Similarly to the case of series of real numbers, we have

(1.3.9) THEOREM. *If $\sum_{n=0}^{\infty} z_n$ and $\sum_{n=0}^{\infty} z'_n$ converge absolutely, then so does their Cauchy product $\sum_{n=0}^{\infty} w_n$, and*

$$\sum_{n=0}^{\infty} w_n = \left(\sum_{n=0}^{\infty} z_n\right)\left(\sum_{n=0}^{\infty} z'_n\right).$$

A formal product $\prod_{n=0}^{\infty} z_n$ for a sequence $\{z_n\}_{n=0}^{\infty}$ is called an *infinite product.* An infinite product $\prod_{n=0}^{\infty} z_n$ is said to be convergent if the following conditions are satisfied:

(1.3.10) i) There is an $n_0 \in \mathbf{N}$ such that $a_n \neq 0$ for all $n \geqq n_0$.

ii) The sequence $\left\{\prod_{n=n_0}^{n_0+m} z_n\right\}_{m=0}^{\infty}$ converges to a *non-zero* number.

In this case, we set

$$\prod_{n=0}^{\infty} z_n = \left(\prod_{n=0}^{n_0-1} z_n\right)\left(\lim_{m\to\infty}\prod_{n=n_0}^{n_0+m} z_n\right),$$

which is called the limit of the infinite product $\prod_{n=0}^{\infty} z_n$. It is clear that the limit, if it exists, does not depend on the choice of n_0.

(1.3.11) THEOREM. i) *If an infinite product* $\prod_{n=0}^{\infty} z_n$ *converges, then* $\lim_{n\to\infty} z_n = 1$.

ii) *An infinite product* $\prod_{n=0}^{\infty} z_n$ *converges if and only if for an arbitrary* $\epsilon > 0$ *there is an* $n_0 \in \mathbf{N}$ *such that*

$$\left|\prod_{n=n_1}^{n_2} z_n - 1\right| < \epsilon \qquad \textit{for all } n_2 \geqq n_1 \geqq n_0. \tag{1.3.12}$$

PROOF. i) Take n_0 as in (1.3.10), i), and put $a = \lim_{m\to\infty}\prod_{n=n_0}^{m} z_n$. Then $a \neq 0$ and

$$z_m = \frac{\prod_{n=n_0}^{m} z_n}{\prod_{n=n_0}^{m-1} z_n} \to \frac{a}{a} = 1 \qquad (m\to\infty).$$

ii) We first show the "only if" part. Take n_0 as above. There is an $M > 0$ such that

$$\frac{1}{M} \leqq \left|\prod_{n=n_0}^{n_0+m} z_n\right| \leqq M, \qquad m \geqq 0. \tag{1.3.13}$$

It follows from Theorem (1.3.5) that for any $\epsilon > 0$ there is an n_1 such that

$$\left|\prod_{n=n_0}^{n_0+m_2} z_n - \prod_{n=n_0}^{n_0+m_1} z_n\right| < \epsilon, \qquad m_2 \geqq m_1 \geqq n_1.$$

Therefore

$$\left|\prod_{n=n_0+m_1+1}^{n_0+m_2} z_n - 1\right| < \frac{\epsilon}{\prod_{n=0}^{n_0+m_1}|z_n|} \leqq M\epsilon.$$

Since $\epsilon > 0$ is arbitrary, the claim is proved.

We next show the "if" part. We may assume by the condition that (1.3.13) holds (putting $\epsilon = 1/2$, we may take $M = 2$). For an arbitrary $\epsilon > 0$ there is an n_0' $(> n_0)$ such that

$$\left|\prod_{n=n_1}^{n_2} z_n - 1\right| < \epsilon \qquad \text{for all } n_2 \geqq n_1 \geqq n_0'.$$

It follows from this and (1.3.13) that

$$\left|\prod_{n=n_0}^{n_2} z_n - \prod_{n=n_0}^{n_1} z_n\right| = \left|\prod_{n=n_0}^{n_1-1} z_n\right|\left|\prod_{n=n_1}^{n_2} z_n - 1\right| \leqq M\epsilon.$$

Hence $\left\{\prod_{n=n_0}^{n_0+m} z_n\right\}_{m=0}^{\infty}$ gives rise to a Cauchy sequence, and so it converges by Theorem (1.3.5). By (1.3.13) the limit is not zero. $\square$

We will discuss infinite products again in the next chapter, after defining the exponential and logarithmic functions for complex numbers.

Problems

1. Write $\sqrt{i}, \sqrt{1+i}, \sqrt{1-\sqrt{3}i}$ in the form of $x+iy$.
2. In general, put $\sqrt{x+iy}=u+iv$. Express u and v as functions of x and y.
3. To a complex number $z=x+iy$ we assign a real matrix $\phi(z)=\left(\begin{smallmatrix} x & y \\ -y & x \end{smallmatrix}\right)$. Let $\phi(z_1)+\phi(z_2)$ and $\phi(z_1)\phi(z_2)$ stand for the addition and product as matrices. Then show the following.

$$\phi(z_1 \pm z_2) = \phi(z_1) \pm \phi(z_2),$$
$$\phi(z_1)\phi(z_2) = \phi(z_1 z_2),$$
$$\phi(z)^{-1} = \phi(z^{-1}) \quad (z \neq 0).$$

4. Take n points $\mathrm{P}_j = \cos\frac{2\pi j}{n} + i\sin\frac{2\pi j}{n}$, $n \in N$, $0 \leqq j \leqq n-1$, on the unit circle. Let $\overline{\mathrm{P}_0\mathrm{P}}_j$ be the distance between P_0 and P_j. Show that $\prod_{j=1}^{n-1} \overline{\mathrm{P}_0\mathrm{P}}_j = n$.
5. Let $0<r<1$, and $\theta_n \in \mathbf{R}$, $n=0,1,2,\ldots$. Show that

$$\sum_{n=0}^{\infty} r^n(\cos\theta_n + i\sin\theta_n)$$

converges.

6. Define a sequence $\{z_n\}_{n=0}^{\infty}$ by $z_{n+1}-z_n = a(z_n - z_{n-1})$ with $0<|a|<1$. Write the limit $\lim z_n$ in terms of z_0 and z_1.
7. Let $E_\alpha \subset \mathbf{C}, \alpha \in \Gamma$, be compact subsets such that $E_{\alpha_1} \cap \cdots \cap E_{\alpha_k} \neq \emptyset$ for any finitely many $E_{\alpha_1}, \ldots, E_{\alpha_k}$. Show that $\bigcap_{\alpha\in\Gamma} E_\alpha \neq \emptyset$.
8. Let $B \subset A$ be a discrete subset in A $(\subset \mathbf{C})$. Show that for any compact subset $K \subset A$, $K \cap B$ is a finite set.
9. Show that the ring $R(r_1, r_2) = \{z \in \mathbf{C}; r_1 < |z| < r_2\}$ $(0 \leqq r_1 < r_2)$ is a domain.
10. Let $D \subset \mathbf{C}$ be a domain, and $E \subset D$ be a discrete subset. Show that $D \setminus E$ is also a domain.

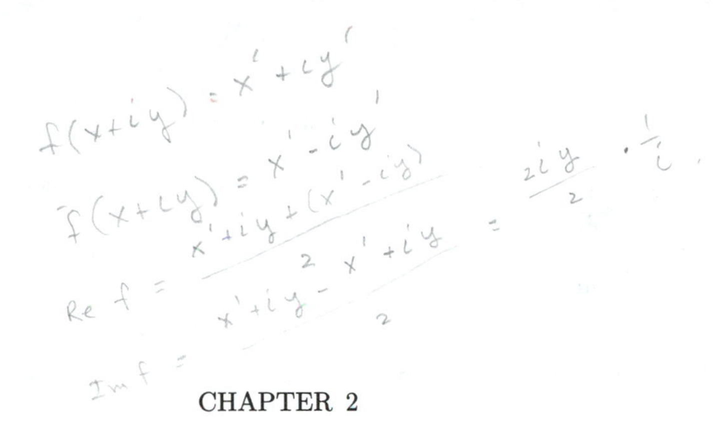

CHAPTER 2

Complex Functions

In this chapter we use the properties of complex numbers described in the previous chapter to deal with complex functions, power series and analytic functions, and to study their properties. By making use of power series we define elementary functions such as exponential functions, trigonometric functions, etc. Then we define the Riemann sphere and introduce linear transformations.

The material presented in this chapter may be said to form the foundation of complex functions theory.

2.1. Complex Functions

Let $A \subset \mathbf{C}$, and let $f : A \longrightarrow \mathbf{C}$ be a mapping. We call f a *complex function* defined on A. We set

$$\overline{f}(z) = \overline{f(z)}, \qquad z \in A,$$
$$\mathrm{Re} f = \frac{f + \overline{f}}{2},$$
$$\mathrm{Im} f = \frac{f - \overline{f}}{2i}.$$

We call $\mathrm{Re} f$ (resp., $\mathrm{Im} f$) the real (resp., imaginary) part of f.

The function f is said to be *bounded* if there is a number $M > 0$ such that $|f(z)| < M$ for all $z \in A$. Let $z_0 \in \mathbf{C}$ be an accumulation point of A. Then we say that f has *limit* $\alpha \in \mathbf{C}$ at z_0, if for an arbitrary $\epsilon > 0$ there is a $\delta > 0$ such that

$$|f(z) - \alpha| < \delta, \qquad z \in (\Delta(z_0; \delta) \setminus \{z_0\}) \cap A,$$

and write

$$\alpha = \lim_{z \longrightarrow z_0} f(z).$$

EXERCISE 1. A function $f(z)$ has limit α at z_0 if and only if for an arbitrary sequence $\{z_n\}_{n=0}^{\infty}$ of $A \setminus \{z_0\}$ converging to z_0, $\lim_{n \to \infty} f(z_n) = \alpha$.

In particular, if $z_0 \in A$ and $f(z_0) = \lim_{z \to z_0} f(z)$, f is said to be *continuous* at z_0. If f is continuous at all points of A, f is said to be continuous on A. Let g be another function on A. If f and g are continuous at z_0 (or on A) and $a, b \in \mathbf{C}$, then

$$af(z) + bg(z), \qquad f(z)g(z)$$

are continuous there; moreover, if $g(z_0) \neq 0$, then $f(z)/g(z)$ is continuous at z_0, too.

EXERCISE 2. i) $f(z)$ is continuous at z_0 if and only if $\mathrm{Re} f(z)$ and $\mathrm{Im} f(z)$ are continuous at z_0.

ii) If $f(z)$ is continuous at z_0, then so is $|f(z)|$.

We say that a complex function $f : A \to \mathbf{C}$ is *uniformly continuous* if for an arbitrary $\epsilon > 0$ there is a $\delta > 0$ such that

$$z, w \in A, \qquad |z - w| < \delta \Longrightarrow |f(z) - f(w)| < \epsilon.$$

(2.1.1) THEOREM. i) *A continuous complex function f on a compact set A is bounded, and $|f(z)|$ attains the maximum.*

ii) *A continuous complex function f on a compact set A is uniformly continuous.*

PROOF. By definition, for arbitrary $\epsilon > 0$ and $a \in A$ there is a $\delta(a) > 0$ such that

$$z \in \Delta(a, \delta(a)) \cap A \Longrightarrow |f(z) - f(a)| < \epsilon.$$

Since $\{\Delta(a; \delta(a)/2); a \in A\}$ is an open covering of A, there are finitely many $a_1, \ldots, a_l \in A$ by Theorem (1.3.6) such that

$$A \subset \bigcup_{j=1}^{l} \Delta(a_j, \delta_j/2), \qquad \delta_j = \delta(a_j).$$

Therefore, for an arbitrary $z \in A$ there is some j with $\Delta(a_j, \delta_j/2) \ni z$, so that

$$\begin{aligned} |f(z)| &= |f(z) - f(a_j) + f(a_j)| < \epsilon + |f(a_j)| \\ &\leqq \epsilon + \max_{1 \leqq j \leqq l} |f(a_j)|. \end{aligned}$$

Thus f is bounded. Put

$$M = \sup\{|f(z)|; z \in A\} < +\infty.$$

Take a sequence $\{z_n\}_{n=0}^{\infty}$ such that $\lim_{n \to \infty} |f(z_n)| = M$. Since A is compact, we may assume by taking a subsequence that $\{z_n\}_{n=0}^{\infty}$ converges to $z_0 \in A$. By the continuity of f

$$|f(z_0)| = \lim_{n \to \infty} |f(z_n)| = M.$$

Thus i) follows.

To show ii), we put $\delta_0 = \min_{1 \leqq j \leqq l} \delta_j$. We arbitrarily take $z, w \in A$ with $|z - w| < \delta_0/2$. Then $\Delta(a_j; \delta_j/2) \ni z$ for some j, and so $w \in \Delta(a_j; \delta_j)$. Hence

$$|f(z) - f(w)| \leqq |f(z) - f(a_j)| + |f(w) - f(a_j)| < 2\epsilon.$$

This shows that f is uniformly continuous on A. $\square$

EXERCISE 3. Let $f(z) = z^2$ and $0 < \epsilon < 2$. Determine the largest δ such that for $z, w \in \Delta(1)$

$$|z - w| < \delta \Longrightarrow |f(z) - f(w)| < \epsilon.$$

2.2. Sequences of Complex Functions

Let $\{f_n\}_{n=0}^{\infty}$ be an infinite sequence of complex functions defined on A. We simply call it a *sequence of complex functions*. We say that $\{f_n\}_{n=0}^{\infty}$ *converges* at $z \in A$ if the sequence $\{f_n(z)\}_{n=0}^{\infty}$ of numbers is convergent. If $\{f_n(z)\}_{n=0}^{\infty}$ converges for all $z \in A$, then $\{f_n\}_{n=0}^{\infty}$ is said to converge on A, and the function $f(z) = \lim_{n\to\infty} f_n(z)$ in $z \in A$ is called the *limit function* of $\{f_n\}_{n=0}^{\infty}$. We write

$$f = \lim_{n\to\infty} f_n, \quad \text{or} \quad f_n \to f \quad (n \to \infty).$$

Moreover, we say that $\{f_n\}_{n=0}^{\infty}$ *converges uniformly* to f if for an arbitrary $\epsilon > 0$ there is a number n_0 such that for all $n \geqq n_0$ and $z \in A$

$$|f_n(z) - f(z)| < \epsilon.$$

(2.2.1) THEOREM. i) *A sequence $\{f_n\}_{n=0}^{\infty}$ of complex functions on A converges uniformly if and only if for an arbitrary $\epsilon > 0$ there is a number n_0 such that for all $m, n \geqq n_0$ and $z \in A$*

$$|f_m(z) - f_n(z)| < \epsilon.$$

ii) *If a sequence $\{f_n\}_{n=0}^{\infty}$ of continuous complex functions f_n on A converges uniformly, then the limit function $f = \lim_{n\to\infty} f_n$ is continuous.*

PROOF. i) This is easy and is left to the reader.

ii) Take an arbitrary $\epsilon > 0$. By assumption there is a number n_0 such that

$$|f(z) - f_{n_0}(z)| < \epsilon, \qquad z \in A.$$

Take a point $a \in A$, and fix it. Since f_{n_0} is continuous at a, there is a $\delta > 0$ such that

$$|f_{n_0}(z) - f_{n_0}(a)| < \epsilon, \quad z \in \Delta(a; \delta).$$

It follows from the above two equations that for $z \in \Delta(a; \delta)$

$$|f(z) - f(a)| \leqq |f(z) - f_{n_0}(z)| + |f_{n_0}(z) - f_{n_0}(a)| + |f_{n_0}(a) - f(a)| < 3\epsilon.$$

Thus f is continuous at a. $\square$

A sequence $\{f_n\}_{n=0}^{\infty}$ of complex functions on A is said to be *uniformly bounded* if there is an $M > 0$ such that for all n and $z \in A$

$$|f_n(z)| \leqq M.$$

A bounded sequence of complex numbers necessarily has a convergent subsequence (Theorem (1.3.4)). If a sequence of complex functions on A is uniformly bounded, then does it have a convergent subsequence? The answer is "No" in general (see problem 3 at the end of this chapter for an example). Furthermore, here we consider sequences of continuous complex functions, and require the limit functions to be continuous. With this in mind we introduce the following new concept:

A family $\mathcal{F}$ of complex functions on A is said to be *equicontinuous* if for an arbitrary $\epsilon > 0$ there is a $\delta > 0$ such that for all $f \in \mathcal{F}$

$$z, w \in A,\ |z - w| < \delta \Longrightarrow |f(z) - f(w)| < \epsilon. \tag{2.2.2}$$

In this case, every $f \in \mathcal{F}$ is, of course, uniformly continuous. The following theorem is called the *Ascoli-Arzelà theorem* and is widely used in all fields of analysis.

(2.2.3) THEOREM. *If a sequence $\{f_n\}_{n=0}^{\infty}$ of complex functions on a compact set A is uniformly bounded and equicontinuous, then $\{f_n\}_{n=0}^{\infty}$ has a uniformly convergent subsequence.*

PROOF. If A is a finite set, then the claim is clear, so that A may be assumed to be infinite. We take a subset $E = \{z_\nu; \nu = 1, 2, \dots\}$ of A such that $\overline{E} = A$. We may take such E by the following procedure: By Theorem (1.3.6) A is bounded, and so there is an $M > 0$ such that $|\mathrm{Re}\, z| \leqq M$, $|\mathrm{Im}\, z| \leqq M$ for all $z \in A$. Set

$$F_1 = \{z \in \mathbf{C}; |\mathrm{Re}\, z| \leqq M,\ |\mathrm{Im}\, z| \leqq M\}.$$

Dividing each side of the closed square F_1 into two equal parts, we have four closed squares $F_{2,1}, \dots, F_{2,4}$. Repeating this process n times, we have 4^{n-1} squares $F_{n,j}$, $1 \leqq j \leqq 4^{n-1}$, and

$$\bigcup_{j=1}^{4^{n-1}} F_{n,j} = F_1 \supset A.$$

Taking one point $z_{n,j} \in A \cap F_{n,j}$ for non-empty $A \cap F_{n,j}$, we set $E_n = \{z_{n,j}\}_j$, which is a finite set. Then set $E = \bigcup_{n=1}^{\infty} E_n$. Then it is clear that E satisfies the required property.

Now, for $z_1 \in E$ the sequence $\{f_n(z_1)\}_{n=0}^{\infty}$ is bounded, and so by Theorem (1.3.4) it has a convergent subsequence $\{f_{n(1)_\nu}(z_1)\}_{\nu=0}^{\infty}$. In the same way, $\{f_{n(1)_\nu}(z_2)\}_{\nu=0}^{\infty}$ has a convergent subsequence $\{f_{n(2)_\nu}(z_2)\}_{\nu=0}^{\infty}$. Repeating this process, we obtain a subsequence $\{f_{n(k)_\nu}(z)\}_{\nu=0}^{\infty}$ of $\{f_n(z)\}_{n=0}^{\infty}$ converging at $z_1, \dots, z_k$ for $k = 1, 2, \dots$. Thus we get a subsequence $\{f_{n(\nu+1)_\nu}\}_{\nu=0}^{\infty}$ of $\{f_n\}_{n=0}^{\infty}$

which converges on E. We write this subsequence as $\{g_\mu\}_{\mu=0}^\infty$ for the sake of simplicity. Given $\epsilon > 0$, we choose a $\delta > 0$ so that (2.2.2) holds. By Theorem (1.3.6) A is covered by finitely many $\Delta(a_j;\delta)$ with $a_j \in A$ and $1 \leqq j \leqq l$. Since $\overline{E} = A$, there is a point $z_{\nu(j)} \in \Delta(a_j;\delta) \cap E$ for every j. There is a number μ_0 such that for all $\mu, \mu' \geqq \mu_0$

$$(2.2.4) \qquad |g_\mu(z_{\nu(j)}) - g_{\mu'}(z_{\nu(j)})| < \epsilon, \qquad 1 \leqq j \leqq l.$$

Now, let $a \in A$ be an arbitrary point. Then $\Delta(a_j;\delta) \ni a$ for some j. It follows from (2.2.2) and (2.2.4) that for the above $\mu, \mu' \geqq \mu_0$

$$\begin{aligned} |g_\mu(a) - g_{\mu'}(a)| \leqq & |g_\mu(a) - g_\mu(z_{\nu(j)})| + |g_\mu(z_{\nu(j)}) - g_{\mu'}(z_{\nu(j)})| \\ & + |g_{\mu'}(z_{\nu(j)}) - g_{\mu'}(a)| < 3\epsilon. \end{aligned}$$

Therefore we infer from Theorem (2.2.1), i) that $\{g_\mu\}_{\mu=0}^\infty$ converges uniformly on A. □

Let D be an open subset of $\mathbf{C}$ and $\{f_n\}_{n=0}^\infty$ be a sequence of complex functions on D. We say that $\{f_n\}_{n=0}^\infty$ converges uniformly on compact subsets if it converges uniformly on every compact subset of D.

(2.2.5) THEOREM. i) *A sequence $\{f_n\}_{n=0}^\infty$ of complex functions on D converges uniformly on every compact subset if and only if for every $a \in D$ there is an $r > 0$ such that $\overline{\Delta(a;r)} \subset D$ and $\{f_n\}_{n=0}^\infty$ converges uniformly on $\overline{\Delta(a;r)}$.*

ii) *The limit function of a sequence of continuous complex functions on D which converges uniformly on compact subsets is continuous.*

PROOF. i) The "only if" part is clear. We show the "if" part. Let K be a compact subset of D. By assumption, there is an $r_a > 0$ for every $a \in K$ such that $\overline{\Delta(a;r_a)} \subset D$ and $\{f_n\}_{n=0}^\infty$ converges uniformly on $\overline{\Delta(a;r_a)}$. Since $\{\Delta(a;r_a)\}_{a\in K}$ is an open covering of K, there are finitely many $a_1, \dots, a_l \in K$ such that $K \subset \bigcup_{j=1}^l \Delta(a_j;r_j)$ with $r_j = r_{a_j}$. Since $\{f_n\}_{n=0}^\infty$ converges uniformly on $\bigcup_{j=1}^l \Delta(a_j;r_j)$, it also converges uniformly on K.

ii) To show continuity we may restrict the complex functions to a disk neighborhood $\Delta(z;r) \Subset D$ of a given point $z \in D$. The result then follows immediately from Theorem (2.2.1), ii). □

We say that a sequence $\{f_n\}_{n=0}^\infty$ of complex functions on D is uniformly bounded (resp., equicontinuous) on compact subsets if it is uniformly bounded (resp., equicontinuous) on every fixed compact subset of D.

(2.2.6) THEOREM. *If a sequence $\{f_n\}_{n=0}^\infty$ of complex functions on D is uniformly bounded and equicontinuous on compact subsets of D, then it contains a subsequence converging uniformly on compact subsets of D.*

We prepare the following for the proof.

(2.2.7) LEMMA. *There are relatively compact subsets $U_n, n = 1, 2, \ldots$, of D satisfying*

i) $\overline{U}_n \Subset U_{n+1}, n = 1, 2, \ldots$;

ii) *for any compact subset $K \Subset D$, there is an $n \in \mathbf{N}$ with $K \subset U_n$.*

PROOF. If $\partial D = \emptyset$, then $D = \mathbf{C}$, and we may put $U_n = \Delta(n), n = 1, 2, \ldots$. Assume that $\partial D \neq \emptyset$. We define the *boundary distance* by

$$d(z; \partial D) = \inf\{|z - w|; w \in \partial D\}, \quad z \in D. \tag{2.2.8}$$

Take a sequence $w_n \in \partial D, n = 0, 1, 2, \ldots$, so that

$$|z - w_n| \to d(z; \partial D) \quad (n \to \infty).$$

Then $\{w_n\}_{n=0}^{\infty}$ is bounded, and so it has a converging subsequence by Theorem (1.3.4). Hence we may assume from the beginning that $\lim w_n = w_0$. Since ∂D is closed, $w_0 \in \partial D$. Therefore we have

$$d(z; \partial D) = |z - w_0| = \min\{|z - w|; w \in \partial D\} > 0. \tag{2.2.9}$$

For $z' \in D$

$$|z' - w_0| - |z - w_0| \leqq |z' - z|,$$

and hence $d(z'; \partial D) - d(z; \partial D) \leqq |z' - z|$. By the symmetry we get

$$|d(z'; \partial D) - d(z; \partial D)| \leqq |z' - z|. \tag{2.2.10}$$

In particular, $d(z; \partial D)$ is continuous. For $r > 0$ the set defined by

$$D_r = \{z \in D; d(z; \partial D) > r\}$$

is open.

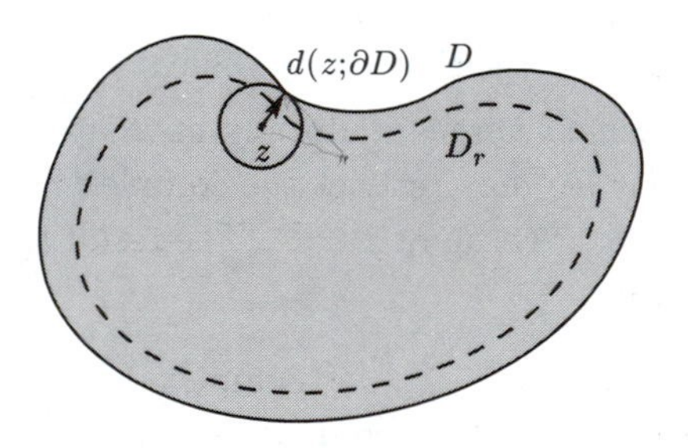

FIGURE 11

Put

$$U_n = D_{1/n} \cap \Delta(n), \qquad n = 1, 2, \ldots.$$

Then $\overline{U}_n$ are compact, and $D = \bigcup_{n=1}^{\infty} U_n$. Thus ii) follows from this and Theorem (1.3.6), iii). □

PROOF OF THEOREM (2.2.6). Take $U_n, n = 1, 2, \ldots$, as in Lemma (2.2.7). By Theorem (2.2.3) there is a subsequence $\{f_{n(1)\nu}\}_{\nu=0}^{\infty}$ of $\{f_n\}_{n=0}^{\infty}$ which converges uniformly on $\overline{U}_1$. In the same way, we take a subsequence $\{f_{n(2)\nu}\}_{\nu=0}^{\infty}$ of $\{f_{n(1)\nu}\}_{\nu=0}^{\infty}$ which converges uniformly on $\overline{U}_2$. Inductively, we take a subsequence $\{f_{n(\mu)\nu}\}_{\nu=0}^{\infty}$ of $\{f_{n(\mu-1)\nu}\}_{\nu=0}^{\infty}$ which converges uniformly on $\overline{U}_{\mu}$ $(\mu = 1, 2, \ldots)$. Then the subsequence $\{f_{n(\nu+1)\nu}\}_{\nu=0}^{\infty}$ of $\{f_n\}$ converges uniformly on every $\overline{U}_{\mu}$. Hence, $\{f_{n(\nu+1)\nu}\}_{\nu=0}^{\infty}$ converges uniformly on compact subsets of D. □

EXERCISE 1. Show that the functions $f_n(z) = z^n, z \in \overline{\Delta(1)}, n = 1, 2, \ldots$, are not equicontinuous.

EXERCISE 2. Show that if $\{f_n(z)\}_{n=1}^{\infty}$ is uniformly bounded and equicontinuous on a subset $A \subset \mathbf{C}$, then so are $g_n(z) = \frac{1}{n}\sum_{k=1}^{n} f_k(z), n = 1, 2, \ldots$.

2.3. Series of Functions

In what follows, functions mean complex functions. Let $\{f_n\}_{n=0}^{\infty}$ be a sequence of functions defined on $A \subset \mathbf{C}$. Then the formal sum $\sum_{n=0}^{\infty} f_n$ is called a *series of functions*. For a number N $(= 0, 1, \cdots)$ the sum $\sum_{n=0}^{N} f_n$ is called an N-*partial sum* or simply a partial sum. If the sequence $\{s_N(z)\}_{N=0}^{\infty}$ converges at a point $z \in A$, the series $\sum_{n=0}^{\infty} f_n$ of functions is said to converge at z, and we write

$$\lim_{N\to\infty} s_N(z) = \sum_{n=0}^{\infty} f_n(z).$$

If $\{s_N\}_{N=0}^{\infty}$ converges at all points of A, we say that $\sum_{n=0}^{\infty} f_n$ converges on A and write the limit function as

$$\lim_{N\to\infty} s_N = \sum_{n=0}^{\infty} f_n.$$

If two series, $\sum_{n=0}^{\infty} f_n$ and $\sum_{n=0}^{\infty} g_n$, of functions on A converge at a point $z \in A$ (resp., on A), then for constants $a, b \in \mathbf{C}$

$$\sum_{n=0}^{\infty}(af_n(z) + bg_n(z)) = a\sum_{n=0}^{\infty} f_n(z) + b\sum_{n=0}^{\infty} g_n(z)$$

at z (resp., on A). If $\sum_{n=0}^{\infty} |f_n(z)|$ converges at z (resp., on A), $\sum_{n=0}^{\infty} f_n$ is said to *converge absolutely* at z (resp., on A). The following is clear.

(2.3.1) Theorem (1.3.8), iii) and iv) hold for a series $\sum_{n=0}^{\infty} f_n$ of functions at $z \in A$ or on A.

We say that $\sum_{n=0}^{\infty} f_n$ converges uniformly on A if $\{s_N\}_{N=0}^{\infty}$ converges uniformly on A.

(2.3.2) THEOREM. *A series $\sum_{n=0}^{\infty} f_n$ of functions on A converges uniformly on A if and only if for an arbitrary $\epsilon > 0$ there is a number n_0 such that*

$$\left|\sum_{j=m}^{n} f_j(z)\right| < \epsilon$$

for all $n \geqq m \geqq n_0$ and $z \in A$. Moreover, if the series converges uniformly and all the functions f_n are continuous, then the limit function $\sum_{n=0}^{\infty} f_n$ is continuous.

PROOF. Assume that $\sum_{n=0}^{\infty} f_n = f$ converges uniformly. Then for an arbitrary $\epsilon > 0$ there is a number n_0 such that

$$\left|\sum_{j=0}^{n} f_j(z) - f(z)\right| < \epsilon$$

for all $n \geqq n_0$ and $z \in A$. Take any $n \geqq m > n_0$. Then we have

$$\begin{aligned}\left|\sum_{j=m}^{n} f_j(z)\right| &= \left|\left(\sum_{j=0}^{n} f_j(z) - f(z)\right) - \left(\sum_{j=0}^{m-1} f_j(z) - f(z)\right)\right| \\ &\leqq \left|\sum_{j=0}^{n} f_j(z) - f(z)\right| + \left|\sum_{j=0}^{m-1} f_j(z) - f(z)\right| < 2\epsilon.\end{aligned}$$

This proves the "only if" part.

Now, we show the "if" part. Put $f = \sum_{n=0}^{\infty} f_n(z)$. Take an arbitrary $\epsilon > 0$. By assumption there is a number n_0 such that

$$\left|\sum_{j=0}^{n} f_j(z) - \sum_{j=0}^{m} f_j(z)\right| = \left|\sum_{j=m+1}^{n} f_j(z)\right| < \epsilon$$

for all $n \geqq m \geqq n_0$ and $z \in A$. Letting $n \to \infty$, we have

$$\left|f(z) - \sum_{j=0}^{m} f_j(z)\right| \leqq \epsilon.$$

Thus $\sum_{n=0}^{\infty} f_j$ converges uniformly to f.

Continuity is a direct consequence of Theorem (2.2.1), ii). □

We say that a series $\sum_{n=0}^{\infty} M_n$ with $M_n \geqq 0$ is a *majorant* of a series $\sum_{n=0}^{\infty} f_n$ of functions on A if

$$|f_n(z)| \leqq M_n, \qquad z \in A,\ n = 0, 1, \ldots .$$

The following theorem is called the *majorant test* or *Weierstrass' M-test.* This test is simple but quite useful.

(2.3.3) THEOREM. *Let $\sum_{n=0}^{\infty} M_n$ be a majorant of a series $\sum_{n=0}^{\infty} f_n$ of functions on A. If $\sum_{n=0}^{\infty} M_n$ converges, then $\sum_{n=0}^{\infty} f_n$ converges absolutely and uniformly on A.*

The proof is clear by definition and by Theorem (2.3.2).

EXERCISE 1. Show the above Theorem (2.3.3).

Let $\sum_{n=0}^{\infty} f_n$ be defined over an open set $D \subset \mathbf{C}$. If the sequence $\{s_N\}$ of its partial sums converges uniformly on compact subsets of D, then $\sum_{n=0}^{\infty} f_n$ is said to converge uniformly on compact subsets of D. By Theorem (2.2.5), ii) we have

(2.3.4) THEOREM. *Let $\sum_{n=0}^{\infty} f_n$ be a series of functions on an open set $D \subset \mathbf{C}$. If $\sum_{n=0}^{\infty} f_n$ converges uniformly on compact sets of D, then the limit function is continuous.*

2.4. Power Series

Let $\{a_n\}_{n=0}^{\infty}$ be a sequence and $c \in \mathbf{C}$. Then the series $\sum_{n=0}^{\infty} a_n(z-c)^n$ of functions is called a *power series*. By a translation we can reduce to the case $c = 0$, and so we henceforth mainly assume that $c = 0$. For a power series $\sum_{n=0}^{\infty} a_n z^n$ and $z_0 \in \mathbf{C}$ we consider the following three conditions:

(2.4.1) i) $\sum_{n=0}^{\infty} a_n z_0^n$ converges.
 ii) $\lim_{n\to\infty} a_n z_0^n = 0$.
 iii) $\overline{\lim}_{n\to\infty} |a_n z_0^n| < +\infty$.

Clearly, i) implies ii), and ii) implies iii).

(2.4.2) LEMMA. *If any of (2.4.1) holds, then $\sum_{n=0}^{\infty} a_n z^n$ converges absolutely and uniformly on compact subsets of $\Delta(|z_0|)$.*

PROOF. It suffices to assume that $z_0 \neq 0$ and (2.4.1), iii) holds. Then there is an $M > 0$ such that

$$|a_n z_0^n| \leqq M, \qquad n = 0, 1, \ldots .$$

Take arbitrarily $0 < r < |z_0|$. We are going to show that $\sum_{n=0}^{\infty} |a_n z^n|$ converges uniformly on $\overline{\Delta(r)}$. We have

$$|a_n z^n| \leqq |a_n| r^n \leqq |a_n z_0^n| \frac{r^n}{|z_0|^n} \leqq M \left(\frac{r}{|z_0|} \right)^n .$$

Since $r/|z_0| < 1$, $\sum_{n=0}^{\infty} M(r/|z_0|)^n$ converges. By Theorem (2.3.3) $\sum_{n=0}^{\infty} |a_n z^n|$ converges uniformly on $\overline{\Delta(r)}$. □

Let R ($\leqq +\infty$) be the supremum of those $r \geqq 0$ such that $\sum_{n=0}^{\infty} a_n z^n$ converges. We call R the *radius of convergence* of $\sum_{n=0}^{\infty} a_n z^n$, and $\{z \in \mathbf{C}; |z| = R\}$ the *circle of convergence*. If $R > 0$, we call $\sum_{n=0}^{\infty} a_n z^n$ a *convergent power series*. The limit $\sum_{n=0}^{\infty} a_n z^n$ defines a continuous function on $\Delta(R)$.

(2.4.3) THEOREM. *The radius of convergence of* $\sum_{n=0}^{\infty} a_n z^n$ *is given by*

$$R = \frac{1}{\overline{\lim\limits_{n\to\infty}} \sqrt[n]{|a_n|}},$$

where $1/+\infty = 0$ *and* $1/0 = +\infty$.

PROOF. Suppose that $R = 0$. Then it follows from Lemma (2.4.2) that for any $r > 0$ there are infinitely many n with $|a_n|r^n \geqq 1$. Hence

$$\overline{\lim_{n\to\infty}} \sqrt[n]{|a_n|} \geqq \frac{1}{r}.$$

Letting $r \searrow 0$, we have that $\overline{\lim\limits_{n\to\infty}} \sqrt[n]{|a_n|} = +\infty$.

Suppose that $R > 0$. Take arbitrarily $0 < r < R$. By Lemma (2.4.2) there is a number n_0 such that $|a_n r^n| \leqq 1$ for $n \geqq n_0$. Therefore

$$\overline{\lim_{n\to\infty}} \sqrt[n]{|a_n|} \leqq \frac{1}{r}.$$

Letting $r \nearrow R$, we have that $\overline{\lim\limits_{n\to\infty}} \sqrt[n]{|a_n|} \leqq 1/R$. Next we take $r > 0$ so that $\overline{\lim\limits_{n\to\infty}} \sqrt[n]{|a_n|} < 1/r$. Then there is a number n_0 such that $|a_n|r^n < 1$ for all $n \geqq n_0$. By Lemma (2.4.2), $r \leqq R$. Letting $r \nearrow 1/\overline{\lim\limits_{n\to\infty}} \sqrt[n]{|a_n|}$, we get

$$\frac{1}{\overline{\lim\limits_{n\to\infty}} \sqrt[n]{|a_n|}} \leqq R.$$

Thus $R = 1/\overline{\lim\limits_{n\to\infty}} \sqrt[n]{|a_n|}$. $\square$

(2.4.4) THEOREM. *If* $\lim_{n\to\infty} |a_n|/|a_{n+1}|$ $(\leqq +\infty)$ *exists, then*

$$R = \lim_{n\to\infty} \frac{|a_n|}{|a_{n+1}|}.$$

PROOF. Assume that $0 < \lim_{n\to\infty} |a_n|/|a_{n+1}| \leqq +\infty$. Take arbitrarily $0 < r < \lim_{n\to\infty} |a_n|/|a_{n+1}|$. Then there is a number n_0 such that $r < |a_n|/|a_{n+1}|$ for $n \geqq n_0$. Therefore

$$r^{n-n_0} < \frac{|a_{n_0}|}{|a_{n_0+1}|} \cdot \frac{|a_{n_0+1}|}{|a_{n_0+2}|} \cdots \frac{|a_{n-1}|}{|a_n|} = \frac{|a_{n_0}|}{|a_n|}.$$

Taking the n-th root of both sides, and letting $n \to \infty$, we get

$$r \leqq \frac{1}{\overline{\lim\limits_{n\to\infty}} \sqrt[n]{|a_n|}} = R.$$

We let $r \nearrow \lim_{n\to\infty} |a_n|/|a_{n+1}|$, so that

$$\lim_{n\to\infty} \frac{|a_n|}{|a_{n+1}|} \leqq \frac{1}{\overline{\lim\limits_{n\to\infty}} \sqrt[n]{|a_n|}} = R. \tag{2.4.5}$$

Therefore, if $\lim_{n\to\infty} |a_n|/|a_{n+1}| = +\infty$, $R = +\infty$, and so they are equal. Suppose that $0 \leqq \lim_{n\to\infty} |a_n|/|a_{n+1}| < +\infty$. Take an arbitrary $r > \lim_{n\to\infty} |a_n|/|a_{n+1}|$. Similarly to the above there is a number n_0 such that $r^{n-n_0} > |a_{n_0}|/|a_n|$ for $n \geqq n_0$. It follows that

$$r \geqq \frac{1}{\varliminf_{n\to\infty} \sqrt[n]{|a_n|}} \geqq \frac{1}{\varlimsup_{n\to\infty} \sqrt[n]{|a_n|}} = R.$$

Letting $r \searrow \lim_{n\to\infty} |a_n|/|a_{n+1}|$, we get

$$\lim_{n\to\infty} \frac{|a_n|}{|a_{n+1}|} \geqq \frac{1}{\varliminf_{n\to\infty} \sqrt[n]{|a_n|}} \geqq R. \tag{2.4.6}$$

Thus the desired equality follows from (2.4.5) and (2.4.6). □

By (2.4.5) and (2.4.6) we have

(2.4.7) COROLLARY. *If* $\lim_{n\to\infty} \frac{|a_{n+1}|}{|a_n|}$ $(\leqq +\infty)$ *exists, then so does* $\lim_{n\to\infty} \sqrt[n]{|a_n|}$ $(\leqq +\infty)$, *and*

$$\lim_{n\to\infty} \frac{|a_{n+1}|}{|a_n|} = \lim_{n\to\infty} \sqrt[n]{|a_n|}.$$

For example, the radius of convergence of $\sum_{n=0}^{\infty} z^n$ is $R = \lim_{n\to\infty} 1/1 = 1$, and

$$\sum_{n=0}^{\infty} z^n = \frac{1}{1-z}, \qquad z \in \Delta(1). \tag{2.4.8}$$

This is verified by letting $N \to \infty$ in the sum $\sum_{n=0}^{N} z^n = (1 - z^{N+1})/(1-z)$. Taking the Cauchy product of (2.4.8) with itself and using $\sum_{k=0}^{n} z^k z^{n-k} = (n+1)z^n$, we see that

$$\sum_{n=0}^{\infty} (n+1)z^n = \frac{1}{(1-z)^2}. \tag{2.4.9}$$

The radius of convergence of this power series is $\lim_{n\to\infty}(n+2)/(n+1) = 1$.

Let R be the radius of convergence of a power series $\sum_{n=0}^{\infty} a_n z^n$. Then the series does not converge at any point outside the circle of convergence. On the circle of convergence there are, in general, points where it converges and where it does not converge. For example, the power series

$$f(z) = \sum_{n=0}^{\infty} \frac{(-1)^n}{n+1} z^{n+1} \tag{2.4.10}$$

has radius of convergence 1 and converges at $z = 1$, but not at $z = -1$.

Let f be a function in an open set $D \subset \mathbf{C}$. The function f is said to be *analytic* if for any $z_0 \in D$ there are $r > 0$ with $\Delta(z_0; r) \subset D$ and a power series $\sum_{n=0}^{\infty} a_n(z - z_0)^n$, converging in $\Delta(z_0; r)$, such that

$$f(z) = \sum_{n=0}^{\infty} a_n(z - z_0)^n, \qquad z \in \Delta(z_0; r). \tag{2.4.11}$$

In this case, f is, of course, continuous in D. We call (2.4.11) the *power series expansion* or the *Taylor series* of f about z_0.

(2.4.12) THEOREM. *A function defined by a power series*

$$f(z) = \sum_{n=0}^{\infty} a_n(z - a)^n,$$

which converges in $\Delta(a; r)$ with $a \in \mathbf{C}$, is analytic in $\Delta(a; r)$.

PROOF. Fix an arbitrary point $b \in \Delta(a; r)$. Then for partial sums we have

(2.4.13)

$$\begin{aligned}
\sum_{n=0}^{N} a_n(z - a)^n &= \sum_{n=0}^{N} a_n(z - b + b - a)^n \\
&= \sum_{n=0}^{N} \sum_{m=0}^{n} a_n \binom{n}{m} (b - a)^{n-m} (z - b)^m \\
&= \sum_{n=0}^{N} \left\{ \sum_{m=0}^{n} a_n \binom{n}{m} (b - a)^{n-m} \right\} (z - b)^m \\
&= \sum_{m=0}^{N} c_m^N (z - b)^m,
\end{aligned}$$

where $c_m^N = \sum_{n=m}^{N} a_n \binom{n}{m} (b - a)^{n-m}$.

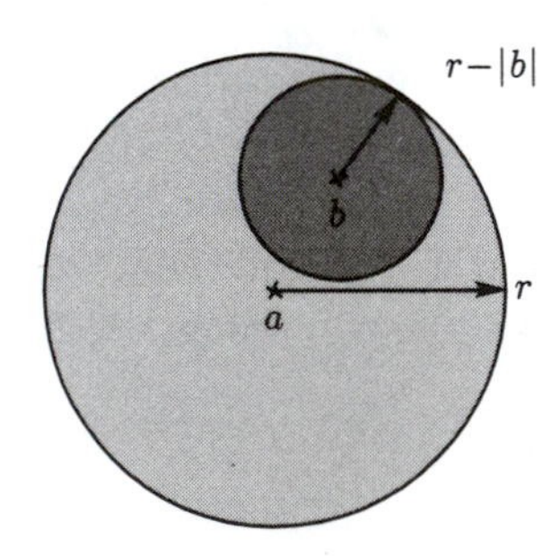

FIGURE 12

In the same way we have

$$\sum_{m=0}^{N} |c_m^N||z-b|^m \leqq \sum_{m=0}^{N}\sum_{n=m}^{N} |a_n|\binom{n}{m}|b-a|^{n-m}|z-b|^m$$
$$= \sum_{n=0}^{N} |a_n|(|b-a|+|z-b|)^n \leqq \sum_{n=0}^{\infty} |a_n|r'^n < \infty,$$

where $|b-a|+|z-b| < r' < r$. The above estimate implies also that for every fixed m the series $\sum_{n=m}^{\infty} a_n\binom{n}{m}(b-a)^{n-m}$ converges absolutely. Thus, setting $c_m = \lim_{N\to\infty} c_m^N$ and $M = \sum_{n=0}^{\infty} |a_n|r'^n$, we have for every fixed N

$$\sum_{m=0}^{N} |c_m||z-b|^m = \lim_{N'\to\infty} \sum_{m=0}^{N} |c_m^{N'}||z-b|^m \leqq M.$$

Hence, $\sum c_m(z-b)^m$ converges absolutely on $\Delta(b; r-|b|)$, and

$$f(z) = \sum_{m=0}^{\infty} c_m(z-b)^m, \qquad z \in \Delta(b; r-|b|). \qquad \square$$

The following theorem is called the *identity theorem.*

(2.4.14) THEOREM. *Let f be an analytic function in D. Let $E \subset D$ have at least one accumulation point in D. If the restriction $f|E$ of f to E is identically 0, then so is f in D.*

PROOF. Let $a \in D$ be an accumulation point of E. As in (2.4.11) we expand f:

$$f(z) = \sum_{n=0}^{\infty} a_n(z-a)^n, \qquad z \in \Delta(a; r).$$

We have that $a_0 = f(a) = 0$. Now we want to show that all $a_n = 0$. Suppose that it is not the case. Put $n_0 = \min\{n \geqq 1; a_n \neq 0\}$, and

$$f(z) = (z-a)^{n_0}\{a_{n_0} + a_{n_0+1}(z-a) + \cdots\}$$
$$= (z-a)^{n_0}\{a_{n_0} + (z-a)g(z)\},$$

where $g(z) = \sum_{n=1}^{\infty} a_{n_0+n}(z-a)^{n-1}$. Since $g(z)$ is continuous on $\Delta(a; r)$, there is a positive $r_1 < r$ such that

$$|(z-a)g(z)| < |a_{n_0}|, \qquad z \in \Delta(a; r_1).$$

It follows that for all $z \in \Delta(a; r_1) \setminus \{a\}$

$$f(z) \neq 0. \tag{2.4.15}$$

Hence $E \cap (\Delta(a; r_1) \setminus \{a\}) = \emptyset$. This contradicts the assumption that a is an accumulation point of E.

Now, let D' denote the set of all those points $z \in D$ such that there is a disc $\Delta(z; r) \subset D$ with $f|\Delta(z; r) \equiv 0$. Clearly, D' is open and contains a, so that

$D' \neq \emptyset$. Let $z \in D$ be an accumulation point of D'. Since $f|D' \equiv 0$, $z \in D'$ by the fact shown above. Therefore $\overline{D'} \cap D = D'$, and so D' is open and closed in D. It follows from (1.2.4) that $D' = D$. □

We deduce from Theorem (2.4.14) that if f is a non-constant analytic function in D, then the set of solutions of the equation $f(z) = w$ with $w \in \mathbf{C}$ is discrete in D. The following theorem is a direct consequence of Theorem (2.4.14), but a fundamental property of analytic functions.

(2.4.16) THEOREM. *Let f and g be analytic functions in D. If $f(z) \cdot g(z) \equiv 0$ in D, then $f(z) \equiv 0$ in D or $g(z) \equiv 0$ in D.*

EXERCISE 1. What is the radius of convergence of $\sum_{n=1}^{\infty} n^{\log n} z^n$?

EXERCISE 2. What is the radius of convergence of $\sum_{n=1}^{\infty} \frac{n!}{n^n} z^n$?

EXERCISE 3. Let $f(z) = \sum_{n=0}^{\infty} a_n z^n$ converge in $\Delta(r)$. Show that $f \equiv 0$ is equivalent to $a_n = 0$ for all $n = 0, 1, \ldots$.

EXERCISE 4. Show Theorem (2.4.16).

2.5. Exponential Functions and Trigonometric Functions

The radius R of convergence of the power series

$$\sum_{n=0}^{\infty} \frac{z^n}{n!} = 1 + \frac{z}{1!} + \frac{z^2}{2!} + \cdots + \frac{z^n}{n!} + \cdots \tag{2.5.1}$$

is $R = \lim_{n\to\infty}\{(n+1)!/n!\} = \lim_{n\to\infty}(n+1) = +\infty$ by Theorem (2.4.4). The analytic function in $\mathbf{C}$ defined by (2.5.1) is called the *exponential function* and denoted by e^z or $\exp z$. When z is a real number, the reader should know that it coincides with the z-th power of the natural logarithm e. Let $z, w \in \mathbf{C}$ be two arbitrary points. Then $\sum_{n=0}^{\infty} \frac{z^n}{n!}$ and $\sum_{n=0}^{\infty} \frac{w^n}{n!}$ converge absolutely. By Theorem (1.3.9) the Cauchy product is

$$\begin{aligned}\sum_{n=0}^{\infty}\left(\sum_{j=0}^{n} \frac{z^j}{j!} \cdot \frac{w^{n-j}}{(n-j)!}\right) &= \sum_{n=0}^{\infty} \frac{1}{n!} \sum_{j=0}^{n} \binom{n}{j} z^j w^{n-j} \\ &= \sum_{n=0}^{\infty} \frac{(z+w)^n}{n!}.\end{aligned}$$

Therefore we have

$$e^{z+w} = e^z \cdot e^w. \tag{2.5.2}$$

In particular, $e^z \cdot e^{-z} = e^0 = 1$, so that $e^z \neq 0$ for all $z \in \mathbf{C}$.

When $z = x \in \mathbf{R}$ and $|x| < 1$, the reader should know that the logarithmic function $\log(1+x)$ is given by (2.4.10); that is,

$$\log(1+x) = \sum_{n=0}^{\infty} \frac{(-1)^n}{n+1} x^{n+1}.$$

Thus we define the logarithmic function for $z \in \Delta(1)$ by

$$\log(1+z) = \sum_{n=0}^{\infty} \frac{(-1)^n}{n+1} z^{n+1}. \tag{2.5.3}$$

The logarithmic function $\log(1+z)$ is analytic in $\Delta(1)$. One may also write

$$\log z = \sum_{n=0}^{\infty} \frac{(-1)^n}{n+1} (z-1)^{n+1}, \qquad z \in \Delta(1;1). \tag{2.5.4}$$

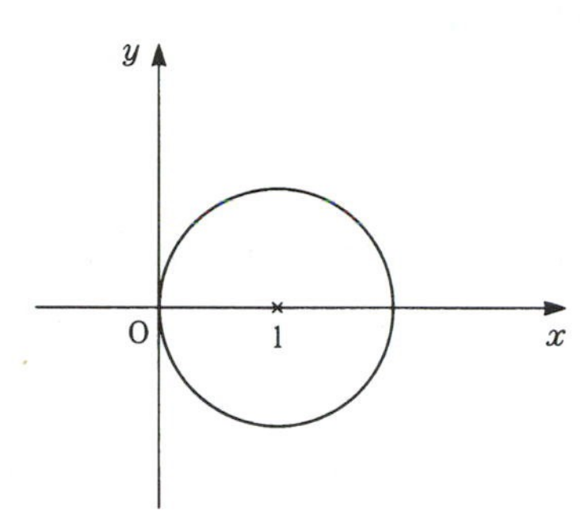

FIGURE 13

We have

$$\begin{aligned} e^{\log x} &= x \qquad (x \in \Delta(1;1) \cap \mathbf{R}), \\ \log e^x &= x \qquad (|e^x - 1| < 1). \end{aligned}$$

Since compositions of analytic functions are again analytic (this can be verified directly, but we will prove it in Remark (3.5.9) of the next chapter), the Identity Theorem (2.4.14) implies that

$$\begin{aligned} e^{\log z} &= z \qquad (z \in \Delta(1;1)), \\ \log e^z &= z \qquad (|e^z - 1| < 1). \end{aligned} \tag{2.5.5}$$

For $z \in \mathbf{C}$ we define *trigonometric functions* by

$$\cos z = \frac{e^{iz} + e^{-iz}}{2}, \qquad \sin z = \frac{e^{iz} - e^{-iz}}{2i}.$$

They both are analytic in $\mathbf{C}$, and the following hold:

$$e^{iz} = \cos z + i \sin z \qquad \text{(Euler's formula)},$$
$$\cos z = \sum_{n=0}^{\infty} \frac{(-1)^n}{(2n)!} z^{2n},$$
$$\sin z = \sum_{n=0}^{\infty} \frac{(-1)^n}{(2n+1)!} z^{2n+1}.$$

By these we define other trigonometric functions, $\tan z = \sin z/\cos z$, $\cot z = \cos z/\sin z$, etc. It follows from (2.5.2) that $e^{iz} \cdot e^{-iz} = 1$, so that the following functional equation holds:

$$\cos^2 z + \sin^2 z = 1. \tag{2.5.6}$$

We also infer from (2.5.2) the following addition formulas of trigonometric functions:

$$\cos(z+w) = \cos z \cos w - \sin z \sin w$$
$$\sin(z+w) = \sin z \cos w + \cos z \sin w.$$

For a moment we assume that z is a real number $x \in \mathbf{R}$, and prove that $\cos x$ and $\sin x$ are so-called periodic functions. By definition, $\cos 0 = 1 > 0$. Since

$$\frac{1}{(2n-2)!} - \frac{4}{(4n)!} > 0, \qquad n \geqq 2,$$

letting $x = 2$ we have

$$\cos 2 = 1 - \frac{2^2}{2!} + \frac{2^4}{4!} - \sum_{m=2}^{\infty} \left(\frac{1}{(4m-2)!} - \frac{4}{(4m)!} \right) 2^{4m-2}$$
$$< 1 - 2 + \frac{2}{3} = -\frac{1}{3}.$$

Therefore $\cos x$ has a zero point in $0 < x < 2$ by the mean value theorem. We show that this is a unique one. For $m \geqq 0$ and $0 < x \leqq 2$ we get

$$\frac{1}{(4m+1)!} - \frac{x^2}{(4m+3)!} > 0,$$
$$\sin x = \sum_{m=0}^{\infty} \left(\frac{1}{(4m+1)!} - \frac{x^2}{(4m+3)!} \right) x^{4m+1} > 0.$$

We take $0 \leqq x < y \leqq 2$. Then $0 < (y-x)/2 \leqq (y+x)/2 < 2$, and the addition formulas imply that

(2.5.7)
$$\begin{aligned}\cos x - \cos y &= \cos\left(\frac{x+y}{2} - \frac{y-x}{2}\right) - \cos\left(\frac{x+y}{2} + \frac{y-x}{2}\right) \\ &= 2\sin\frac{x+y}{2}\sin\frac{y-x}{2} > 0.\end{aligned}$$

Hence $\cos x$ is strictly decreasing in the interval $[0, 2]$. This proves that $\cos x$ has a unique zero point $\alpha_0 \in (0, 2)$. Set

(2.5.8)
$$\pi = 2\alpha_0.$$

We call π the *ratio of the circumference of a circle to its diameter.* It follows from the above arguments that

(2.5.9)
$$\cos\frac{\pi}{2} = 0, \qquad \sin\frac{\pi}{2} = 1.$$

Again making use of the addition formulas we get

(2.5.10)
$$\begin{aligned}&\sin\left(x+\frac{\pi}{2}\right) = \cos x, \quad \cos\left(x+\frac{\pi}{2}\right) = -\sin x, \\ &\sin\,(x+\pi) = -\sin x, \quad \cos\,(x+\pi) = -\cos x.\end{aligned}$$

Thus we obtain the following periodicity:

(2.5.11)
$$\sin(x+2\pi) = \sin x, \qquad \cos(x+2\pi) = \cos x.$$

(2.5.12) THEOREM. *The fundamental period (the positive minimum period) of* $\sin x$ *and* $\cos x$ *is* 2π.

PROOF. It follows from (2.5.10) that $\sin x$ and $\cos x$ have the same periods. We deal with $\sin x$. Since $\sin 0 = 0$, $\sin\alpha = 0$ for any period α. By the Identity Theorem (2.4.14) the periods do not accumulate at 0. Therefore there is the positive minimum period ω, so that $0 < \omega \leqq 2\pi$. If $2\pi/\omega$ is not a natural number,

$$0 < 2\pi - \left[\frac{2\pi}{\omega}\right]\omega < \omega,$$

where $[2\pi/\omega]$ stands for Gauss' symbol. This shows the existence of a positive period less than ω, which contradicts the minimality of ω. Set $p = 2\pi/\omega \in \mathbf{N}$. Since $\sin x > 0$ for $0 < x \leqq \pi/2 (< 2)$ and $\sin(\pi - x) = \sin x$,

(2.5.13)
$$\sin x > 0, \qquad 0 < x < \pi.$$

Suppose that $p > 1$. Since $\sin(x+\pi) = -\sin x$, $p \neq 2$, so that $p \geqq 3$. Thus we see that $0 < \omega \leqq 2\pi/3$, and so $\sin\omega \neq 0$ by (2.5.13). This is again a contradiction. Hence we get $\omega = 2\pi$. □

If $0 \leqq x < y \leqq \pi$, $0 < (y-x)/2 \leqq (y+x)/2 < \pi$. Thus (2.5.7) and (2.5.13) imply that $\cos x > \cos y$. That is, $\cos x$ is strictly decreasing in $[0, \pi]$. Therefore the image of the injective continuous mapping

$$[0, \pi] \ni \theta \to \cos\theta + i\sin\theta \in \mathbf{C}$$

just describes the upper half part ($\operatorname{Im} z \geqq 0$) of the unit circle $\{z \in \mathbf{C}; |z| = 1\}$.

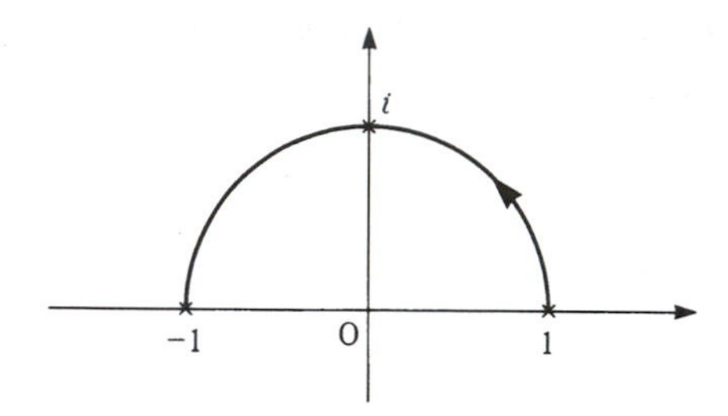

FIGURE 14

Therefore the polar coordinate representation (1.1.6) is written as

$$z = re^{i\theta} = |z|e^{i\arg z}.$$

Setting $w = |w|e^{i\arg w}$, we have

$$zw = |z||w|e^{i(\arg z + \arg w)}.$$

Henceforth we have

$$\arg(zw) = \arg z + \arg w \text{ (modulo } 2\pi),$$

and

$$\log z = \log|z| + i\arg z \tag{2.5.14}$$

for $z \in \Delta(1; 1)$.

EXERCISE 1. Represent $\sqrt[4]{1+\sqrt{3}i}$ in polar coordinates.

EXERCISE 2. Let $z \in \Delta(1), z \neq 0$, and let ℓ be the half line from the origin passing through z. Express in terms of z a point $z' \in \ell$ with $|z'||z| = 1$.

EXERCISE 3. Show that $\tan(z + \pi) = \tan z$ for $z \neq \pi/2 + n\pi$, $n \in \mathbf{Z}$.

2.6. Infinite Products

The convergence of an infinite product $\prod_{n=0}^{\infty} z_n$ is defined by (1.3.10). If this converges, then

$$\lim_{n\to\infty} z_n = 1.$$

Therefore, to investigate the convergence of the infinite product, it is more convenient to replace z_n by $1 + z_n$, and to deal with $\prod_{n=0}^{\infty}(1 + z_n)$. If $\prod_{n=0}^{\infty}(1 + z_n)$

converges, then $\lim z_n = 0$, and there is a number n_0 such that $|z_n| < 1/2$ for all $n \geqq n_0$ and

$$\left| \prod_{j=n_0}^{n} (1+z_j) - 1 \right| < \frac{1}{2}.$$

Therefore the values of $1 + z_n$ $(n \geqq n_0)$, $\prod_{j=n_0}^{n}(1+z_n)$ and $\prod_{j=n_0}^{\infty}(1+z_n)$ are contained in the domain $\Delta(1;1)$ where the logarithmic function log is defined (cf. (2.5.5)), so that

$$\begin{aligned}\sum_{j=n_0}^{n} \log(1+z_j) &= \log\left(\prod_{j=n_0}^{n}(1+z_j)\right)\\ &\to \log \prod_{j=n_0}^{\infty}(1+z_j) \qquad (n\to\infty).\end{aligned}$$

On the other hand, if $|z_n| < 1$ for $n \geqq n_0$ for some n_0 and $\sum_{j=n_0}^{\infty} \log(1+z_j)$ converges, then

$$1 + z_n = e^{\log(1+z_n)} \to e^0 = 1 \qquad (n \to \infty).$$

Hence we have $\lim_{n\to\infty} z_n = 0$ and

$$\begin{aligned}\prod_{j=n_0}^{n}(1+z_n) &= \exp\left(\sum_{j=n_0}^{n}\log(1+z_j)\right)\\ &\to \exp\left(\sum_{j=n_0}^{\infty}\log(1+z_j)\right) \neq 0 \ (n\to\infty).\end{aligned} \tag{2.6.1}$$

Therefore the infinite product $\prod_{n=0}^{\infty}(1+z_n)$ converges. Thus we have proved the following theorem.

(2.6.2) THEOREM. *An infinite product* $\prod_{n=0}^{\infty}(1+z_n)$ *converges if and only if there is a number* n_0 *such that* $|z_n| < 1$ *for* $n \geqq n_0$ *and* $\sum_{n=n_0}^{\infty} \log(1+z_n)$ *converges.*

Given this fact, we define the *absolute convergence* of an infinite product $\prod_{n=0}^{\infty}(1+z_n)$ by the convergence of

$$\sum_{j=n_0}^{\infty} |\log(1+z_j)| < +\infty$$

for some number n_0. It follows from (2.5.3) that

$$
\begin{aligned}
|\log(1+z)| &= \left| z \sum_{n=0}^{\infty} \frac{(-1)^n}{n+1} z^n \right| \\
&= |z| \cdot \left| 1 + \frac{1}{2} \sum_{n=1}^{\infty} \frac{2}{n+1} (-1)^n z^n \right| .
\end{aligned}
$$

Since for $|z| \leqq 1/2$

$$
\left| \frac{1}{2} \sum_{n=1}^{\infty} \frac{2}{n+1} (-1)^n z^n \right| \leqq \frac{1}{4} \sum_{n=0}^{\infty} \left(\frac{1}{2} \right)^n = \frac{1}{2},
$$

we deduce that

$$
\frac{1}{2}|z| \leqq |\log(1+z)| \leqq \frac{3}{2}|z|, \qquad z \in \Delta\left(\tfrac{1}{2}\right). \tag{2.6.3}
$$

As in Theorem (1.3.8), iv), we consider an infinite product $\prod_{n=0}^{\infty}(1 + z_{\lambda(n)})$ of order change of an infinite product $\prod_{n=0}^{\infty}(1 + z_n)$.

(2.6.4) THEOREM. i) *An absolutely convergent infinite product converges.*

ii) *An infinite product of order change of an absolutely convergent infinite product again converges absolutely, and the limit does not change.*

iii) *An infinite product* $\prod_{n=0}^{\infty}(1 + z_n)$ *converges absolutely if and only if* $\sum_{n=0}^{\infty} z_n$ *converges absolutely.*

PROOF. i) and ii) follow easily from (2.6.1) and Theorem (1.3.8), iii), iv). iii) is clear by the definition and (2.6.3). □

Let $\{f_n\}_{n=0}^{\infty}$ be a series of functions defined over a subset $A \subset \mathbf{C}$. Then the formal product $\prod_{n=0}^{\infty} f_n$ is called an *infinite product of functions.* If $\prod_{n=0}^{\infty} f_n(z)$ converges at $z \in A$, then $\prod_{n=0}^{\infty} f_n$ is said to converge at the point z. If the product converges at all points of A, it is said to converge on A, and the limit function is also denoted by $\prod_{n=0}^{\infty} f_n$. We say that $\prod_{n=0}^{\infty} f_n$ converges uniformly on A if the following conditions are satisfied:

(2.6.5) i) There is a number n_0 such that $f_n(z) \neq 0$ for all $n \geqq n_0$ and $z \in A$.

ii) $\left\{ \prod_{n=n_0}^{n_0+m} f_n \right\}_{m=0}^{\infty}$ converges uniformly on A to a non-vanishing function g such that there is a constant $\delta > 0$ with $|g| \geqq \delta$.

If every f_n is continuous in A and $\prod_{n=0}^{\infty} f_n$ converges uniformly on A, then the limit function is continuous there.

(2.6.6) THEOREM. *An infinite product* $\prod_{n=0}^{\infty} f_n$ *of functions on* A *converges uniformly on* A *if and only if for an arbitrary* $\epsilon > 0$ *there is a number* n_0 *such that*

$$
\left| \prod_{j=m}^{n} f_j(z) - 1 \right| < \epsilon
$$

for all $n \geqq m \geqq n_0$ and $z \in A$.

PROOF. Noting that $f_n \to 1$ $(n \to \infty)$ uniformly, we make use of the same arguments as in Theorem (1.3.11), ii). □

For an infinite product $\prod_{n=0}^{\infty} f_n$ of functions on an open set, we define its uniform convergence on compact subsets as in the cases of sequences and series of functions. In this case, if every f_n is continuous, then the limit function is continuous. By the definition and (2.6.3) we have

(2.6.7) THEOREM. *Let $\prod_{n=0}^{\infty}(1+f_n)$ be an infinite product of functions on A. Then $\prod_{n=0}^{\infty}(1+f_n)$ converges absolutely if and only if the series $\sum_{n=0}^{\infty}|f_n|$ converges. Moreover, if $\sum_{n=0}^{\infty}|f_n|$ converges uniformly, then so does $\prod_{n=0}^{\infty}(1+f_n)$.*

EXERCISE 1. Prove the above theorem.

EXERCISE 2. Show that the infinite product $\prod_{n=1}^{\infty}\left(1-\frac{z}{n^2}\right)$ converges uniformly on compact subsets of $\mathbf{C}$.

2.7. Riemann Sphere

A bounded closed set of the complex plane $\mathbf{C}$ is compact, but $\mathbf{C}$ itself is not compact. On the other hand, it might be convenient to make analytic arguments dealing with sequences or sequences of functions if we can choose a convergent subsequence of them. Let us also consider the following simple function:

$$f_0(z) = \frac{1}{z}. \tag{2.7.1}$$

By the definition of complex numbers, f_0 is not defined at $z = 0$, but is continuous outside $\{0\}$, that is, in $\mathbf{C}^* = \mathbf{C} \setminus \{0\}$. The mapping

$$f_0 : \mathbf{C}^* \to \mathbf{C}^* \tag{2.7.2}$$

is injective and surjective, and the inverse function $f_0^{-1} = f_0$ is continuous, too.

EXERCISE 1. Show that f_0 is an analytic function in $\mathbf{C}$.

As $z \to 0$, $f_0(z)$ does not have a limit in $\mathbf{C}$. It might be useful to have a compact space that contains $\mathbf{C}$ and such that $\lim_{z\to 0} f_0(z)$ is defined. The following concept of the *Riemann sphere* was invented with this in mind.

Let (T_1, T_2, T_3) be the standard coordinate system of the real 3-dimensional euclidean space $\mathbf{R}^3$. Let $\mathbf{S}$ be the unit sphere with center at the origin:

$$\mathbf{S} = \{\mathrm{P} = (T_1, T_2, T_3) \in \mathbf{R}^3; T_1^2 + T_2^2 + T_3^2 = 1\}.$$

We identify $\mathbf{C}$ with the hyperplane $\{T_3 = 0\}$ of $\mathbf{R}^3$ through

$$\mathbf{C} \ni z = x + iy \longleftrightarrow (x, y, 0) \in \mathbf{R}^3.$$

The point $\mathrm{N} = (0,0,1) \in \mathbf{S}$ is called the *north pole* and $\mathrm{S} = (0,0,-1) \in \mathbf{S}$ the *south pole*. Take a point $\mathrm{P} = (T_1, T_2, T_3) \neq \mathrm{N}$. The line passing through

N and P intersects $\mathbf{C}$ at a unique point $z = x + iy = (x, y, 0)$. The mapping $\mathrm{P} \to z$ is called the *stereographic projection* (from the north pole N). By simple calculations we have

$$z = z(\mathrm{P}) = \frac{T_1}{1 - T_3} + i\frac{T_2}{1 - T_3}, \tag{2.7.3}$$

$$T_1 = \frac{z + \bar{z}}{1 + |z|^2}, \quad T_2 = \frac{z - \bar{z}}{i(1 + |z|^2)}, \quad T_3 = \frac{|z|^2 - 1}{|z|^2 + 1}.$$

Henceforth $\mathbf{C}$ is identified with $\mathbf{S} \setminus \{\mathrm{N}\}$, and z (resp., P) is continuous in P (resp., z). Through this identification, the topology of $\mathbf{C}$ is the same as $\mathbf{S} \setminus \{\mathrm{N}\}$ (that is, convergent sequences of $\mathbf{C}$ converge as sequences of $\mathbf{S} \setminus \{\mathrm{N}\}$, the converse holds, too, and the limits are equal to those corresponding points).

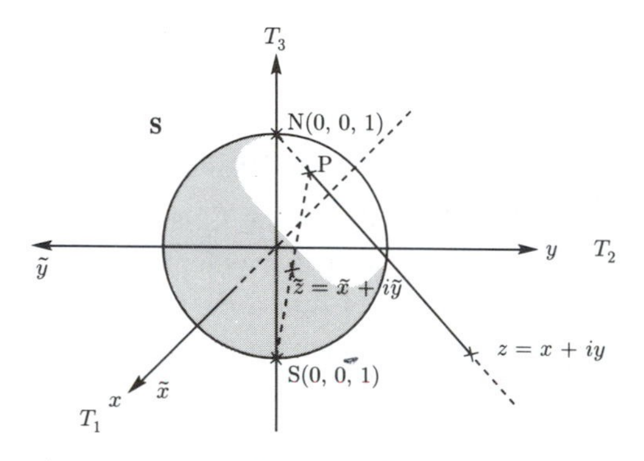

FIGURE 15

The sphere $\mathbf{S}$ is a bounded closed subset of $\mathbf{R}^3$, and hence compact. We write $\mathrm{N} = \infty$:

$$\mathbf{S} = \mathbf{C} \cup \{\infty\}.$$

The point ∞ is called the *infinity*, or the *point at infinity*. Let $\{z_n\}_{n=0}^{\infty}$ be a sequence of $\mathbf{C}$. If it is bounded, it contains a subsequence converging to a point of $\mathbf{C}$. If it is unbounded, we can take a subsequence $\{z_{n_\nu}\}_{\nu=0}^{\infty}$ of $\{z_n\}_{n=0}^{\infty}$ such that $\lim_{\nu \to \infty} |z_{n_\nu}| = +\infty$. This $\{z_{n_\nu}\}_{\nu=0}^{\infty}$ converges to N in $\mathbf{S}$. A priori, we have the complex coordinate z in the subset $\mathbf{C}$ of $\mathbf{S}$. We want to introduce a complex coordinate in a neighborhood of ∞ as well. We consider the stereographic projection from S similar to the one above. Take a point $\mathrm{P} = (T_1, T_2, T_3) \in \mathbf{S} \setminus \{\mathrm{S}\}$, and let $(x', y', 0)$ be the point at which the line passing through S and P intersects the hyperplane $T_3 = 0$. In the same way as (2.7.3) we get

$$x' = \frac{T_1}{1 + T_3}, \qquad y' = \frac{T_2}{1 + T_3}. \tag{2.7.4}$$

Take a new complex plane $\mathbf{C}$ with complex coordinate $\tilde{z} = \tilde{x} + i\tilde{y}$, and set

$$\tilde{x} = x', \qquad \tilde{y} = -y'.$$

(This is natural because of the orientation of the stereographic projection.) We assign P the complex number

$$\tilde{z} = \tilde{z}(\mathrm{P}) = \frac{T_1}{1+T_3} + i\frac{-T_2}{1+T_3}.$$

By computation we obtain

$$z(\mathrm{P})\tilde{z}(\mathrm{P}) = 1, \qquad \mathrm{P} \in \mathbf{S} \setminus \{\mathrm{N}, \mathrm{S}\}. \tag{2.7.5}$$

It follows that $\tilde{z}(\infty) = 0$. The function f_0 in (2.7.2) is extended over $\mathbf{S}$ to a continuous mapping into $\mathbf{S}$ by setting $f_0(0) = \infty$ and $f_0(\infty) = 0$. We use the complex coordinate $\tilde{z}$ in neighborhoods of ∞. A neighborhood $\{\tilde{z} \in \mathbf{C}; |\tilde{z}| < r\}$ with $r > 0$ is called a *disk neighborhood* of ∞. The subspace $\mathbf{S}\setminus\{\mathrm{N},\mathrm{S}\}$ is identified with $\mathbf{C}^*$, which carries two complex coordinates z and $\tilde{z}$, and by (2.7.2) both coordinates give the same topology. The space $\mathbf{S}$, whose subspace $\mathbf{C}\setminus\{\mathrm{N}\}$ (resp., $\mathbf{C} \setminus \{\mathrm{S}\}$) is assigned the complex coordinate z (resp., $\tilde{z}$) is called the *Riemann sphere*, and denoted by $\widehat{\mathbf{C}}$.

The statements on the topology of $\mathbf{C}$ such as the convergence of sequences, accumulation points, closures, closed sets, open sets, connectedness, domains, continuous functions and mappings, etc., continue to hold on $\widehat{\mathbf{C}}$. For instance, let $E_j \subset \widehat{\mathbf{C}}$, $j = 1, 2$, be subsets, and let $f : E_1 \to E_2$ be a mapping. Then f is said to be continuous if for an arbitrary point $\mathrm{P} \in E_1$ and an arbitrary sequence $\{\mathrm{P}_\nu\}_{\nu=0}^{\infty}$ in E_1 converging to P, $\{f(\mathrm{P}_\nu)\}_{\nu=0}^{\infty}$ converges to $f(\mathrm{P})$, i.e., $\lim_{\nu\to\infty} f(\mathrm{P}_\nu) = f(\mathrm{P})$. In particular, if f is continuous, injective, and surjective, and if the inverse $f^{-1} : E_2 \to E_1$ is also continuous, then f is called a *homeomorphism*. For example, f_0 in (2.7.2) is a homeomorphism.

Since the composition of analytic functions is analytic (see Remark (3.5.9) in the next chapter), it follows from (2.7.5) that if a function in an open subset of $\widehat{\mathbf{C}} \setminus \{0, \infty\}$ is analytic in z, it is analytic in $\tilde{z}$, and the converse holds, too. Therefore the notion of analyticity is well defined on an arbitrary open subset of $\widehat{\mathbf{C}}$.

A *circle* of $\widehat{\mathbf{C}}$ stands for a circle, or a line of $\mathbf{C}$ (plus the point at infinity ∞ to be precise). Here a line of $\mathbf{C}$ is considered as a circle passing through ∞.

EXERCISE 2. Let P (resp., P′) be the point of $\mathbf{S}$ corresponding to $z \in \mathbf{C}$ (resp., $z' \in \mathbf{C}$) by the stereographic projection from N. Express the distance $d(\mathrm{P}, \mathrm{P}')$ between P and P′ in $\mathbf{R}^3$ in terms of z and z'. Moreover, express the distance $d(\mathrm{P}, \mathrm{N})$ in terms of z.

EXERCISE 3. Show that under the stereographic projection circles in $\widehat{\mathbf{C}}$ correspond to circles in $\mathbf{S}$ which are intersections of $\mathbf{S}$ and hyperplanes in $\mathbf{R}^3$, and vise versa. (This fact is referred to as the correspondence of circle to circle under stereographic projection.)

2.8. Linear Transformations

A mapping f from $\widehat{\mathbf{C}}$ onto itself written as

$$f(z) = \frac{az+b}{cz+d} \tag{2.8.1}$$

is called a linear fractional transformation, a *linear transformation*, or a *Möbius transformation*. Here the coefficients $a, b, c, d \in \mathbf{C}$ must satisfy

$$ad - bc \neq 0.$$

If $c = 0$, $f(\infty) = \infty$, and otherwise

$$f(\infty) = \frac{a}{c}, \qquad f\left(-\frac{d}{c}\right) = \infty.$$

The mapping f has the inverse f^{-1}:

$$f^{-1}(w) = \frac{dw - b}{-cw + a}$$

which is a linear transformation. In particular, $f : \widehat{\mathbf{C}} \to \widehat{\mathbf{C}}$ is a homeomorphism. The multiplication of $a, b, c,$ and d by a common non-zero number does not change the linear transformation $f(z)$, and hence we may assume that

$$ad - bc = 1. \tag{2.8.2}$$

The set of all 2×2 complex matrices $\left(\begin{smallmatrix} a & b \\ c & d \end{smallmatrix}\right)$ satisfying (2.8.2) is denoted by $SL(2, \mathbf{C})$.

In general, a set G endowed with an operation $a \cdot b$ $(a, b \in G)$ satisfying the following conditions is called a *group*:

i) Two arbitrary elements $a, b \in G$ determine a third element $a \cdot b \in G$, and $(a \cdot b) \cdot c = a \cdot (b \cdot c)$ holds for any three elements $a, b, c \in G$.
ii) There is an element $e \in G$ such that $a \cdot e = e \cdot a = a$ for all $a \in G$. (This e is unique and is called the unit element.)
iii) For an arbitrary $a \in G$ there is an element $b \in G$ such that $ab = ba = e$. (This b is unique; it is called the inverse element of a and is denoted by a^{-1}.)

EXERCISE 1. Show the uniqueness asserted in ii) and iii).

The set $SL(2, \mathbf{C})$ forms a group under matrix multiplication, and is called the *special linear group*. By a direct computation we see that the composition of two linear transformations is also a linear transformation. It is easily seen that this composition satisfies the above conditions so that the set $\mathrm{Aut}(\widehat{\mathbf{C}})$ of all linear transformations is a group. The unit element of $\mathrm{Aut}(\widehat{\mathbf{C}})$ is the identity mapping.

(2.8.3) THEOREM. *The mapping*

$$\Phi : \begin{pmatrix} a & b \\ c & d \end{pmatrix} \in SL(2, \mathbf{C}) \to \frac{az+b}{cz+d} \in \mathrm{Aut}(\widehat{\mathbf{C}})$$

is a surjective group homomorphism (i.e., $\Phi(\alpha \cdot \beta) = \Phi(\alpha) \cdot \Phi(\beta)$*). The kernel* Ker Φ $(= \Phi^{-1}$*(the unit element of* $\widehat{\mathbf{C}}$*)) is as follows:*

$$\mathrm{Ker}\ \Phi = \left\{ \pm \begin{pmatrix} 1 & 0 \\ 0 & 1 \end{pmatrix} \right\}.$$

PROOF. The first half follows from a direct computation. For the latter half we use the identity

$$\frac{az+b}{cz+d} = z$$

and (2.8.2). Then we have $b = c = 0$, $ad = 1$, and $a/d = 1$. Therefore $a = d = 1$ or $a = d = -1$. □

Identifying $\alpha \in SL(2, \mathbf{C})$ and $-\alpha \in SL(2, \mathbf{C})$, we have the following group:

$$SL(2, \mathbf{C}) / \left\{ \pm \begin{pmatrix} 1 & 0 \\ 0 & 1 \end{pmatrix} \right\}.$$

By Theorem (2.8.3) the group $\mathrm{Aut}(\widehat{\mathbf{C}})$ is written as

$$\mathrm{Aut}(\widehat{\mathbf{C}}) = SL(2, \mathbf{C}) / \left\{ \pm \begin{pmatrix} 1 & 0 \\ 0 & 1 \end{pmatrix} \right\}.$$

The right-hand side of this equality is often denoted by $PSL(2, \mathbf{C})$, and called the *projective special linear group.*

(2.8.4) THEOREM. *For two triples,* (z_1, z_2, z_3) *and* (w_1, w_2, w_3)*, of distinct points of* $\widehat{\mathbf{C}}$*, there is a unique linear transformation* f *such that* $f(z_i) = w_i$*,* $i = 1, 2, 3$.

PROOF. We first show existence. If $z_1 = \infty$, we set $f_1(z) = 1/z$; otherwise, we set $f_1(z) = z - z_1$. Then $f_1(z_1) = 0$. If $f_1(z_3) = \infty$, we set $f_2(z) = z$; otherwise we set $f_2(z) = z/(z - f_1(z_3))$. Then $f_2 \circ f_1(z_1) = 0$, and $f_2 \circ f_1(z_3) = \infty$. Moreover, we set $f_3(z) = z/f_2 \circ f_1(z_2)$ and $f = f_3 \circ f_2 \circ f_1$. Then f is a linear transformation satisfying $f(z_1) = 0$, $f(z_2) = 1$, and $f(z_3) = \infty$, Similarly, there is a linear transformation g satisfying $g(w_1) = 0$, $g(w_2) = 1$, and $g(w_3) = \infty$. Then $g^{-1} \circ f$ is the required linear transformation.

For uniqueness it suffices to show that if a linear transformation f satisfies $f(0) = 0$, $f(1) = 1$, and $f(\infty) = \infty$, then $f(z) = z$. This is easy. □

If $c = 0$, the linear transformation (2.8.1) gives rise to

$$f(z) = \frac{a}{d} z + \frac{b}{d};$$

otherwise,

$$f(z) = \frac{a}{c} - \frac{ad - bc}{c^2} \cdot \frac{1}{z + d/c}.$$

Therefore an arbitrary linear transformation is represented by a composition of the following linear transformations of three kinds:

$$
\begin{aligned}
&\text{i)} \quad f(z) = z + b, && b \in \mathbf{C} && \text{(translation)}.\\
(2.8.5) \qquad &\text{ii)} \quad f(z) = az, && a \in \mathbf{C}^* && \text{(non-zero multiplication)}.\\
&\text{iii)} \quad f(z) = \frac{1}{z} && && \text{(inversion)}.
\end{aligned}
$$

A circle of $\widehat{\mathbf{C}}$ is written by

$$
(2.8.6) \qquad \frac{|z - z_1|}{|z - z_2|} = k
$$

with distinct $z_1, z_2 \in \mathbf{C}$ and $k > 0$, which is the so-called Appollonius' circle.

EXERCISE 2. Show that if $k = 1$, equation (2.8.6) presents a line passing through $(z_1 + z_2)/2$ and perpendicular to the line passing through z_1 and z_2, and that if $k \neq 1$ it presents a circle of $\mathbf{C}$ with center at $(z_1 - k^2 z_2)/(1 - k^2)$ and with radius $|(z_1 - z_2)k/(1 - k^2)|$.

Substituting a linear transformation $z = f^{-1}(w)$ of (2.8.5), i)$\sim$iii) into (2.8.6), we obtain

$$
(2.8.7) \qquad \frac{|w - f(z_1)|}{|w - f(z_2)|} = k' \quad (> 0).
$$

This defines a circle. Thus circles of $\widehat{\mathbf{C}}$ are mapped to circles of $\widehat{\mathbf{C}}$ by linear transformations. This property of linear transformations is called the *correspondence of circle to circle.* Take a circle $C(a; r)$. If two points $z_1, z_2 \in \mathbf{C}$ satisfy

$$
(2.8.8) \qquad (z_1 - a)\overline{(z_2 - a)} = r^2,
$$

then z_1 is called the *reflection* of z_2 with respect to the circle $C(a; r)$, and vise versa. We also say that z_1 and z_2 are *mutual reflections* with respect to $C(a; r)$. We define the center a and ∞ to be mutual reflections. In particular, a point of $C(a; r)$ is the reflection of itself with respect to $C(a; r)$. The points z_1 and z_2 in (2.8.6) are mutual reflections with respect to the circle defined by (2.8.6). This is verified by Exercise 2 and a direct computation.

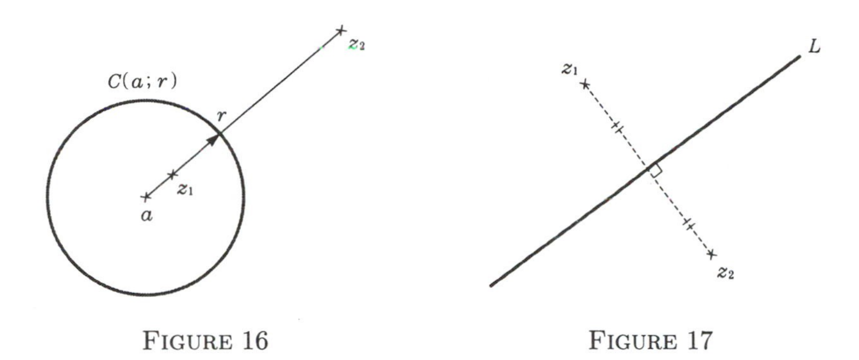

FIGURE 16 FIGURE 17

A line L of $\mathbf{C}$ was defined to be a circle of $\widehat{\mathbf{C}}$. If two points $z_1, z_2 \in \mathbf{C}$ are mirror symmetric with respect to L, then they are said to be *mutual reflections* with respect to L. The points on L, including ∞, are reflections of themselves with respect to L. Therefore one deduces the following *reflection principle.*

(2.8.9) THEOREM. *Let f be a linear transformation, C a circle of $\widehat{\mathbf{C}}$, and z_1 the reflection of z_2 with respect to C. Then $f(z_1)$ and $f(z_2)$ are mutual reflections with respect to $f(C)$.*

Note that two points $z_1, z_2 \in \mathbf{C} \setminus L$ are mutual reflections with respect to a line L of $\mathbf{C}$ if and only if L is defined by (2.8.6) with $k = 1$. The same holds for a circle $C(a; r)$. That is, two points $z_1, z_2 \in \mathbf{C} \setminus C(a; r)$ are mutual reflections with respect to a circle $C(a; r)$ if and only if $C(a; r)$ is defined by (2.8.6) with suitable $a \in \mathbf{C}$ and $r > 0$.

In general, a subset F of a group G is called a *subgroup* if F itself is a group under the operation of G. The unit element is always shared by all subgroups of G. For example, $\{\pm \left(\begin{smallmatrix} 1 & 0 \\ 0 & 1 \end{smallmatrix}\right)\}$ is a subgroup of $SL(2, \mathbf{C})$.

If a linear transformation f satisfies $f(\Delta(1)) = \Delta(1)$, f is said to preserve $\Delta(1)$. We denote the set of all those f by $\mathrm{Aut}(\Delta(1))$, which is a subgroup of $\mathrm{Aut}(\widehat{\mathbf{C}})$.

We determine the type of $f \in \mathrm{Aut}(\Delta(1))$. For $a \in \Delta(1)$ we look for $\phi_a \in \mathrm{Aut}(\Delta(1))$ such that $\phi_a(C(0; 1)) = C(0; 1)$, $\phi_a(\Delta(1)) = \Delta(1)$, and $\phi_a(a) = 0$. The reflection of a with respect to $C(0; 1)$ is $1/\bar{a}$, and so $\phi_a(1/\bar{a}) = \infty$. Hence for a candidate we set

$$\phi_a(z) = \frac{z - a}{-\bar{a}z + 1}. \tag{2.8.10}$$

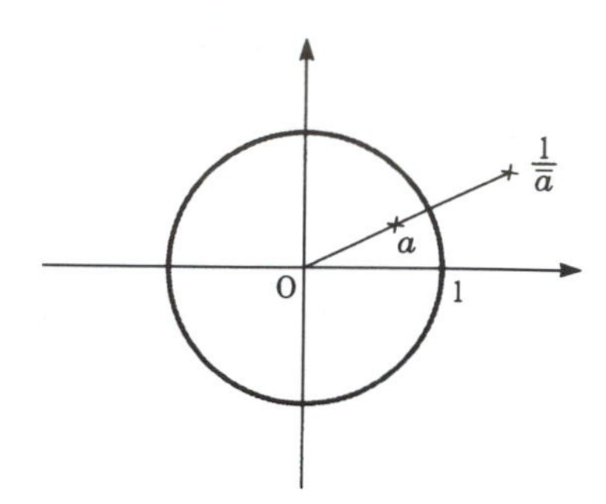

FIGURE 18

In fact, $\phi_a(a) = 0$, and, if $|z| = 1$,

$$\left|\frac{z-a}{-\bar{a}z+1}\right| = \frac{|z-a|}{|\bar{a}-\bar{z}|} = 1.$$

Furthermore, we see that

$$|z| < 1 \Longleftrightarrow 1 - \frac{|z-a|^2}{|\bar{a}z-1|^2} = \frac{(1-|a|^2)(1-|z|^2)}{|\bar{a}z-1|^2} > 0.$$

It follows that $\phi_a(\Delta(1)) = \Delta(1)$. Note that $\phi_a^{-1} = \phi_{-a}$. Let a linear transformation f preserve $\Delta(1)$. Then $g = f \circ \phi_{f^{-1}(0)}^{-1}$ is a linear transformation, and $g(0) = 0$. Since the reflection point of 0 with respect to $C(0;1)$ is ∞, $g(z) = \alpha z$ with some $\alpha \in \mathbf{C}^*$. Since $g(C(0;1)) = C(0;1)$, $|\alpha| = 1$, so that $\alpha = e^{i\theta}$ $(\theta \in \mathbf{R})$. Therefore

$$f(z) = e^{i\theta}\frac{z-a}{-\bar{a}z+1}, \qquad a = f^{-1}(0).$$

(2.8.11) THEOREM. i) *A linear transformation* $f \in \mathrm{Aut}(\Delta(1))$ *is written as*

$$f(z) = e^{i\theta}\frac{z-a}{-\bar{a}z+1}, \qquad a \in \Delta(1),\ \theta \in \mathbf{R}.$$

ii) *For two arbitrary points* $\alpha, \beta \in \Delta(1)$, *there is an* $f \in \mathrm{Aut}(\Delta(1))$ *such that* $f(\alpha) = \beta$.

PROOF. The first half is already proved. The latter half follows from $f = \phi_\beta^{-1} \circ \phi_\alpha$. □

Property ii) of the above theorem is referred to by saying $\mathrm{Aut}(\Delta(1))$ acts *transitively* on $\Delta(1)$.

Set

$$\mathbf{H} = \{z \in \mathbf{C}; \mathrm{Im}\, z > 0\}.$$

$\mathbf{H}$ is called the *upper half plane*. Next we find a linear transformation $\psi \in \mathrm{Aut}(\widehat{\mathbf{C}})$ such that $\psi(\mathbf{H}) = \Delta(1)$.

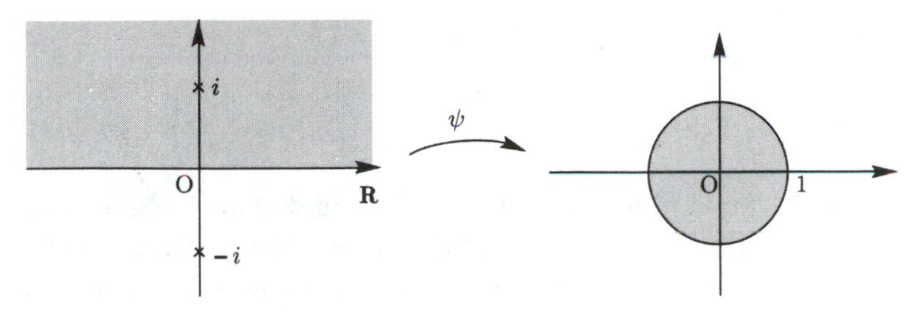

FIGURE 19

We set a condition, $\psi(i) = 0$. Since $\psi(\mathbf{R} \cup \{\infty\}) = C(0;1)$, it follows from Theorem (2.8.9) that $\psi(-i) = \infty$. Hence for a candidate we set

$$\psi(z) = \frac{z-i}{z+i}. \tag{2.8.12}$$

By easy computations $|\psi(z)| = 1$ for $z \in \mathbf{R}$, and $|\psi(z)| < 1$ if and only if $z \in \mathbf{H}$.

Let $\mathrm{Aut}(\mathbf{H})$ denote the set of all linear transformations preserving $\mathbf{H}$. Then $\mathrm{Aut}(\mathbf{H})$ is a subgroup of $\mathrm{Aut}(\widehat{\mathbf{C}})$. For an element $f \in \mathrm{Aut}(\mathbf{H})$ there is a $g \in \mathrm{Aut}(\Delta(1))$ such that $f = \psi^{-1} \circ g \circ \psi$. From this and Theorem (2.8.11) we deduce that $\mathrm{Aut}(\mathbf{H})$ acts transitively on $\mathbf{H}$. Since $\{f(0), f(1), f(\infty)\} \subset \mathbf{R} \cup \{\infty\}$, we see that

$$f(z) = \frac{az+b}{cz+d}, \qquad ad - bc = 1,$$

where $a, b, c, d \in \mathbf{R}$. Hence $f(\mathbf{R} \cup \{\infty\}) = \mathbf{R} \cup \{\infty\}$ and

$$\mathrm{Im}\ z > 0 \Longleftrightarrow \mathrm{Im}\ f(z) = \frac{1}{2i}\left\{\frac{az+b}{cz+d} - \frac{a\bar{z}+b}{c\bar{z}+d}\right\} = \frac{\mathrm{Im}\ z}{|cz+d|^2} > 0.$$

It follows that $f(\mathbf{H}) = \mathbf{H}$. Let $SL(2;\mathbf{R})$ denote the set of all real matrices of $SL(2,\mathbf{C})$. We hence obtain the following theorem.

(2.8.13) THEOREM. i) *The mapping*

$$\Phi : \left(\begin{smallmatrix} a & b \\ c & d \end{smallmatrix}\right) \in SL(2,\mathbf{R}) \to f(z) = \frac{az+b}{cz+d} \in \mathrm{Aut}(\mathbf{H})$$

is a surjective group homomorphism, and $\mathrm{Ker}\ \Phi = \{\pm \left(\begin{smallmatrix} 1 & 0 \\ 0 & 1 \end{smallmatrix}\right)\}$.

ii) $\mathrm{Aut}(\mathbf{H})$ *acts transitively on* $\mathbf{H}$.

We have the following expression:

$$\mathrm{Aut}(\mathbf{H}) = SL(2,\mathbf{R}) / \left\{ \pm \begin{pmatrix} 1 & 0 \\ 0 & 1 \end{pmatrix} \right\}.$$

The right-hand side of the above equation is denoted by $PSL(2, \mathbf{R})$. The group $SL(2, \mathbf{R})$ carries subgroups such as

$$SL(2, \mathbf{Z}) = \left\{ \begin{pmatrix} a & b \\ c & d \end{pmatrix}; \ a, b, c, d \in \mathbf{Z}, \ ad - bc = 1 \right\},$$

$$\Gamma(n) = \left\{ \begin{pmatrix} a & b \\ c & d \end{pmatrix} \in SL(2, \mathbf{Z}); \ \begin{pmatrix} a & b \\ c & d \end{pmatrix} \equiv \begin{pmatrix} 1 & 0 \\ 0 & 1 \end{pmatrix} \pmod{n} \right\} \quad (n \geqq 2).$$

Here $\left(\begin{smallmatrix} a & b \\ c & d \end{smallmatrix}\right) \equiv \left(\begin{smallmatrix} 1 & 0 \\ 0 & 1 \end{smallmatrix}\right) \pmod{n}$ means that $a-1, b, c,$ and $d-1$ are integral multiples of n. The subgroup $\Gamma(n)$ or $\Gamma(n)/\left\{\pm \left(\begin{smallmatrix} 1 & 0 \\ 0 & 1 \end{smallmatrix}\right)\right\}$ is called the *principal congruence subgroup*, and in particular $\Gamma(1) = SL(2, \mathbf{Z})$ or $PSL(2, \mathbf{Z})$ the *modular group*. These groups play important roles in the theory of elliptic functions and the theory of automorphic forms.

EXERCISE 3. Show that $\Gamma(n)$ is a subgroup of $SL(2, \mathbf{R})$.

Problems

1. Compute the limits $\displaystyle\lim_{z \to 0} \frac{\sin z}{z}$, $\displaystyle\lim_{z \to 0} \frac{e^z - 1}{z}$, and $\displaystyle\lim_{z \to 0} \frac{\log(1+z)}{z}$.

2. Let $f(z) = (z^n - 1)/(z - 1)$. Find the supremum and the infimum of $|f(z)|$ on $\Delta(1)$.

3. Let $f_n(x) = \sin nx$, $n = 1, 2, \ldots$, be a sequence of functions of the real variable $x \in [0, 2\pi]$. Then $\{f_n(x)\}_{n=1}^{\infty}$ is uniformly bounded. Prove that any subsequence of $\{f_n(x)\}_{n=1}^{\infty}$ does not converges on $[0, 2\pi]$. (This requires a knowledge of Lebesgue integration. If the reader finds it too difficult, see the answer in the back of the book.)

4. Find the radii of convergence of the following power series:

i) $\displaystyle\sum_{n=0}^{\infty} n^p z^n \quad (p > 0)$. ii) $\displaystyle\sum_{n=0}^{\infty} q^{n^2} z^n \quad (|q| < 1)$.

iii) $\displaystyle\sum_{n=0}^{\infty} (n!)^{1/n} z^n$. iv) $\displaystyle 1 + \sum_{n=0}^{\infty} z^n \prod_{\nu=0}^{n-1} \frac{(\alpha+\nu)(\beta+\nu)}{(1+\nu)(\gamma+\nu)}$.
(Gauss' hypergeometric function)

5. (*Stolz' domain and Abel's continuity theorem*) Let $\zeta \in C(0; 1)$, and let $G(\zeta; \tau)$ denote a subdomain of $\Delta(1)$ hedged by two line segments passing through ζ and forming an angle $0 < \tau < \pi/2$ with the line passing through 0 and ζ. We call $G(\zeta; \tau)$ *Stolz' domain* with vertex ζ.

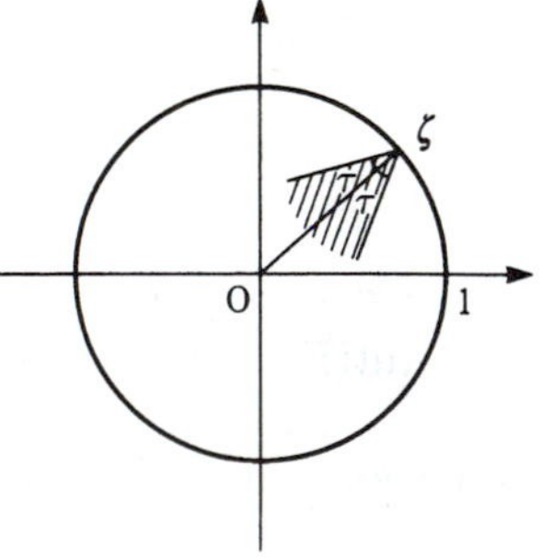

(a) Show that

$$G(\zeta; \tau) \cap \Delta(\zeta; \cos\tau) \subset \left\{ z \in \Delta(1); \frac{|z - \zeta|}{1 - |z|} < \frac{2}{\cos\tau} \right\} \cap \Delta(\zeta; \cos\tau).$$

On the other hand, for any given $K > 1$, take $\tau = \cos^{-1}\frac{1}{2K} \in (0, \pi/2)$. Then, show that

$$\left\{z \in \Delta(1); \frac{|z-\zeta|}{1-|z|} < K\right\} \cap \Delta(\zeta; 2\cos\tau) \subset G(\zeta;\tau) \cap \Delta(\zeta; 2\cos\tau).$$

(b) Let $f(z) = \sum_{n=0}^{\infty} a_n z^n$ be a power series converging on $\Delta(1)$. Prove that if $\sum_{n=0}^{\infty} a_n \zeta^n$ converges, then

$$\lim_{\substack{z\to\zeta \\ z\in G(\zeta;\tau)}} f(z) = \sum_{n=0}^{\infty} a_n \zeta^n.$$

6. Show that $\mathrm{Aut}(\Delta(1))$ is equicontinuous on compact subsets of $\Delta(1)$ as a family of functions. (The uniform boundedness is clear.)

7. Show that $\tan z = \dfrac{\sin z}{\cos z}$ is a function on $\mathbf{C} \setminus \{2n + \pi/2; n \in \mathbf{Z}\}$ with fundamental period π.

8. Show that if $\alpha > 1$, $\displaystyle\prod_{n=1}^{\infty}\left(1 - \frac{z}{n^\alpha}\right)$ converges uniformly on compact subsets of $\mathbf{C}$.

9. Show that $\displaystyle\prod_{n=1}^{\infty}(1 - z^n)$ converges absolutely and uniformly on compact subsets of $\Delta(1)$.

10. Show that $\displaystyle\prod_{n=0}^{\infty}(1 + z^{2n}) = \frac{1}{1-z}$.

11. The *cross ratio* of four points $z_1, \ldots, z_4$ of $\widehat{\mathbf{C}}$ is defined by

$$(z_1, z_2, z_3, z_4) = \frac{(z_1 - z_3)(z_2 - z_4)}{(z_1 - z_4)(z_2 - z_3)}.$$

Here, if some $z_j = \infty$, the above right hand side is supposed to represent the limit as $z_j \to \infty$. Let $f \in \mathrm{Aut}(\widehat{\mathbf{C}})$ be a linear transformation and $z_1, z_2, z_3 \in \widehat{\mathbf{C}}$ such that $f(z_2) = 1, f(z_3) = 0$, and $f(z_4) = \infty$. Show that

$$f(z) = (z, z_2, z_3, z_4).$$

12. Show that for any linear transformation f and four points $z_1, \ldots, z_4$ of $\widehat{\mathbf{C}}$

$$(z_1, z_2, z_3, z_4) = (f(z_1), f(z_2), f(z_3), f(z_4))$$

(the invariance of the cross ratio).

13. Let $f(z) = (az + b)/(cz + d) \not\equiv z$, $ad - bc = 1$, be a linear transformation. Prove that if $a + d = \pm 2$, then f has one *fixed point* z (a point z such that $f(z) = z$); otherwise, f has two fixed points.

14. Let α and β be the above fixed points of f. Show that if $\alpha \neq \beta$, $w = f(z)$ is given by

$$\frac{w-\alpha}{w-\beta} = Ke^{i\theta}\frac{z-\alpha}{z-\beta}, \qquad K > 0,\ \theta \in \mathbf{R}.$$

The linear transformation f is said to be *hyperbolic* if $e^{i\theta} = 1$, *elliptic* if $K = 1$, and *loxodromic* otherwise.

15. Let α and β be as above. Suppose that $\alpha = \beta$. Show that $w = f(z)$ is given by

$$\frac{1}{w-\alpha} = \frac{1}{z-\alpha} + \gamma \qquad (\alpha = \beta \neq \infty),$$
$$w = z + \gamma \qquad (\alpha = \beta = \infty).$$

In this case, f is said to be *parabolic*.

16. Let $f(z)$ be as in problem 13. Show that if $a+d$ is real, then f is hyperbolic, elliptic, or parabolic according to whether $|a+d| > 2, < 2$, or $= 2$; show that if $a+d$ is not real, f is loxodromic.

17. Find a general form for linear transformations preserving $\Delta(R)$, $0 < R < \infty$.

18. i) Find a linear transformation which maps $0, 1, \infty$ to $i, 1+i, 2+i$.
ii) Find a linear transformation which maps $-1, i, 1$ to $-2, i, 2$.

CHAPTER 3

Holomorphic Functions

In this chapter we first define holomorphic functions, and then prove Cauchy's integral formula, which is the most fundamental and important theorem in complex analysis. One sees the close relationship between the topological structure of a domain and complex analysis on it.

3.1. Complex Derivatives

Let D be an open set of $\mathbf{C}$, and let f be a function on D. We say that f is *complex differentiable* at a point $a \in D$ if there is a number $\alpha \in \mathbf{C}$ such that for an arbitrary $\epsilon > 0$ there is a $\delta > 0$ with

$$\left| \frac{f(a+h) - f(a)}{h} - \alpha \right| < \epsilon \tag{3.1.1}$$

for all $h \in \Delta(\delta) \setminus \{0\}$. We call α the *complex derivative* of f at a, and denote it by $f'(a) = \frac{df}{dz}(a)$. The expression (3.1.1) is equivalent to

$$f(a+h) = f(a) + \alpha h + o(h), \qquad \alpha = f'(a), \tag{3.1.2}$$

where $o(h)$ is a term such that $\lim_{h \to 0} o(h)/h = 0$. If f is complex differentiable at a, then f is continuous at a. If f is complex differentiable at every point of D, f is said to be *holomorphic*. In this case the function $f' : z \in D \to f'(z) \in \mathbf{C}$ is called the *derived function* or the *derivative* of f, and is also denoted by df/dz.

If f is holomorphic, so is $1/f$ outside the set $\{f = 0\}$, and

$$\left(\frac{1}{f}\right)' = -\frac{f'}{f^2}.$$

When f_1 and f_2 are holomorphic functions, and a_1 and a_2 are constants, the following linearity is clear:

$$(a_1 f_1 + a_2 f_2)' = a_1 f_1' + a_2 f_2'.$$

We have the Leibniz formula:

$$(f_1 \cdot f_2)' = f_1' \cdot f_2 + f_1 \cdot f_2'.$$

EXERCISE 1. Show that for $n \in \mathbf{Z}$, $(z^n)' = nz^{n-1}$.

We set $z = x + iy$, and consider $f(z)$ as a function $f(x, y)$ in x and y. Set

$$f(x,y) = u(x,y) + iv(x,y),$$

where $u(x,y)$ (resp., $v(x,y)$) is the real (resp., imaginary) part of $f(x,y)$. Assume that f is holomorphic. Then by (3.1.2) $u(x,y)$ and $v(x,y)$ are totally differentiable. Letting $h \to 0$ with real h and with purely imaginary h, we have

$$\frac{\partial u}{\partial x} + i\frac{\partial v}{\partial x} = \frac{\partial v}{\partial y} - i\frac{\partial u}{\partial y}.$$

Therefore

$$\frac{\partial u}{\partial x} = \frac{\partial v}{\partial y}, \qquad \frac{\partial u}{\partial y} = -\frac{\partial v}{\partial x}. \tag{3.1.3}$$

These are called the *Cauchy-Riemann equations.*

EXERCISE 2. Let $u(x,y)$ and $v(x,y)$ be real valued, totally differentiable functions satisfying the Cauchy-Riemann equations (3.1.3). Show that the complex valued function $f(x,y) = u(x,y) + iv(x,y)$ is holomorphic as a function of $z = x + iy$.

Partial differential operators $\partial_z = \partial/\partial z$ and $\bar{\partial}_z = \partial/\partial \bar{z}$ are defined by

$$\partial_z f = \frac{\partial f}{\partial z} = \frac{1}{2}\left(\frac{\partial f}{\partial x} + \frac{1}{i}\frac{\partial f}{\partial y}\right),$$
$$\bar{\partial}_z f = \frac{\partial f}{\partial \bar{z}} = \frac{1}{2}\left(\frac{\partial f}{\partial x} - \frac{1}{i}\frac{\partial f}{\partial y}\right).$$

Then (3.1.3) is equivalent to

$$\bar{\partial}_z f = 0. \tag{3.1.4}$$

This is called the $\bar{\partial}$-equation (d-bar-equation). For a holomorphic function f,

$$f' = \partial_z f. \tag{3.1.5}$$

Let $\psi : I \to D$ be a continuously differentiable curve. For $t, t' \in I$ with $t \neq t'$

$$\frac{f \circ \psi(t') - f \circ \psi(t)}{t' - t} = f' \circ \psi(t)\frac{\psi(t') - \psi(t)}{t' - t} + o\left(\frac{\psi(t') - \psi(t)}{t' - t}\right).$$

Since $\lim_{t' \to t}(\psi(t') - \psi(t))/(t' - t) = \psi'(t)$,

$$\frac{d}{dt} f \circ \psi(t) = f' \circ \psi(t) \cdot \psi'(t). \tag{3.1.6}$$

(3.1.7) THEOREM. *If a holomorphic function f on a domain D satisfies $f' \equiv 0$, then f is constant.*

PROOF. If we set $f(z) = f(x,y) = u(x,y) + iv(x,y)$, we have by (3.1.5) or (3.1.3)

$$\frac{\partial u}{\partial x} = \frac{\partial u}{\partial y} = \frac{\partial v}{\partial x} = -\frac{\partial v}{\partial y} \equiv 0.$$

Since D is connected, u and v are constant. □

(3.1.8) THEOREM. *Let f be a holomorphic function on a domain D. Then, if* $\mathrm{Re} f$ *or* $\mathrm{Im} f$ *is constant, so is f.*

PROOF. It is sufficient to deal with the case where $\mathrm{Re} f$ is constant. Setting $f = u + iv$, we have $\partial u/\partial x = \partial u/\partial y \equiv 0$. It follows from (3.1.3) that $\partial v/\partial x = \partial v/\partial y \equiv 0$. Hence Theorem (3.1.7) implies the constancy of f. □

Let f be a holomorphic function on a domain D. Let D' be a domain of $\mathbf{C}$ containing the image $f(D)$. Let g be a holomorphic function on D'. Fix $z_0 \in D$ arbitrarily, and set $w_0 = f(z_0)$. Then for a small h

$$\begin{aligned} g(f(z_0+h)) - g(f(z_0)) &= g(w_0 + hf'(z_0) + o(h)) - g(w_0) \\ &= g'(w_0)(hf'(z_0) + o(h)) + o(hf'(z_0) + o(h)) \\ &= g'(w_0)f'(z_0)h + o(h). \end{aligned}$$

Therefore

$$\lim_{h\to 0} \frac{g(f(z_0+h)) - g(f(z_0))}{h} = g'(f(z_0))f'(z_0).$$

Thus we have the following:

(3.1.9) THEOREM. *The composite $g \circ f$ of two holomorphic functions f and g is again holomorphic, and*

$$(g \circ f)'(z) = g'(f(z))f'(z).$$

We take a convergent power series about $a \in \mathbf{C}$:

$$f(z) = \sum_{n=0}^{\infty} a_n (z-a)^n.$$

Let $R > 0$ be its radius of convergence.

(3.1.10) THEOREM. *The above $f(z)$ is a holomorphic function on $\Delta(a;R)$, and*

$$f'(z) = \sum_{n=1}^{\infty} n a_n (z-a)^{n-1}.$$

Here the radius of convergence of the right side is R, too.

PROOF. Without loss of generality we may assume $a = 0$. Since $\lim_{n\to\infty} \sqrt[n]{n+1} = 1$,

$$\varlimsup_{n\to\infty} \sqrt[n]{|a_n|} = \varlimsup_{n\to\infty} \sqrt[n-1]{|a_n|} = \varlimsup_{n\to\infty} \sqrt[n]{|a_{n+1}|}$$
$$= \varlimsup_{n\to\infty} \sqrt[n]{(n+1)|a_n|}.$$

Thus the radius of convergence of

$$g(z) = \sum_{n=1}^{\infty} n a_n z^{n-1}$$

is R. Take an arbitrary point $z \in \Delta(R)$. Fix r with $|z| < r < R$. By Lemma (2.4.2) there is an $M > 0$ such that

$$\text{(3.1.11)} \qquad |a_n r^n| \leqq M, \qquad n = 1, 2, \ldots.$$

For $h \in \Delta(r - |z|) \setminus \{0\}$ we have

$$\left|\frac{f(z+h) - f(z)}{h} - g(z)\right| = \left|\sum_{n=1}^{\infty} a_n \left\{\frac{(z+h)^n - z^n}{h} - nz^{n-1}\right\}\right|$$
$$= \left|\sum_{n=2}^{\infty} a_n \sum_{\nu=2}^{n} \binom{n}{\nu} z^{\nu} h^{n-\nu-1}\right| \leqq \sum_{n=2}^{\infty} |a_n| \sum_{\nu=2}^{n} \binom{n}{\nu} |z|^{\nu} |h|^{n-\nu-1}$$
$$= \sum_{n=1}^{\infty} |a_n| \left\{\frac{(|z| + |h|)^n - z^n}{|h|} - n|z|^{n-1}\right\}$$
$$\leqq M \sum_{n=1}^{\infty} \left[\frac{1}{|h|}\left\{\left(\frac{|z|+|h|}{r}\right)^n - \left(\frac{|z|}{r}\right)^n\right\} - \frac{n}{r}\left(\frac{|z|}{r}\right)^{n-1}\right].$$

Here we used (3.1.11) in the last estimate. It follows from (2.4.8) and (2.4.9) that

$$\left|\frac{f(z+h) - f(z)}{h} - g(z)\right|$$
$$\leqq M\left\{\frac{1}{|h|}\left(\frac{|z|+|h|}{r - (|z|+|h|)} - \frac{|z|}{r - |z|}\right) - \frac{r}{(r-|z|)^2}\right\}$$
$$= \frac{Mr|h|}{(r-|z|)^2(r - (|z|+|h|))} \to 0 \qquad (h \to 0). \qquad \square$$

(3.1.12) COROLLARY. *An analytic function is holomorphic.*

We mention that the converse of the above statement holds, and will be proved in Theorem (3.5.7).

It follows from Theorem (3.1.10) that a function $f(z)$ expressed by a convergent power series about a is complex differentiable arbitrarily many times, and the coefficients a_n are determined by the n-th derivatives $f^{(n)}(a)$ of $f(z)$:

$$\text{(3.1.13)} \qquad a_n = \frac{1}{n!} f^{(n)}(a).$$

Hence we see that if a function is expanded to a power series, the coefficients of the power series are uniquely determined.

Let $f(z) = u(x,y) + iv(x,y)$ be a holomorphic function of $z = x + iy \in D$. The Jacobian $Jf(x,y)$ of

$$f : (x,y) \in D \to (u(x,y), v(x,y)) \in \mathbf{R}$$

is given by

$$Jf(x,y) = \begin{vmatrix} \frac{\partial u}{\partial x}(x,y) & \frac{\partial u}{\partial y}(x,y) \\ \frac{\partial v}{\partial x}(x,y) & \frac{\partial v}{\partial y}(x,y) \end{vmatrix}.$$

The Cauchy-Riemann equations (3.1.3) imply

$$Jf(x,y) = |f'(z)|^2 \qquad (z = x + iy). \tag{3.1.14}$$

Therefore $Jf(x,y) \geqq 0$ everywhere. If $f'(z) \neq 0$, then $Jf(x,y) > 0$, so that f gives rise to an orientation preserving mapping from (x,y)-plane to (u,v)-plane. That is, to vectors $X_j = (a_{j1}, a_{j2})$, $j = 1,2$, at z we assign vectors

$$U_j = \left(a_{j1}\frac{\partial u}{\partial x} + a_{j2}\frac{\partial u}{\partial y}, a_{j1}\frac{\partial v}{\partial x} + a_{j2}\frac{\partial v}{\partial y} \right), \ j = 1,2,$$

at $f(z)$. If $\{X_1, X_2\}$ is a right-hand system ($\det \binom{X_1}{X_2} > 0$), then so is $\{U_1, U_2\}$.

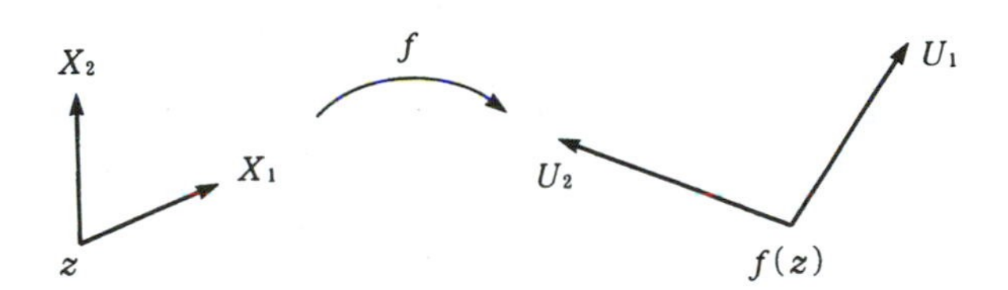

FIGURE 20

Now let $z_0 \in D$, and assume $f'(z_0) \neq 0$. For the sake of simplicity, we assume that $z_0 = 0$ and $f(z_0) = 0$. Let $C_j(\phi_j : [0, t_j] \to D)$, $j = 1,2$, be two curves such that the limit $\lim_{t \to +0} \phi_j(t)/t = a_j$ exists. Take $\theta_j = \arg a_j$, $j = 1,2$, so that $0 \leqq \theta_2 - \theta_1 < 2\pi$. Then we call $\theta = \theta_2 - \theta_1$ the angle pinched by C_1 and C_2.

(3.1.15) THEOREM. *Let the notation be as above. Then the angle pinched by $f(C_1)$ and $f(C_2)$ is equal to that pinched by C_1 and C_2.*

PROOF. Since f is holomorphic, it follows from (3.1.6) that

$$\lim_{t \to +0} \frac{f(\phi_j(t))}{t} = f'(0)a_j = |f'(0)| \cdot |a_j| e^{i(\theta_j + \arg f'(0))}.$$

Thus the angle pinched by $f(C_1)$ and $f(C_2)$ is $\theta_2 - \theta_1$. $\square$

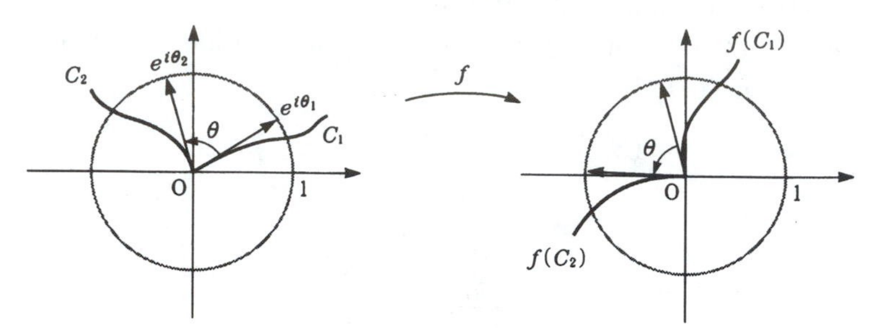

FIGURE 21

The property of f in the above theorem is called *conformality*. Hence, an injective holomorphic function f with nowhere vanishing f' is called a *conformal mapping*.

EXERCISE 3. Let $u(x,y)$ and $v(x,y)$ be totally differentiable real functions. Set $f(x,y) = u(x,y) + iv(x,y)$. Show that if $Jf(x,y) \neq 0$ and f is conformal, then f is holomorphic.

EXERCISE 4. Show the following:

$$\frac{d}{dz}e^z = e^z, \qquad z \in \mathbf{C}.$$

$$\frac{d}{dz}\log(1+z) = \frac{1}{1+z}, \qquad z \in \Delta(1).$$

EXERCISE 5. Let $f(z) = f(x,y)$ be a continuously differentiable function on a domain $D \subset \mathbf{C}$. Let K be a compact subset of D. For $z \in K$, set

$$f(z+h) = f(z) + \frac{\partial f}{\partial z}(z)h + \frac{\partial f}{\partial \bar{z}}(z)\bar{h} + o(h).$$

Show that $o(h)/h \to 0$ uniformly in $z \in K$ as $h \to 0$; i.e., for an arbitrary $\epsilon > 0$, there is a $\delta > 0$ such that $|o(h)/h| < \epsilon$ for all $0 < |h| < \delta$ and $z \in K$.

3.2. Curvilinear Integrals

Let $C(\phi : [T_0, T_1] \to \mathbf{C})$ be a curve. We define the *length* $L(C)$ of C. Let $(d) : T_0 = t_0 \leqq t_1 \leqq \cdots \leqq t_l = T_1$ be a *partition* of the interval $[T_0, T_1]$. Set

$$L(C;(d)) = \sum_{j=1}^{l} |\phi(t_j) - \phi(t_{j-1})|.$$

We call each t_j a *partition point*, and define the width $|(d)|$ by

$$|(d)| = \max\{t_j - t_{j-1}; 1 \leqq j \leqq l\}.$$

Let (d') be a partition of $[T_0, T_1]$ such that every partition point of (d) is a partition point of (d'). Then the partition (d') is called a *refinement* of (d), and

$$L(C;(d)) \leqq L(C;(d')).$$

We define

$$L(C) = \sup\{L(C;(d)); (d) \text{ is an arbitrary partition of } [T_0, T_1]\} \leqq +\infty. \tag{3.2.1}$$

Note that $L(C)$ is invariant under the change of parameter of C.

EXERCISE 1. Prove this fact.

When $L(C) < \infty$, C is said to have a *finite length.* For a general curve $C(\phi : I \to \mathbf{C})$, we take a bounded closed interval $J \subset I$. Let $\phi|J$ be the restriction of ϕ to J, and let $C|J$ denote the restriction of the curve C to J given by $\phi|J$. Define

$$L(C) = \sup\{L(C|J); J \subset I\} \leqq +\infty.$$

Assume that $\phi : [T_0, T_1] \to \mathbf{C}$ is a piecewise continuously differentiable curve. Take a partition $(d') : T_0 = t'_0 < t'_1 < \cdots < t'_l = T_1$ so that $\phi|[t'_{j-1}, t'_j]$ are continuously differentiable. Let $(d) : T_0 = t_0 \leqq t_1 \leqq \cdots \leqq t_l = T_1$ be a refinement of (d'). Setting $\phi(t) = \phi_1(t) + i\phi_2(t)$, we have by the mean value theorem

(3.2.2)

$$\begin{aligned}
|\phi(t_j) - \phi(t_{j-1})| &= \sqrt{(\phi(t_j) - \phi(t_{j-1}))^2 + (\phi_2(t_j) - \phi_2(t_{j-1}))^2} \\
&= \sqrt{(\phi'_1(t_{j-1} + \theta_1(t_j - t_{j-1})))^2 + (\phi'_2(t_{j-1} + \theta_2(t_j - t_{j-1})))^2}\,(t_j - t_{j-1}) \\
&\quad (0 < \theta_1 < 1, 0 < \theta_2 < 1) \\
&= \sqrt{(\phi'_1(t_{j-1}))^2 + (\phi'_2(t_{j-1}))^2}\,(t_j - t_{j-1}) + o(1)(t_j - t_{j-1}).
\end{aligned}$$

Since $\phi'_1(t)$ and $\phi'_2(t)$ are continuous on every closed interval between the partition points of (d'), $o(1) \to 0$ uniformly in j as $|(d)| \to 0$. Thus we have the Riemann integral

$$L(C) = \int_{T_0}^{T_1} \sqrt{(\phi'_1(t))^2 + (\phi'_2(t))^2}\, dt. \tag{3.2.3}$$

EXERCISE 2. Expressing the half circle C of radius $r > 0$ as $\phi(t) = \cos t + i \sin t$, $0 \leqq t \leqq \pi$, confirm that $L(C) = \pi$.

(3.2.4) LEMMA. *Let $C(\phi : [T_0, T_1] \to \mathbf{C})$ be a curve with finite length.*

i) *The real function*

$$\psi : t \in [T_0, T_1] \to L(C|[T_0, t]) \in [0, L(C)]$$

is monotone increasing, and continuous.

ii) *For* $s \in [0, L(C)]$ *we take* $t \in \psi^{-1}(s)$, *and set* $\tilde{\phi}(s) = \phi(t)$. *Then* $\tilde{\phi}(s)$ *is independent of the choice of* $t \in \psi^{-1}(s)$, *and satisfies*

$$|\tilde{\phi}(s) - \tilde{\phi}(s')| \leqq |s - s'| \tag{3.2.5}$$

for $s, s' \in [0, L(C)]$.

PROOF. i) The monotone increasingness is clear. Assume that ψ is not continuous at a point $t_0 \in [T_0, T_1]$. Then one of the following must hold:

$$\lim_{t \to t_0 - 0} L(C|[T_0, t]) < L(C|[T_0, t_0]),$$
$$\lim_{t \to t_0 + 0} L(C|[T_0, t]) < L(C|[T_0, t_0]).$$

Since the argument is the same in either case, we assume the first. We have $t_0 > T_0$, and generally for $t \in [T_0, T_1]$

$$L(C|[T_0, T_1]) = L(C|[T_0, t]) + L(C|[t, T_1]). \tag{3.2.6}$$

Set

$$\epsilon_0 = L(C|[T_0, t_0]) - \lim_{t \to t_0 - 0} L(C|[T_0, t]) > 0.$$

By definition there are infinitely many $t_j < \tilde{t}_j < t_{j+1} < \tilde{t}_{j+1} < t_0$, $j = 1, 2, \ldots$, such that

$$L(C|[t_j, \tilde{t}_j]) > \frac{\epsilon_0}{2}.$$

It follows from (3.2.6) that $L(C|[T_0, t_0]) = +\infty$. This is a contradiction.

ii) It is clear that $\tilde{\phi}(s)$ is defined independently of the choice of $t \in \psi^{-1}(s)$. The definition of the length of curves immediately implies (3.2.5). □

An inequality of the type (3.2.5) with right side replaced by $K|s - s'|$, where K is a positive constant, is called *Lipschitz' condition* for $\tilde{\phi}$.

In general, $\tilde{\phi}$ is not a parameter change of ϕ, but it follows that $\phi(t) = \tilde{\phi}(\psi(t))$. In particular, if $C(\phi)$ is a simple curve, then $\tilde{\phi}$ is a parameter change of ϕ. The parameter s of $\tilde{\phi}(s)$ is called the *length parameter*, and $\tilde{\phi}$ is called the *length parametrization*.

REMARK. From now on we include the length parametrization of a curve with finite length in parameter changes.

EXERCISE 3. Show that a curve $C(\phi)$ in $\mathbf{C}$ has finite length if and only if there is a change of parameter of $C(\phi)$ satisfying Lipschitz' condition.

Let $D \subset \mathbf{C}$ be a domain and $C(\phi : I \to D)$ be a curve with finite length. Let f be a continuous function on D. Let $I = [T_0, T_1]$, and take a partition $(d) : T_0 = t_0 \leqq t_1 \leqq \cdots \leqq t_l = T_1$. For $\xi_j \in [t_{j-1}, t_j]$, $1 \leqq j \leqq l$, we set

$$S((d); (\xi_j)) = \sum_{j=1}^{l} f(\phi(\xi_j))(\phi(t_j) - \phi(t_{j-1})). \tag{3.2.7}$$

We first take (d) and ξ_j as follows:

$$(d_n): \quad t_{nj} = T_0 + \frac{T_1 - T_0}{n} j, \qquad \xi_{nj} = t_{nj}.$$

Since $f \circ \phi(t)$ is uniformly continuous in $t \in I$, for an arbitrary $\epsilon > 0$ there is a number n such that

$$|t - t'| \leqq \frac{T_1 - T_0}{n} \quad \Longrightarrow \quad |f \circ \phi(t) - f \circ \phi(t')| < \epsilon.$$

Let $|(d)| < (T_1 - T_0)/2n$ and ξ_j be as above. If a point t_{nj} lies in the interior of $[t_{\nu-1}, t_\nu]$, we add t_{nj} to (d) as a partition point, and divide $[t_{\nu-1}, t_\nu] = [t_{\nu-1}, t_{nj}] \cup [t_{nj}, t_\nu]$ into two parts. We use the same $\xi_\nu \in [t_{\nu-1}, t_\nu]$ for the intervals $[t_{\nu-1}, t_{nj}]$ and $[t_{nj}, t_\nu]$.

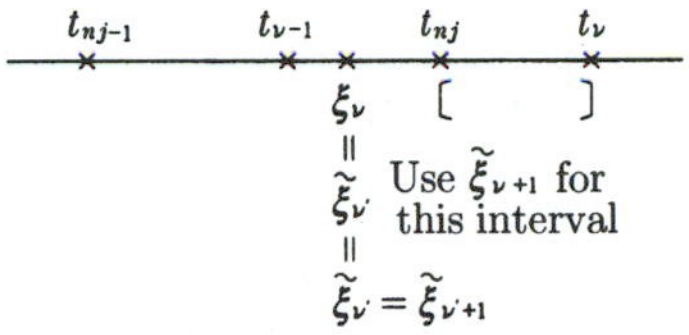

FIGURE 22

Thus we have a refinement $(\tilde{d})$ of (d) and $(\tilde{\xi}_\nu)$. Let $t_{nj-1} = t_{j0} \leqq \cdots \leqq t_{jk(j)} = t_{nj}$ be the partition points of $(\tilde{d})$ contained in $[t_{nj-1}, t_{nj}]$. Write $\tilde{\xi}_{j\mu}$ for $\tilde{\xi}_\nu$ associated to $[t_{j\mu-1}, t_{j\mu}]$, $1 \leqq \mu \leqq k(j)$. Then

$$S((d); (\xi_j)) = \sum_{j=1}^{n} \sum_{\mu=1}^{k(j)} f \circ \phi(\tilde{\xi}_{j\mu})(\phi(t_{j\mu}) - \phi(t_{j\mu} - 1)).$$

It follows that

$$\begin{aligned}
&|S((d_n); (t_{nj})) - S((d); (\xi_j))| \\
&= \left| \sum_{j=1}^{n} \left\{ f \circ \phi(t_{nj})(\phi(t_{nj}) - \phi(t_{nj-1})) - \sum_{\mu=1}^{k(j)} f \circ \phi(\tilde{\xi}_{j\mu})(\phi(t_{j\mu}) - \phi(t_{j\mu} - 1)) \right\} \right| \\
&= \left| \sum_{j=1}^{n} \sum_{\mu=1}^{k(j)} (f \circ \phi(t_{nj}) - f \circ \phi(\tilde{\xi}_{j\mu}))(\phi(t_{j\mu}) - \phi(t_{j\mu} - 1)) \right|.
\end{aligned}$$

By the choice of $\tilde{\xi}_{j\mu}$, $|t_{nj} - \tilde{\xi}_{j\mu}| \leqq (T_1 - T_0)/n$. It follows that for all j

$$|f \circ \phi(t_{nj}) - f \circ \phi(\tilde{\xi}_{j\mu})| < \epsilon.$$

Therefore we have

$$|S((d_n); (t_{nj})) - S((d); (\xi_j))| < L(C)\epsilon.$$

For arbitrary $m_1, m_2 > 2n$

$$\begin{aligned}
&|S((d_{m_1});(t_{m_1 j})) - S((d_{m_2});(t_{m_2 j}))| \\
&\leqq |S((d_{m_1});(t_{m_1 j})) - S((d_n);(t_{nj}))| \\
&+ |S((d_{m_2});(t_{m_2 j})) - S((d_n);(t_{nj}))| < 2L(C)\epsilon.
\end{aligned}$$

Thus we see that $\{S((d_n);(t_{nj}))\}_{n=1}^{\infty}$ is a Cauchy sequence; we denote its limit by σ. Then it follows that

$$|(d)| < \frac{T_1 - T_0}{2n} \implies |S((d);(\xi_j)) - \sigma| \leqq L(C)\epsilon.$$

That is,

$$\sigma = \lim_{|(d)| \to 0} S((d);(\xi_j)).$$

We call σ the *curvilinear integral* of f along C. The limit σ is invariant under the change of parameter of C. We write

$$\int_C f(z)dz = \sigma = \lim_{|(d)| \to 0} S((d);(\xi_j)). \tag{3.2.8}$$

If f_1 and f_2 are continuous functions on D, and if a_1 and a_2 are constants, then

$$\int_C (a_1 f_1(z) + a_2 f_2(z))dz = a_1 \int_C f_1(z)dz + a_2 \int_C f_2(z)dz.$$

This might be clear. Taking the absolute value of the right side of (3.2.7), we set

$$\begin{aligned}
S &= \inf_{\delta > 0} \sup \left\{ \sum_{j=1}^{l} |f(\phi(\xi_j))| \cdot |\phi(t_j) - \phi(t_{j-1})|; |(d)| < \delta \right\} \\
&\leqq \max\{|f \circ \phi|\} \cdot L(C) < \infty.
\end{aligned}$$

In fact, we show that

$$S = \lim_{|(d)| \to 0} \sum_{j=1}^{l} |f \circ \phi(\xi_j)| \cdot |\phi(t_j) - \phi(t_{j-1})|. \tag{3.2.9}$$

For an arbitrary $\epsilon > 0$ we take a $\delta_0 > 0$ so that

$$|t - t'| < \delta_0 \implies \begin{cases} |\phi(t) - \phi(t')| < \epsilon, \\ |f \circ \phi(t) - f \circ \phi(t')| < \epsilon. \end{cases}$$

Letting $\delta_0 > 0$ be smaller if necessary, we have for $|(d)| < \delta_0$

$$S + \epsilon > \sum_{j=1}^{l} |f \circ \phi(\xi_j)| \cdot |\phi(t_j) - \phi(t_{j-1})|.$$

On the other hand, it follows from the definition that there are $(\Delta): T_0 = \tau_0 < \tau_1 < \cdots < \tau_n = T_1$ and $\zeta_k \in [\tau_{k-1}, \tau_k]$ with $|(\Delta)| < \delta_0$ such that

$$S - \epsilon < \sum_{k=1}^{n} |f \circ \phi(\zeta_k)| \cdot |\phi(\tau_k) - \phi(\tau_{k-1})|.$$

Define $\delta_1 = \min\{\tau_k - \tau_{k-1}; 1 \leqq k \leqq n\}$. Then $0 < \delta_1 < \delta_0$. We take a partition (d) of $[T_0, T_1]$ and (ξ_j) with $|(d)| < \delta_1$. Let $(\tilde{d})$ be the partition formed by all partition points of (d) and (Δ).

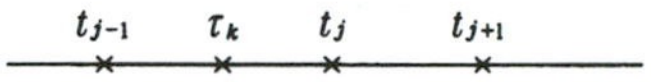

t_{j-1} τ_k t_j t_{j+1}

FIGURE 23

If an interval $[\tilde{t}_{i-1}, \tilde{t}_i]$ contains a ξ_j, we set $\tilde{\xi}_i = \xi_j$; otherwise, we take $\tilde{\xi}_i \in [\tilde{t}_{i-1}, \tilde{t}_i]$. So, we have $(\tilde{\xi}_i)$, and by the choice we get

$$\left| \sum_{i=1}^{\tilde{n}} |f \circ \phi(\tilde{\xi}_i)| \cdot |\phi(\tilde{t}_i) - \phi(\tilde{t}_{i-1})| - \sum_{j=1}^{l} |f \circ \phi(\xi_j)| \cdot |\phi(t_j) - \phi(t_{j-1})| \right| \\ \leqq 3n \max\{|f \circ \phi|\} \cdot \epsilon;$$

$$\sum_{i=1}^{\tilde{n}} |f \circ \phi(\tilde{\xi}_i)| \cdot |\phi(\tilde{t}_i) - \phi(\tilde{t}_{i-1})| - \sum_{k=1}^{n} |f \circ \phi(\zeta_k)| \cdot |\phi(\tau_k) - \phi(\tau_{k-1})| \\ > -\epsilon \sum_{i=1}^{\tilde{n}} |\phi(\tilde{t}_i) - \phi(\tilde{t}_{i-1})| \geqq -\epsilon L(C).$$

Summarizing the above, we obtain

$$S + \epsilon > \sum_{j=1}^{l} |f \circ \phi(\xi_j)| \cdot |\phi(t_j) - \phi(t_{j-1})| \\ > S - \epsilon \{L(C) + 3n \max\{|f \circ \phi|\} + 1\}.$$

Hence we have shown (3.2.9).

We write

$$S = \int_C |f(z)||dz|.$$

It follows that

$$\left| \int_C f(z) dz \right| \leqq \int_C |f(z)||dz|. \tag{3.2.10}$$

Let f_n, $n = 0, 1, \ldots$, be continuous functions on D. If $\{f_n\}_{n=0}^{\infty}$ or $\sum_{n=0}^{\infty} f_n$ converges uniformly on compact subsets of D, then by (3.2.10) we have respectively

$$\int_C \lim_{n\to\infty} f_n(z)dz = \lim_{n\to\infty} \int_C f_n(z)dz, \tag{3.2.11}$$

$$\int_C \sum_{n=0}^{\infty} f_n(z)dz = \sum_{n=0}^{\infty} \int_C f_n(z)dz.$$

Let $\phi(t) = \phi_1(t) + i\phi_2(t)$ be piecewise continuously differentiable. As in (3.2.2) it follows from (3.2.7) that

$$\begin{aligned} S((d);(\xi_j)) &= \sum_{j=1}^{l} f\circ\phi(\xi_j)\{\phi_1'(t_{j-1}+\theta_{1j}(t_j-t_{j-1})) \\ &\qquad + i\phi_2'(t_{j-1}+\theta_{2j}(t_j-t_{j-1}))\}(t_j-t_{j-1}) \\ &\qquad (0<\theta_{1j}<1, \quad 0<\theta_{2j}<1) \\ &= \sum_{j=1}^{l} f\circ\phi(\xi_j)(\phi_1'(\xi_j)+i\phi_2'(\xi_j))(t_j-t_{j-1}) \\ &\qquad + o(1)(t_j-t_{j-1}). \end{aligned}$$

Here, as $|(d)| \to 0$, $o(1) \to 0$ uniformly. Hence we get

$$\int_C f(z)dz = \int_{T_0}^{T_1} f\circ\phi(t)\phi'(t)dt. \tag{3.2.12}$$

In the same way, we deduce that

$$\int_C |f(z)||dz| = \int_{T_0}^{T_1} |f\circ\phi(t)||\phi'(t)|dt.$$

It is clear that $L(-C) = L(C)$, and for the curvilinear integral we have

$$\int_{-C} f(z)dz = -\int_C f(z)dz. \tag{3.2.13}$$

For curves C_1 and C_2 in D such that the terminal point of C_1 is the initial point of C_2

$$\int_{C_1+C_2} f(z)dz = \int_{C_1} f(z)dz + \int_{C_2} f(z)dz. \tag{3.2.14}$$

We deal only with a general curve $C(\phi : I \to D)$ such that $L(C|J) < \infty$ for every bounded closed interval $J \subset I$. Let f be a continuous function on D. Let $\{J_n\}_{n=0}^{\infty}$ be an increasing sequence of bounded closed intervals such that $J_n \subset \overset{\circ}{J}_{n+1} \subset I$ with the interior $\overset{\circ}{J}_{n+1}$ of J_{n+1}, and $\bigcup_{n=0}^{\infty} J_n = I$. Assume that the sequence

$$\int_{J_n} f(z)dz, \quad n = 0, 1, \ldots ,$$

converges, and that the limit is independent of the choice of such $\{J_n\}_{n=0}^{\infty}$. Then we write for the limit

$$\int_C f(z)dz,$$

and thus the curvilinear integral of f along C is defined.

Let $f : D \to \mathbf{C}$ be a holomorphic function, and assume that the derivative f' is continuous. (The continuity of f' will be proved in Remark (3.5.9), but is assumed here for a moment.) Let $C(\phi : [T_0, T_1] \to D)$ be a curve with finite length. We compute the length $L(C'(f \circ \phi))$ of the curve $C'(f \circ \phi)$. Let $(d) : T_0 = t_0 < t_1 \cdots < t_n = T_1$ be a partition of $[T_0, T_1]$. For sufficiently small $|(d)|$ we have by Exercise 5, §1

$$\begin{aligned}
&\sum_{j=1}^{n} |f \circ \phi(t_j) - f \circ \phi(t_{j-1})| \\
&= \sum_{j=1}^{n} |f' \circ \phi(t_{j-1})(\phi(t_j) - \phi(t_{j-1})) + o(|(\phi(t_j) - \phi(t_{j-1}))|)| \\
&= \sum_{j=1}^{n} |f' \circ \phi(t_{j-1})| \cdot |\phi(t_j) - \phi(t_{j-1})| + o(1) \sum_{j=1}^{n} |(\phi(t_j) - \phi(t_{j-1})|.
\end{aligned}$$

Let $|(d)| \to 0$. Then $o(1) \to 0$, and so

$$L(C') = \int_C |f'(z)||dz|. \tag{3.2.15}$$

When ϕ is piecewise continuously differentiable,

$$L(C') = \int_{T_0}^{T_1} |f' \circ \phi(t)||\phi'(t)|dt. \tag{3.2.16}$$

Let g be a continuous function on a domain D' containing $f(D)$. In the same way as above, we deduce that

$$\begin{aligned}
&\sum_{j=1}^{n} g \circ f \circ \phi(t_{j-1})(f \circ \phi(t_j) - f \circ \phi(t_{j-1})) \\
&\quad = \sum_{j=1}^{n} (g \circ f \circ \phi(t_{j-1}) \cdot f' \circ \phi(t_{j-1}) + o(1))(\phi(t_j) - f \circ \phi(t_{j-1})).
\end{aligned}$$

Letting $|(d)| \to 0$, we obtain

$$\int_{C'} g(w)dw = \int_C g \circ f(z) \cdot f'(z)dz. \tag{3.2.17}$$

When ϕ is piecewise continuously differentiable,

$$\int_{C'} g(w)dw = \int_{T_0}^{T_1} g \circ f \circ \phi(t) \cdot f' \circ \phi(t) \cdot \phi'(t)dt. \tag{3.2.18}$$

For a function f on D a holomorphic function F such that $F' = f$ is called a *primitive function* of f. It follows from Theorem (3.1.7) that if the primitive function exists, it is unique up to a constant term.

EXERCISE 4. Show that a primitive function of z^n, $n \in \mathbf{Z} \setminus \{-1\}$, is of the form $z^{n+1}/(n+1) + \text{constant}$.

(3.2.19) THEOREM. *Let f be a continuous function on D, and let F be a primitive function of f. Let C be a curve with finite length with initial point z_0 and terminal point z_1. Then*

$$\int_C f(z)dz = F(z_1) - F(z_0).$$

In particular, if C is a closed curve, $\int_C f(z)dz = 0$.

PROOF. Let C be given by $\phi : [0,1] \to D$. As in the proof of (3.2.15), we have for a partition $(d) : T_0 = t_0 < t_1 \cdots < t_n = T_1$ with small $|(d)|$

$$F \circ \phi(t_j) - F \circ \phi(t_{j-1}) = (f \circ \phi(t_{j-1}) + o(1))(\phi(t_j) - \phi(t_{j-1})).$$

Therefore

$$\begin{aligned} F(z_1) - F(z_0) = &\sum_{j=1}^{n} f \circ \phi(t_{j-1})(\phi(t_j) - \phi(t_{j-1})) \\ &+ o(1) \sum_{j=1}^{n} (\phi(t_j) - \phi(t_{j-1})). \end{aligned}$$

As $|(d)| \to 0$, we get the desired equation. □

EXERCISE 5. Integrate the function $f(z) = \bar{z}$ along the circle $C(0; r)$ (with anti-clockwise orientation).

EXERCISE 6. Integrate the function $f(z) = 1/z$ along the circle $C(0; r)$.

EXERCISE 7. Integrate the function $f(z) = \operatorname{Re} z = (z + \bar{z})/2$ along the contour of the square $\{z = x + iy; |x| \leqq 1, |y| \leqq 1\}$ with anti-clockwise orientation.

3.3. Homotopy of Curves

We summarize the necessary facts on homotopy of curves. Let $D \subset \widehat{\mathbf{C}}$ be a domain of the Riemann sphere $\widehat{\mathbf{C}}$. Let $\phi_1 : [T_0, T_1] \to D$ and $\phi_2 : [S_0, S_1] \to D$

be curves with the same initial points and the same terminal points.

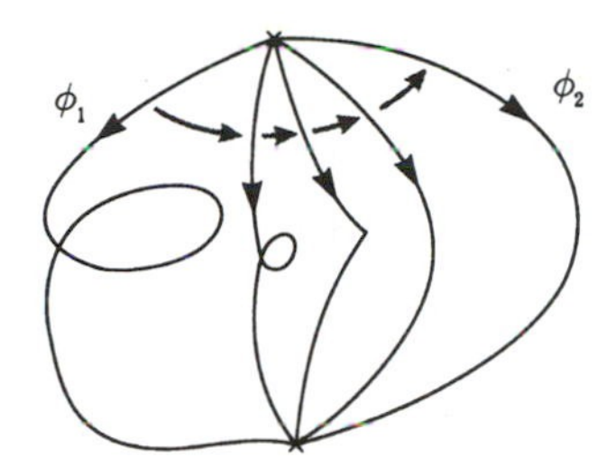

FIGURE 24

We say that ϕ_1 and ϕ_2 are *homotopic* (more precisely, *homotopic in* D), or that ϕ_1 is homotopic to ϕ_2 (in D), if for $\Phi_1(t) = \phi_1(T_0 + t(T_1 - T_0))$ and $\Phi_2(t) = \phi_2(S_0 + t(S_1 - S_0))$ there is a continuous mapping $\Phi : [0,1] \times [0,1] \to D$ satisfying

i) $\Phi(0, s) \equiv \Phi_1(0) = \Phi_2(0)$, $\Phi(1, s) \equiv \Phi_1(1) = \Phi_2(1)$, $0 \leqq s \leqq 1$.
ii) $\Phi(t, 0) = \Phi_1(t)$, $\Phi(t, 1) = \Phi_2(t)$, $0 \leqq t \leqq 1$.

The mapping Φ is called a *homotopy* connecting Φ_1 and Φ_2. The relation of being homotopic is an equivalence relation.

(3.3.1) LEMMA. *Let $\phi_1 : [T_0, T_1] \to D$ and $\phi_2 : [S_0, S_1] \to D$ have the same initial point, and be a change of parameter of each other. Then ϕ_1 and ϕ_2 are homotopic.*

PROOF. We may assume that $[T_0, T_1] = [S_0, S_1] = [0, 1]$. Then there is a strictly monotone increasing continuous function $\tau : [0,1] \to [0,1]$ such that $\phi_1(t) = \phi_2(\tau(t))$, $0 \leqq t \leqq 1$. Set

$$\Phi(t, s) = \phi_1((1-s)\tau(t) + st), \qquad 0 \leqq t \leqq 1, \quad 0 \leqq s \leqq 1.$$

Then Φ is a homotopy connecting ϕ_1 and ϕ_2. □

Two curves C_1 and C_2 in D are said to be homotopic if C_j are given by $\phi_j : [0,1] \to D$, $j = 1, 2$, such that ϕ_1 and ϕ_2 are homotopic. We see by Lemma (3.3.1) that this is an equivalence relation. We denote by $\{C\}$ the equivalence class of a curve C in D, and call it the *homotopy class* of C. When C is a closed curve and is homotopic to a constant curve, C is said to be *homotopic to a point.*

If every closed curve in D is homotopic to a point, D is said to be *simply connected.*

EXERCISE 1. We say that a domain $D \subset \mathbf{C}$ is *star-shaped,* if there is a point $z_0 \in D$ such that for every point $z \in D$ the line segment connecting z and z_0 is contained in D. In this case, show that D is simply connected. In particular, a convex domain and $\mathbf{C} \setminus [0, +\infty) = \{z \in \mathbf{C};\ z \text{ is not a real non-negative number}\}$ are star-shaped. Where can z_0 be taken for these domains?

If C_1 and C_2 have the same initial and terminal points, we set

$$C_1 - C_2 = C_1 + (-C_2),$$

which is a closed curve.

(3.3.2) LEMMA. *For an arbitrary curve C in D, $C - C$ is homotopic to a point.*

PROOF. Let C be given by $\phi[0,1] \to D$. Define $\tilde{\phi} : [0,2] \to D$ by

$$\tilde{\phi}|[0,1] : [0,1] \ni t \to \phi(t) \in D,$$
$$\tilde{\phi}|[1,2] : [1,2] \ni t \to \phi(2-t) \in D,$$

which gives the curve $C - C$. Define $\Phi : [0,2] \times [0,1] \to D$ by

$$\Phi|([0,1] \times [0,1])(t,s) = \phi(st),$$
$$\Phi|([1,2] \times [0,1])(t,s) = \phi(2s - st).$$

Then Φ is continuous, $\Phi(t,1) = \tilde{\phi}(t)$, and $\Phi(t,0) \equiv \phi(0)$. Hence, $C - C$ is homotopic to a point. □

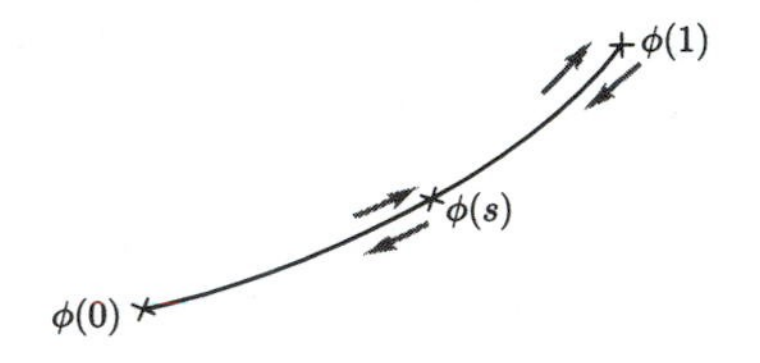

FIGURE 25

(3.3.3) THEOREM. *Let C_0, C_1 be two curves in D which are homotopic. Then $C_0 - C_1$ is homotopic to a point.*

PROOF. Let $\Phi : [0,1] \times [0,1] \to D$ be a homotopy connecting C_0 and C_1. Let C_s be curves given by $\Phi(\cdot, s)$. Then $z_0 \equiv \Phi(0,s)$ (resp., $z_1 \equiv \Phi(1,s)$) is the initial (resp., terminal) point of every C_s. We define a continuous mapping $\Psi : [0,2] \times [0,1/2] \to D$ by

$$(\Psi|[0,1] \times [0,1/2])(t,s) = \Phi(t,s),$$
$$(\Psi|[1,2] \times [0,1/2])(t,s) = \Phi(2-t, 1-s).$$

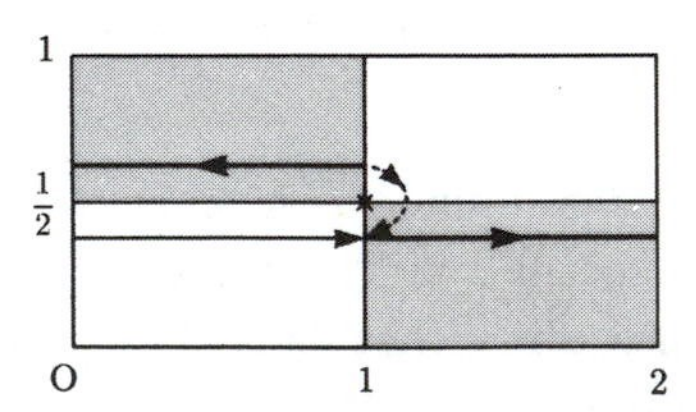

FIGURE 26

It follows that $C_0 - C_1$ is homotopic to $C_{1/2} - C_{1/2}$. Then by Lemma (3.3.2) $C_0 - C_1$ is homotopic to a point. □

By making use of the complex coordinate $\tilde{z}$ $(= 1/z)$ about $\infty \in \widehat{\mathbf{C}}$, the curves connecting $z_0 \in \mathbf{C}^*$ and ∞ given by

$$\tilde{z} = t\tilde{z}_0, \qquad 0 \leqq t \leqq 1,$$
$$\tilde{z} = (1-t)\tilde{z}_0, \qquad 0 \leqq t \leqq 1,$$

where $\tilde{z}_0 = 1/z_0$, are called *line segments* connecting z_0 and ∞.

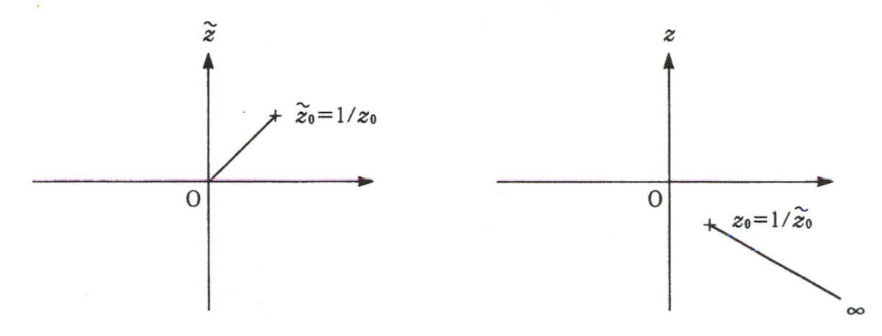

FIGURE 27

A *piecewise linear curve* in $\widehat{\mathbf{C}}$ is defined as a finite sum of line segments in $\mathbf{C}$ and line segments connecting points of $\mathbf{C}^*$ and ∞. A constant curve is considered as a special case of a piecewise linear curve.

(3.3.4) LEMMA. *Every curve $C(\phi : [0,1] \to D)$ is homotopic to a piecewise linear curve in D.*

PROOF. By making use of the uniform continuity of ϕ, we may take a partition $0 = t_0 < t_1 < \cdots < t_n = 1$ satisfying the following:

(3.3.5) For every $j, 1 \leqq j \leqq n$, there is a disk neighborhood Δ_j in $\mathbf{C}$ or a disk neighborhood $\tilde{\Delta}_j$ of ∞ such that $\phi([t_{j-1}, t_j]) \subset \Delta_j$ or $\phi([t_{j-1}, t_j]) \subset \tilde{\Delta}_j$. Here, when $\phi([t_{j-1}, t_j]) \subset \tilde{\Delta}_j$, $\phi(t_{j-1})$ or $\phi(t_j)$ is assumed to be ∞.

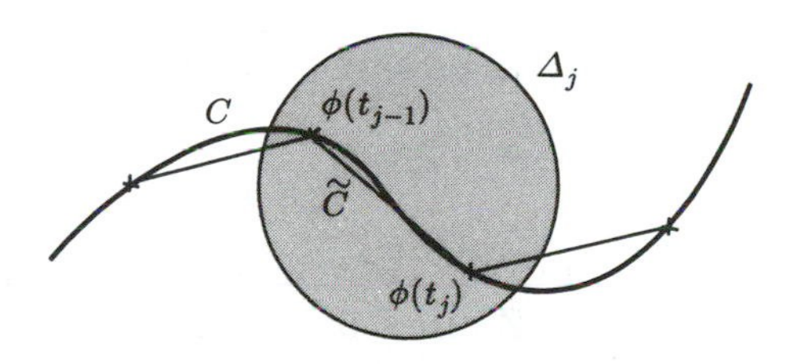

FIGURE 28

Connecting $\phi(t_0)$, $\phi(t_1)$, ..., $\phi(t_n)$ in order by line segments, we have a piecewise linear curve $\tilde{C}$. We show that C and $\tilde{C}$ are homotopic. It is sufficient to show that $C|[t_{j-1}, t_j]$ and the line segment from $\phi(t_{j-1})$ to $\phi(t_j)$ are homotopic. If $C|[t_{j-1}, t_j] \subset \Delta_j$, we set

$$\Phi(t,s) = (1-s)\phi(t) + s\left\{\frac{t_j - t}{t_j - t_{j-1}}\phi(t_{j-1}) + \frac{t - t_{j-1}}{t_j - t_{j-1}}\phi(t_j)\right\},$$
$$t_{j-1} \leqq t \leqq t_j, \quad 0 \leqq s \leqq 1,$$

which gives rise to a required homotopy. If $C|[t_{j-1}, t_j] \subset \tilde{\Delta}_j$, we form a homotopy Φ by making use of $\tilde{z}$ in the same way as above. □

(3.3.6) THEOREM. *There are at most countably many homotopy classes of curves connecting two arbitrarily given points of D.*

PROOF. Let C be a curve connecting the given points of D. We use the same notation as in the proof of Lemma (3.3.4). If $\phi(t_j)$ with $1 \leqq j \leqq n-1$ is not ∞, we take a rational point z_j sufficiently close to $\phi(t_j)$ so that the piecewise linear curve C' connecting $z_0 = \phi(0), z_1, \ldots, z_n = \phi(1)$ in order by line segments is homotopic to C.

The set of rational points is countable, and hence so is the set of such curves C'. Therefore the set of homotopy classes is at most countable. □

The following lemma will be used in the next section.

(3.3.7) LEMMA. *Let $\Phi : [0,1] \times [0,1] \to D$ be a continuous mapping. Then there are a $\delta > 0$, a partition $0 = t_0 < t_1 \cdots < t_n = 1$, and disks Δ_j, $1 \leqq j \leqq n$, in D (some of them may be disks about ∞) such that $\Phi([t_{j-1}, t_j] \times [0, \delta]) \subset \Delta_j$.*

PROOF. Applying (3.3.5) to $\Phi(t, 0)$, we may take $\delta > 0$ so that

$$\Phi([t_{j-1}, t_j] \times [0, \delta]) \subset \Delta_j, \qquad 1 \leqq j \leqq n. \qquad \square$$

The next theorem is called *Jordan's theorem.* We will not use it, but it will be helpful to know that such a theorem holds.

(3.3.8) THEOREM. *A Jordan closed curve in* **C** *divides* **C** *into two domains. One of them is a bounded simply connected domain.*

EXERCISE 2. Show that if a closed curve C in a domain D is homotopic to a point, then so is nC with $n \in \mathbf{Z}$. Here $0C$ is considered as a constant curve mapped to the initial point of C.

EXERCISE 3. Let $D = \mathbf{C} \setminus \{0, 1\}$. Set $C_1 = C(0; 1/2)$ and $C_2 = C(1; 1/2)$, let the initial (=terminal) points of them be $1/2$, and let the orientation be anti-clockwise. Show that $C_1 + C_2$ is homotopic to $C_2 + C_1$.

3

8	7			6				
			8	5			7	
	1	4			2	9		
9				4				2
2	4						1	7
1				8				5
		9	6			3	8	
	8			3	5			
				1			5	9

4

	6		5	4			2	
		7				4		
4			8			6		
3	9	1		8				2
2								3
7				9		1	4	8
		9			6			1
		5				3		
	7			3	8		5	

1

2				4	1	6	5	
			9	6	3		1	
	1						9	
							8	4
9	4						7	6
8	6							
	2						4	
	7		4	5	6			
	8	1	7	9				3

2

4	9		6					
8	7				9			5
		5				6	2	
	2		8	6		5	3	4
6	4	3		7	2		9	
	8	7						
5			7				8	6
					8		5	1

3.4. Cauchy's Integral Theorem

Let D be a domain of $\mathbf{C}$, and let f be a holomorphic function on D. Let C_0 denote the perimeter of a closed triangle E_0 with vertices, z_j, $1 \leqq j \leqq 3$, which is a closed piecewise linear curve formed by the sides of E_0

(3.4.1) LEMMA. *If* $E_0 \subset D$, *then* $\displaystyle\int_{C_0} f(z)dz = 0.$

PROOF. Let C_0 be endowed with the anti-clockwise orientation, and let z_1, z_2 and z_3 be ordered anti-clockwise, too. Let C_j be the side from z_j to z_{j+1} $(1 \leqq j \leqq 3,\ z_4 = z_1)$.

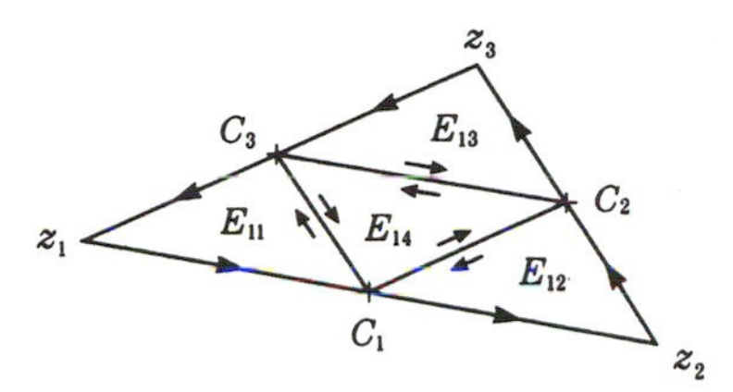

FIGURE 29

Then

$$C_0 = C_1 + C_2 + C_3.$$

Taking the middle point of each C_j and connecting them by line segments, we divide E_0 into four closed triangles, E_{11}, E_{12}, E_{13}, E_{14}. Let C_{1kl}, $1 \leqq l \leqq 3$, denote the sides of E_{1k} as in the case of C_j. The closed piecewise linear curves

$$C_{1k} = C_{1k1} + C_{1k2} + C_{1k3}, \qquad 1 \leqq k \leqq 4,$$

form the perimeter of E_{1k}. It follows from (3.2.13) and (3.2.14) that

$$\int_{C_0} f(z)dz = \sum_{k=1}^{4} \int_{C_{1k}} f(z)dz. \tag{3.4.2}$$

Repeating this process n times, we have 4^n closed triangles E_{nk} $(1 \leqq k \leqq 4^n)$ and their perimeter C_{nk}, and as in (3.4.2) we obtain

$$\int_{C_0} f(z)dz = \sum_{k=1}^{4^n} \int_{C_{nk}} f(z)dz. \tag{3.4.3}$$

Now we assume that

$$\alpha = \left| \int_{C_0} f(z)dz \right| > 0.$$

Then it can be deduced from (3.4.2) that for some C_{1k_1}

$$\left| \int_{C_{1k_1}} f(z)dz \right| \geqq \frac{\alpha}{4}.$$

Applying the same argument to E_{1k_1} and C_{1k_1}, we have E_{2k_2} and C_{2k_2} such that $E_{2k_2} \subset E_{1k_1}$ and

$$\left| \int_{C_{2k_2}} f(z)dz \right| \geqq \frac{\alpha}{4^2}.$$

We have inductively $E_{nk_n} \subset E_{n-1k_{n-1}}$ with perimeter C_{nk_n}, so that

$$\left| \int_{C_{nk_n}} f(z)dz \right| \geqq \frac{\alpha}{4^n}.$$

For the sake of simplicity we write $E'_n = E_{nk_n}$ and $C'_n = C_{nk_n}$, and so

$$\left| \int_{C'_n} f(z)dz \right| \geqq \frac{\alpha}{4^n}, \qquad n = 1, 2, \dots . \tag{3.4.4}$$

Letting $M = \max\{L(C_j); 1 \leqq j \leqq 3\}$, we have

$$L(C'_n) \leqq \frac{3}{2^n} M. \tag{3.4.5}$$

The diameter R_n of E'_n ($R_n = \max\{|z-w|; z, w \in E'_n\}$) satisfies

$$R_n = \frac{M}{2^n}. \tag{3.4.6}$$

Since E'_n, $n = 1, 2, \dots$, are decreasing compact sets, $\bigcap_{n=1}^{\infty} E'_n \neq \emptyset$ (cf. Exercise 7, Chapter 1). Let $a \in \bigcap_{n=1}^{\infty} E'_n$. It follows from (3.4.6) that $R_n \to 0$ ($n \to \infty$), and so $\bigcap_{n=1}^{\infty} E'_n = \{a\}$. Since f is complex differentiable at a, for an arbitrary $\epsilon > 0$ there is a $\delta > 0$ such that

$$|f(z) - \{f(a) + f'(a)(z-a)\}| \leqq \epsilon |z-a|, \qquad z \in \Delta(a; \delta). \tag{3.4.7}$$

Take n_0 so that $M/2^{n_0} < \delta$. Then, for $n \geqq n_0$ we have by (3.4.5)~(3.4.7) and (3.2.10)

$$\begin{aligned} &\left| \int_{C'_n} f(z) - \int_{C'_n} \{f(a) + f'(a)(z-a)\} dz \right| \\ &\leqq \epsilon \int_{C'_n} |z-a||dz| \leqq \epsilon \frac{M}{2^n} L(C'_n) \leqq \epsilon \frac{3M^2}{4^n}. \end{aligned} \tag{3.4.8}$$

The function $\{f(a) + f'(a)(z-a)\}$ has a primitive function

$$f(a)z + \frac{f'(a)}{2}(z-a)^2,$$

and so Theorem (3.2.19) implies

$$\int_{C'_n} \{f(a) + f'(a)(z-a)\} dz = 0.$$

It follows from (3.4.8) that

$$\left| \int_{C'_n} f(z)dz \right| \leqq \epsilon \frac{3}{4^n} M^2.$$

This with (3.4.4) yields $\alpha/4^n < 3\epsilon M^2/4^n$, and hence $\alpha < 3M^2\epsilon$. Since $\epsilon > 0$ is arbitrary, we have a contradiction. □

REMARK. A good point of the above proof is that it is not necessary to assume the continuity nor the integrability of the derivative $f'(z)$ of $f(z)$.

(3.4.9) LEMMA. *On an arbitrary disk $\Delta(a;r) \subset D$, f has a primitive function.*

PROOF. For a point $z \in \Delta(a;r)$ we define a curve C_z by

$$\phi : t \in [0,1] \to a + t(z-a) \in \Delta(a;r),$$

and set

$$F(z) = \int_{C_z} f(\zeta)d\zeta.$$

We show that $F(z)$ is a primitive function of $f(z)$.

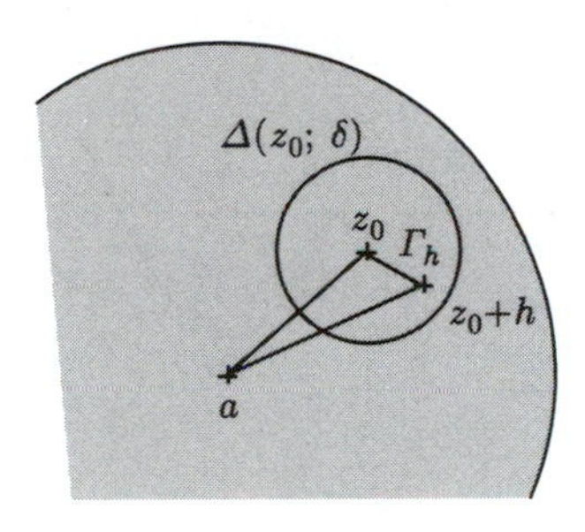

FIGURE 30

Fix an arbitrary point $z_0 \in \Delta(a;r)$. For an arbitrary $\epsilon > 0$ there is a $0 < \delta < r - |a|$ such that

$$|f(z_0 + h) - f(z_0)| < \epsilon, \qquad |h| < \delta. \tag{3.4.10}$$

Let Γ_h denote the line segment from z_0 to $z_0 + h$. It follows from Lemma (3.4.1) and (3.4.10) that

$$\begin{aligned}
&\left| \frac{F(z_0+h) - F(z_0)}{h} - f(z_0) \right| \\
&= \left| \frac{1}{h} \int_{\Gamma_h} f(z)dz - f(z_0) \right| = \left| \frac{1}{h} \int_{\Gamma_h} (f(z) - f(z_0))dz \right| \\
&\leqq \frac{1}{|h|} \int_{\Gamma_h} |f(z) - f(z_0)||dz| < \frac{\epsilon}{|h|} L(\Gamma_h) = \epsilon
\end{aligned}$$

Therefore F is holomorphic, and $F' = f$. □

Let $C(\varphi : [0;1] \to D)$ be a curve. As in the proof of Lemma (3.3.4), there is a partition $(d) : 0 = t_0 < \cdots < t_n = 1$ of $[0,1]$ fine enough so that (3.3.5) holds. Then the piecewise linear curve $\tilde{C}(d)$ connecting $\varphi(t_0), \ldots,$ and $\varphi(t_n)$ in order by line segments is homotopic to C.

(3.4.11) LEMMA. *The curvilinear integral* $\int_{\tilde{C}(d)} f(z)dz$ *is independent of* (d), *provided only that* (3.3.5) *is satisfied.*

PROOF. It is sufficient to prove that for a refinement (d') of (d)

$$\int_{\tilde{C}(d)} f(z)dz = \int_{\tilde{C}(d')} f(z)dz. \tag{3.4.12}$$

For we see by (3.4.12) that for two partitions (d_1) and (d_2)

$$\int_{\tilde{C}(d_1)} f(z)dz = \int_{\tilde{C}(d_3)} f(z)dz = \int_{\tilde{C}(d_2)} f(z)dz,$$

where (d_3) is the partition formed by all partition points of (d_1) and (d_2). To show (3.4.12) it is sufficient to deal with the case where one partition point t'_j is added to (t_j, t_{j+1}).

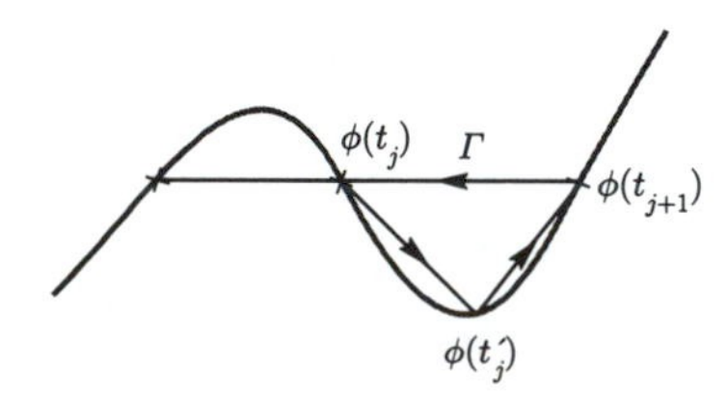

FIGURE 31

In this case we denote by E the triangle with vertices $\varphi(t_j)$, $\varphi(t'_j)$, $\varphi(t_{j+1})$, and by Γ the perimeter of E. Lemma (3.4.1) implies

$$\int_{\tilde{C}(d')} f(z)dz - \int_{\tilde{C}(d)} f(z)dz = \int_{\Gamma} f(z)dz = 0. \qquad \square$$

By Lemma (3.4.11) we can define the *curvilinear integral* of f along C by

$$\int_C f(z)dz = \int_{\tilde{C}(d)} f(z)dz, \tag{3.4.13}$$

where (d) satisfies (3.3.5). The next theorem is called *Cauchy's integral theorem:*

(3.4.14) THEOREM. *Let f be a holomorphic function on D, and let C be a closed curve in D which is homotopic to a point. Then*

$$\int_C f(z)dz = 0.$$

PROOF. Let C be given by $\varphi : [0,1] \to D$, and set $z_0 = \varphi(0) = \varphi(1)$. There is a homotopy

$$\Phi : [0,1] \times [0,1] \to D$$

connecting φ and z_0. Then $\varphi(t) = \Phi(t,0)$, and $\Phi(t,1) \equiv z_0$. Denoting by C_s the curve given by $\Phi_s(t) = \Phi(t,s)$, we define

$$\lambda = \sup\left\{ s \in [0,1]; \int_{C_{s'}} f(z)dz = \int_C f(z)dz, \quad 0 \leqq s' \leqq s \right\}.$$

As in Lemma (3.3.7) we take $\delta > 0$ and $0 = t_0 < \cdots < t_n = 1$. Let Γ_j denote the piecewise linear curve connecting $\Phi(t_j, 0)$, $\Phi(t_j, s)$, $\Phi(t_{j+1}, s)$, $\Phi(t_{j+1}, 0)$, and $\Phi(t_j, 0)$ in that order.

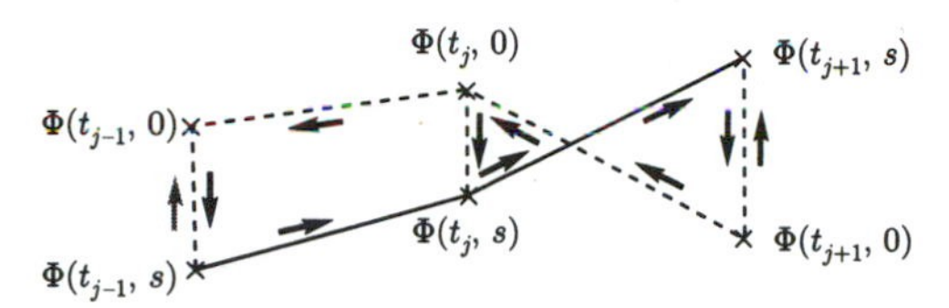

FIGURE 32

We infer from the definition of Γ_j, Lemma (3.4.9), and Theorem (3.2.19) that

$$\int_{\Gamma_j} f(z)dz = 0.$$

The line segment connecting $\Phi(t_j, 0)$ and $\Phi(t_j, s)$ has the opposite orientation in Γ_{j-1} and Γ_j, so that

$$\int_{C_s} f(z)dz - \int_C f(z)dz = \sum_{j=0}^{n-1} \int_{\Gamma_j} f(z)dz = 0.$$

Thus $\lambda \geqq \delta$. If $\lambda < 1$, there would be a small $\delta' > 0$ such that, as in Lemma (3.3.7), $\Phi([t_{j-1}, t_j] \times [\lambda - \delta', \lambda + \delta'])$ is contained in a disk. Then

$$\int_{C_{\lambda+\delta'}} f(z)dz = \int_{C_{\lambda-\delta'}} f(z)dz = \int_C f(z)dz.$$

Therefore $\lambda \geqq \lambda + \delta'$, which is a contradiction. We have $\lambda = 1$. Take $r > 0$ so that $\Delta(z_0; r) \subset D$. There is a $\delta'' > 0$ such that $C_s \subset \Delta(z_0; r)$ for all $1 - \delta'' \leqq s < 1$. Take $\tilde{C}_s(d) \subset \Delta(z_0; r)$ as in (3.3.5). We infer from Lemma (3.4.9) and Theorem (3.2.19) that

$$\int_{C_s} f(z)dz = \int_{\tilde{C}_s(d)} f(z)dz = 0.$$

Hence $\int_C f(z)dz = 0$. $\square$

By Cauchy's integral theorem we find that the curvilinear integral along a curve C in D,

$$\int_C f(z)dz,$$

depends only on the homotopy class of C in D. In fact, if C_1 and C_2 are mutually homotopic, then Theorem (3.3.3) implies that $C_1 - C_2$ is homotopic to a point, and that

$$\int_{C_1} f(z)dz - \int_{C_2} f(z)dz = \int_{C_1 - C_2} f(z)dz = 0.$$

In particular, if D is simply connected, the curvilinear integral $\int_C f(z)dz$ is determined only by the initial point z_0 and the terminal point z_1 of C. Thus we write

$$\int_{z_0}^{z_1} f(z)dz = \int_C f(z)dz.$$

Fixing $z_0 \in D$, we define

$$F(z) = \int_{z_0}^{z_1} f(z)dz.$$

We see by Lemma (3.4.9) that F is holomorphic and a primitive function of f. We call $\int_{z_0}^{z_1} f(z)dz$ the *indefinite integral* of f. Therefore we obtain the following:

(3.4.15) THEOREM. *A holomorphic function f on a simply connected domain D has a primitive function F; moreover, for a fixed point $z_0 \in D$*

$$F(z) = F(z_0) + \int_{z_0}^{z} f(z)dz.$$

EXERCISE 1. Let $f(z) = 1/(z+i)^2$, and $R > 0$. Show that $\int_{[0,R]} f(z)dz$ and $\int_{[-R,0]} f(z)dz$ converge as $R \to +\infty$, and that

$$\int_{\mathbf{R}} f(z) = \lim_{R \to +\infty} \int_{[-R,R]} f(z)dz = 0.$$

EXERCISE 2. On $\Delta(1)$, find a primitive function of $f(z) = (z^2-1)^{-2}$ in power series.

3.5. Cauchy's Integral Formula

Let $C(a;r)$ be the perimeter circle of the disk $\Delta(a;r)$ given by

$$\phi : \theta \in [0, 2\pi] \to a + te^{i\theta} \in C(a;r). \tag{3.5.1}$$

(3.5.2) THEOREM. $\displaystyle\int_{C(a;r)} \frac{1}{z-a} dz = 2\pi i.$

PROOF. Set $z = \phi(\theta) = a + re^{i\theta}$. Since $\phi'(\theta) = rie^{i\theta}$, we have by (3.2.12)

$$\int_{C(a;r)} \frac{1}{z-a} dz = \int_0^{2\pi} \frac{rie^{i\theta}}{re^{i\theta}} d\theta = 2\pi i. \qquad \square$$

Let $D \subset \mathbf{C}$ be a domain, and $a \in D$. We define

$$d(a; \partial D) = \inf\{|a - w|; w \in \partial D\} \leqq +\infty.$$

If $d(a; \partial D) < +\infty$, there is a $w \in \partial D$ with $d(a; \partial D) = |a - w|$. The next theorem is called *Cauchy's integral formula*:

(3.5.3) THEOREM. *For an arbitrary $a \in D$ and $0 < r < d(a; \partial D)$*

$$f(z) = \frac{1}{2\pi i} \int_{C(a;r)} \frac{f(\zeta)}{\zeta - z} d\zeta, \qquad z \in \Delta(a; r).$$

PROOF. For the sake of simplicity we may let $a = 0$. Let $z = |z|e^{i\theta_0} \in \Delta(r)$. Take a $\delta > 0$ with $\overline{\Delta(z; \delta)} \subset \Delta(r)$. Let $C_1 = C(0; r)$, let C_2 be the line segment from $z + \delta e^{i\theta_0}$ to $re^{i\theta_0}$, and let $C_3 = C(z; \delta)$. We may assume that the initial and terminal point of C_1 is $re^{i\theta_0}$, and that of C_3 is $z + \delta e^{i\theta_0}$. Set $C = C_1 - C_2 - C_3 + C_2$. Then C is homotopic to the curve $\tilde{C}$ given in Figure 33.

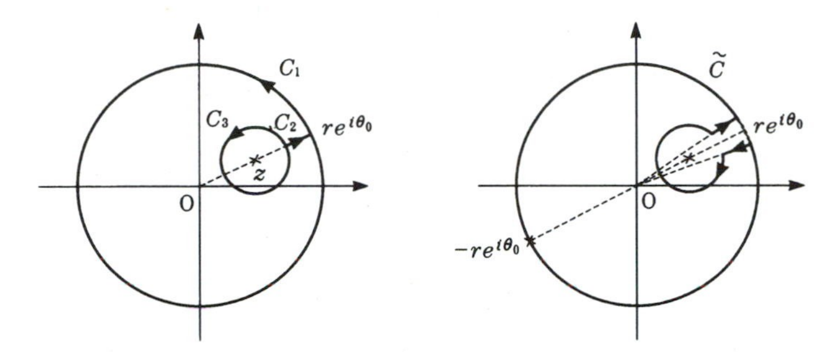

FIGURE 33

Let $\tilde{C}$ be given by $\tilde{\phi} : [0, 1] \to D$, and let

$$\Phi : (t, s) \in [0, 1] \times [0, 1] \to (1 - s)\tilde{\phi}(t) - sre^{i\theta_0} \in D \setminus \{z\}$$

be a homotopy connecting $\tilde{C}$ and the point $-re^{i\theta_0}$ in $D \setminus \{z\}$. Hence C is homotopic to a point in $D \setminus \{z\}$. Since the function $f(\zeta)/(\zeta - z)$ is holomorphic in $D \setminus \{z\}$, it follows from Theorem (3.4.14) that

$$\int_C \frac{f(\zeta)}{\zeta - z} d\zeta = 0.$$

One obtains

$$\int_{C_1} \frac{f(\zeta)}{\zeta - z} d\zeta = \int_{C_3} \frac{f(\zeta)}{\zeta - z} d\zeta.$$

Since

$$\begin{aligned} \int_{C_3} \frac{f(\zeta)}{\zeta - z} d\zeta &= \int_0^{2\pi} \frac{f(z + \delta e^{i\theta})}{(z + \delta e^{i\theta}) - z} \delta i e^{i\theta} d\theta \\ &= \int_0^{2\pi} f(z + \delta e^{i\theta}) i d\theta \to 2\pi i f(z) \qquad (\delta \to 0), \end{aligned}$$

one gets

$$\frac{1}{2\pi i}\int_{C_1}\frac{f(\zeta)}{\zeta - z}d\zeta = f(z). \qquad \square$$

Fix $z \in \Delta(a;r)$. As $h \to 0$, the following convergence is uniform in $\zeta \in C(a;r)$:

$$\frac{1}{h}\left\{\frac{f(\zeta)}{\zeta-(z+h)} - \frac{f(\zeta)}{\zeta - z}\right\} \to \frac{f(\zeta)}{(\zeta - z)^2}.$$

Thus we deduce the integral formula for the derivative of f from Theorem (3.5.3):

$$f'(z) = \frac{1}{2\pi i}\int_{C(a;r)}\frac{f(\zeta)}{(\zeta - z)^2}d\zeta, \qquad z \in \Delta(a;r).$$

Repeating this n times, we find that the n-th derivative $f^{(n)}(z)$ is expressed by

(3.5.4)
$$f^{(n)}(z) = \frac{n!}{2\pi i}\int_{C(a;r)}\frac{f(\zeta)}{(\zeta - z)^{n+1}}d\zeta,$$
$$z \in \Delta(a;r), \quad n = 0, 1, \dots .$$

For $z \in \Delta(a;r)$ and $\zeta \in C(a;r)$ one has

$$\frac{1}{\zeta - z} = \frac{1}{\zeta - a}\frac{1}{1 - \frac{z-a}{\zeta - a}} = \frac{1}{\zeta - a}\sum_{n=0}^{\infty}\left(\frac{z-a}{\zeta - a}\right)^n.$$

Here, note that

$$\left|\frac{z-a}{\zeta - a}\right| \leqq \frac{|z-a|}{r} < 1.$$

Therefore the above power series converges uniformly in $\zeta \in C(a;r)$. It follows from (3.2.11) that

(3.5.5)
$$f(z) = \sum_{n=0}^{\infty} a_n (z-a)^n,$$

(3.5.6)
$$a_n = \frac{1}{2\pi i}\int_{C(a;r)}\frac{f(\zeta)}{(\zeta - a)^{n+1}}d\zeta$$
$$= \frac{1}{2\pi r^n}\int_0^{2\pi} f(a + re^{i\theta})e^{-in\theta}d\theta.$$

Since the right side of (3.5.5) converges for $z \in \Delta(a;r)$, its radius R of convergence is not less than r. Letting $r \nearrow d(a;\partial D)$, we get $R \geqq d(a;\partial D)$. Hence we have the following:

(3.5.7) THEOREM. *A holomorphic function f on D is analytic, and about an arbitrary point $a \in D$ $f(z)$ is expanded to a power series*

$$f(z) = \sum_{n=0}^{\infty} a_n (z-a)^n, \qquad z \in \Delta(a; d(a;\partial D)),$$

whose radius of convergence is not less than $d(a;\partial D)$.

One sees by (3.5.6) and (3.1.13) that

$$f^{(n)}(a) = \frac{n!}{2\pi i}\int_{C(a;r)} \frac{f(\zeta)}{(\zeta - a)^{n+1}} d\zeta = n! a_n. \tag{3.5.8}$$

(3.5.9) REMARK. We see by Theorem (3.5.7) and Corollary (3.1.12) that holomorphic functions and analytic functions are the same. Therefore theorems for analytic functions also hold for holomorphic functions. It follows from (3.5.4) and also from Theorem (3.1.10) that holomorphic functions are complex differentiable arbitrarily many times, so that the derivative of a holomorphic function is again holomorphic. The identity Theorem (2.4.14) is valid for holomorphic functions. Conversely, the theorems for holomorphic functions hold for analytic functions. Thus the composite of analytic functions is analytic (cf. Theorem (3.1.9)).

(3.5.10) THEOREM. *Let f and g be holomorphic functions on D. If $f^{(n)}(a) = g^{(n)}(a)$, $n = 0, 1, \ldots$, at a point $a \in D$, then $f \equiv g$ on D.*

PROOF. Theorem (3.5.7) and (3.5.8) imply that $f \equiv g$ on $\Delta(a; d(a; \partial D))$. One deduces from Theorem (2.4.14) that $f \equiv g$ on D. □

Now we return to the logarithmic function $\log z$, which was defined only for $z \in \Delta(1;1)$ by (2.5.4). Let C be a curve from 0 to $z \neq 0$, not passing through 0, and define

$$F_0(z) = \int_C \frac{1}{\zeta} d\zeta. \tag{3.5.11}$$

If D is a simply connected domain containing 1, but not 0, and if $C \subset D$, then $F_0(z)$ restricted to D gives rise to a holomorphic function on D. A direct computation yields $F_0^{(n)}(1) = \log^{(n)}(1)$, $n = 0, 1, \ldots$, so that by Theorem (3.5.10) $F_0(z) \equiv \log z$ on $\Delta(1;1)$. Therefore

$$e^{F_0(z)} = z, \qquad z \in \Delta(1;1). \tag{3.5.12}$$

Hence we call the function $F_0(z)$ defined by (3.5.11) the logarithmic function, and denote it by $\log z$. The logarithmic function $\log z$ is defined for $z \neq 0$, but not a function such as those we have dealt with before. It is a so-called *multi-valued* function (in contrast, the functions dealt with before are called 1-*valued* functions). One deduces from (3.5.12) and identity Theorem (2.4.14) that

$$e^{\log z} = z. \tag{3.5.13}$$

Therefore $\log z$ is determined modulo $2\pi i$, and

$$\log z = \log|z| + i \arg z.$$

Let $D \subset \mathbf{C}^*$ be a simply connected domain, and fix a point $a \in D$. Take a curve C_1 from 1 to a. For an arbitrary point z we take a curve C_2 in D from a to z. Then

$$\log z = \int_{C_1+C_2} \frac{d\zeta}{\zeta} = \int_{C_1} \frac{d\zeta}{\zeta} + \int_{C_2} \frac{d\zeta}{\zeta}.$$

When C_1 is fixed, this integration is independent of the choice of C_2, and one may write

$$\log z = f(z) = \int_{C_1} \frac{d\zeta}{\zeta} + \int_a^z \frac{d\zeta}{\zeta}.$$

This defines a 1-valued function on D, which is called a *branch* of $\log z$. (The general notion of branch will be given in Chapter 5, §2.) If C_1 is changed, one gets another branch $g(z)$. Since $e^{f(z)} = e^{g(z)}$, there is an $n \in \mathbf{Z}$ such that

$$f(z) - g(z) \equiv 2n\pi i, \qquad z \in D.$$

That is, the difference of two branches of $\log z$ is an integral multiple of $2\pi i$.

Let $a \in \mathbf{C}$, and let C be a curve not passing through a. The *rotation number* of C around a is defined by

$$n(a; C) = \frac{1}{2\pi i} \int_C \frac{d\zeta}{\zeta - a} \in \mathbf{Z}.$$

(3.5.14) THEOREM. *A curve C in $\mathbf{C} \setminus \{a\}$ is homotopic to a point if and only if $n(a; C) = 0$.*

PROOF. We may assume that $a = 0$, and by Lemma (3.3.4) and Theorem (3.4.14) that C is a piecewise linear curve. The "only if" part is clear. Assume that $n(0; C) = 0$. Let C be given by $\varphi : [0,1] \to \mathbf{C}^*$. We may assume $\varphi(0) = 1$. Set

$$\Phi : (t, s) \in [0,1] \times [0,1] \to (1-s)\varphi(t) + \frac{\varphi(t)}{|\varphi(t)|} s \in \mathbf{C}^*.$$

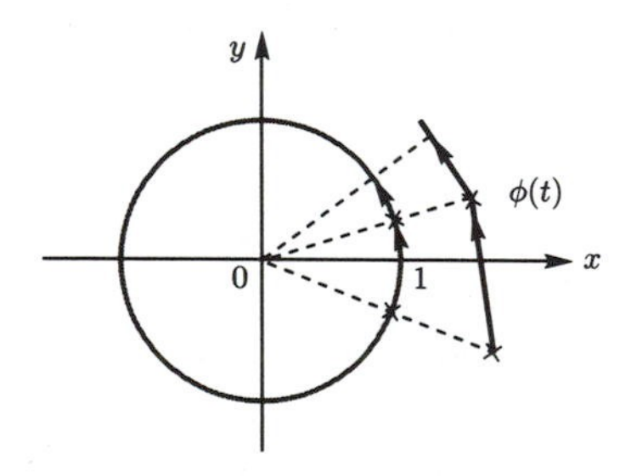

FIGURE 34

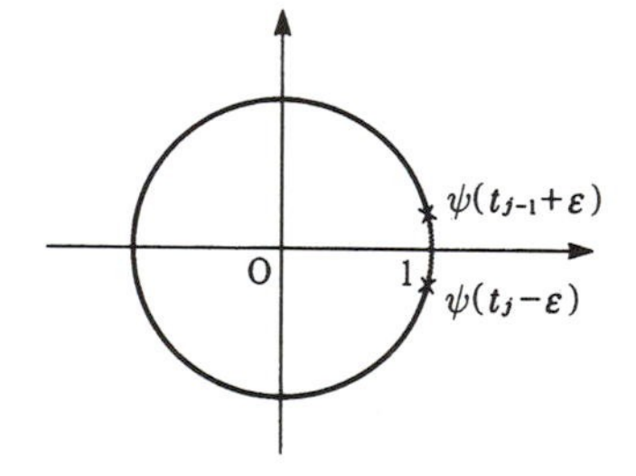

FIGURE 35

Then φ is homotopic to $\tilde{C}(\psi : [0,1] \to C(0;1))$ with $\psi(t) = \Phi(t, 1)$. Since the piecewise linear curve C may be changed within its homotopy class, one may assume that $\operatorname{Im}\psi(t_1) \cdot \operatorname{Im}\psi(t_2)$, $t_1 < t_0 < t_2$, in a neighborhood of t_0 with $\psi(t_0) = 1$. Let $0 = t_0 < t_1 < \cdots < t_l = 1$ be all the points $t \in [0,1]$ such that $\psi(t) = 1$. Let $\epsilon > 0$ be taken small, so that $\psi|[t_{j-1}, t_j]$ does not intersect $\{x \in \mathbf{R}; x \geqq 0\}$. Since $\mathbf{C} \setminus \{x \in \mathbf{R}; x \geqq 0\}$ is simply connected, a branch of $\log z$ is determined. Then

$$\int_{\tilde{C}|[t_{j-1}+\epsilon, t_j-\epsilon]} \frac{d\zeta}{\zeta} = \log \psi(t_j - \epsilon) - \log \psi(t_{j-1} + \epsilon) \\ \to 0, \pm 2\pi i \qquad (\epsilon \to 0).$$

Thus $\int_{\tilde{C}|[t_{j-1}, t_j]} \frac{d\zeta}{\zeta} = 0, \pm 2\pi i$. By the condition we have

$$\sum_{j=1}^{l} \int_{\tilde{C}|[t_{j-1}, t_j]} \frac{d\zeta}{\zeta} = 0. \tag{3.5.15}$$

Let $C_0 = C(0; 1)$ be as defined by (3.5.1). If $\int_{\tilde{C}|[t_{j-1}, t_j]} \frac{d\zeta}{\zeta} = 0$, then $\tilde{C}|[t_{j-1}, t_j]$ is homotopic to a point. If $\int_{\tilde{C}|[t_{j-1}, t_j]} \frac{d\zeta}{\zeta} = 2\pi i$ (resp., $-2\pi i$), then $\tilde{C}|[t_{j-1}, t_j]$ is homotopic to C_0 (resp., $-C_0$). By (3.5.15) the numbers of $\tilde{C}|[t_{j-1}, t_j]$ homotopic to C_0 and to $-C_0$ are equal. Since $C_0 - C_0$ is homotopic to a point, so is $\tilde{C}$. □

From now on, let $D \subset \mathbf{C}$ be a general domain.

(3.5.16) THEOREM. *Let f be a holomorphic function on D. Let C be a curve in D homotopic to a point, and let $z \in D$ not be on C. Then*

$$n(z; C) f(z) = \frac{1}{2\pi i} \int_C \frac{f(\zeta)}{\zeta - z} d\zeta.$$

PROOF. Set

$$g(\zeta) = \frac{f(\zeta)}{\zeta - z} - \frac{f(z)}{\zeta - z}, \qquad \zeta \in D \setminus \{z\}.$$

Expand $f(\zeta)$ in a neighborhood $\Delta(z; r) \subset D$ of z:

$$f(\zeta) = \sum_{n=0}^{\infty} a_n (\zeta - z)^n, \qquad a_0 = f(z).$$

Then one has

$$g(\zeta) = \sum_{n=1}^{\infty} a_n (\zeta - z)^{n-1}, \qquad \zeta \in \Delta(z; r) \setminus \{z\}. \tag{3.5.17}$$

Since this right side converges in $\Delta(z; r)$, $g(\zeta)$ defined by (3.5.17) on $\Delta(z; r)$ is holomorphic on D. Hence

$$\int_C \frac{f(\zeta)}{\zeta - z} d\zeta - \int_C \frac{f(z)}{\zeta - z} d\zeta = \int_C g(\zeta) d\zeta = 0.$$

By the definition of $n(z; C)$ one gets the desired formula. □

The integral formula in the above theorem is called *Cauchy's integral formula*, too. The next gives the inverse of Cauchy's integral Theorem (3.4.14), and is called *Morera's theorem*:

(3.5.18) THEOREM. *Let f be a continuous function on D. If for the perimeter C of an arbitrary closed triangle contained in D*

$$\int_C f(z) = 0,$$

then f is holomorphic.

PROOF. Let $a \in D$. It is sufficient to show that f is holomorphic in $\Delta(a; r_0)$ with $r_0 = d(a; \partial D)$.

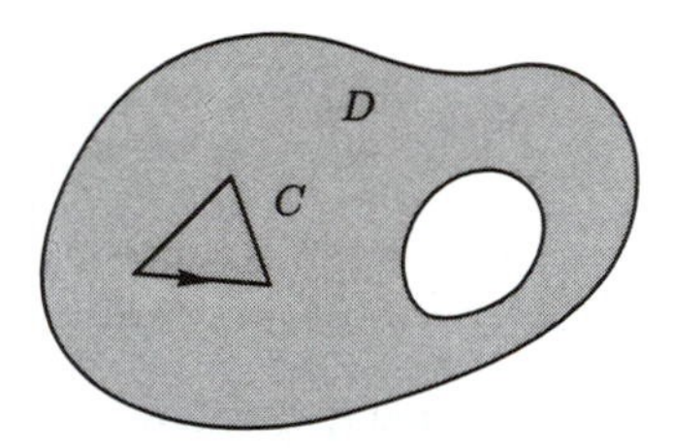

FIGURE 36

For an arbitrary point $z \in \Delta(a; r_0)$ we take the line segment C_z from a to z and set

$$F(z) = \int_{C_z} f(\zeta) d\zeta.$$

By making use of the assumption one sees as in the proof of Lemma (3.4.9) that F is holomorphic. Since $F' = f$, f is holomorphic. □

(3.5.19) THEOREM. *Let $\{f_n\}_{n=0}^{\infty}$ be a sequence of holomorphic functions on D which converges uniformly on compact subsets of D. Then the limit function f is holomorphic and*

$$f^{(k)}(z) = \lim_{n \to \infty} f_n^{(k)}(z), \qquad z \in D, \quad k \in \mathbf{Z}^+.$$

Here the convergence is uniform on compact subsets of D.

PROOF. Let C be a curve as in Theorem (3.5.18). Then

$$\int_C f(z) dz = \int_C \lim f_n(z) dz = \lim \int_C f_n(z) dz = 0.$$

Thus by the same theorem f is holomorphic.

Let $a \in D$ and $0 < r_1 < r < d(a; \partial D)$. It follows from (3.5.4) that for $z \in \overline{\Delta(a; r_1)}$

$$\begin{aligned} f^{(k)}(z) &= \frac{k!}{2\pi i} \int_{C(a;r)} \frac{f(\zeta)}{(\zeta - z)^{k+1}} d\zeta \\ &= \frac{k!}{2\pi i} \int_{C(a;r)} \lim_{n\to\infty} \frac{f_n(\zeta)}{(\zeta - z)^{k+1}} d\zeta \\ &= \lim_{n\to\infty} \frac{k!}{2\pi i} \int_{C(a;r)} \frac{f_n(\zeta)}{(\zeta - z)^{k+1}} d\zeta \\ &= \lim_{n\to\infty} f_n^{(k)}(z). \end{aligned}$$

Here the convergence is uniform in $z \in \overline{\Delta(a; r_1)}$. □

(3.5.20) THEOREM. *Let $f(z) = \sum_{n=0}^{\infty} a_n z^n$ be a holomorphic function on $\Delta(R)$. For $0 < r < R$ we have*

i) $|a_n| \leqq \frac{1}{2\pi r^n} \int_0^{2\pi} |f(re^{i\theta})| d\theta$,

ii) $\sum_{n=0}^{\infty} |a_n|^2 r^{2n} = \frac{1}{2\pi} \int_0^{2\pi} |f(re^{i\theta})^2| d\theta$.

PROOF. i) This follows immediately from (3.5.6).

ii) Noting that $\sum_{n=0}^{\infty} |a_n| r^n$ converges, we have

$$\begin{aligned} \frac{1}{2\pi} \int_0^{2\pi} |f(re^{i\theta})|^2 d\theta &= \frac{1}{2\pi} \int_0^{2\pi} \sum_{n,m=0}^{\infty} a_n \bar{a}_m r^{n+m} e^{i(n-m)\theta} d\theta \\ &= \sum_{n,m=0}^{\infty} a_n \bar{a}_m \frac{r^{n+m}}{2\pi} \int_0^{2\pi} e^{i(n-m)\theta} d\theta. \end{aligned}$$

If $n \neq m$, $\int_0^{2\pi} e^{i(n-m)\theta} d\theta = 0$, and so the desired equality follows. □

The next theorem is called the *maximum principle.*

(3.5.21) THEOREM. *If a holomorphic function f on D takes the maximum at a point of D, then f is constant.*

PROOF. Assume that f takes the maximum at $z_0 \in D$. Let $0 < r < d(z_0; \partial D)$. Expand $f(z) = \sum_{n=0}^{\infty} a_n (z - z_0)^n$, $z \in \Delta(z_0; \partial D)$. Since $|a_0| = |f(z_0)|$, it follows from Theorem (3.5.20), ii) that

$$|a_0|^2 + \sum_{n=1}^{\infty} |a_n|^2 r^{2n} \leqq |a_0|^2.$$

Hence $a_n = 0$, $n \geqq 1$, so that f is constant by Theorem (3.5.10). □

The next is called *Liouville's theorem:*

(3.5.22) THEOREM. *A bounded holomorphic function f on $\mathbf{C}$ is constant.*

PROOF. Expand $f(z)$ about 0 to a power series:

$$f(z) = \sum_{n=0}^{\infty} a_n z^n.$$

Then the radius of convergence is $+\infty$. There is an $M > 0$ such that $|f(z)| \leqq M$ for $z \in \mathbf{C}$. For an arbitrary $R > 0$ one has by Theorem (3.5.20), i)

$$|a_n| \leqq \frac{M}{R^n}, \qquad n = 0, 1, \ldots .$$

Letting $R \to +\infty$, one gets $a_n = 0$ for $n \geqq 1$. Thus f is constant. □

We get the following *fundamental theorem of algebra* due to Gauss as an application of the above theorem.

(3.5.23) THEOREM. *A polynomial $f(z) = \sum_{n=0}^{N} a_n z^n$ necessarily has a zero if it is not constant; i.e., there is a $z_0 \in \mathbf{C}$ with $f(z_0) = 0$.*

PROOF. Suppose that $f(z)$ does not have a zero on $\mathbf{C}$. Then $g(z) = 1/f(z)$ is holomorphic on $\mathbf{C}$. We may assume $a_N \neq 0$, $N \geqq 1$. Taking a sufficiently large R_0, we have for $|z| \geqq R_0$

$$\left| a_N + \frac{a_{N-1}}{z} + \cdots + \frac{a_0}{z^N} \right| \geqq |a_N| - \left| \frac{a_{N-1}}{z} + \cdots + \frac{a_0}{z^N} \right| \geqq \frac{1}{2}|a_N|.$$

Thus, for $|z| \geqq R_0$, $|g(z)| \leqq 2/|a_N||z|^N \leqq 2/|a_N|R_0^N$. Since $g(z)$ is bounded on $\overline{\Delta(R_0)}$, g is bounded on $\mathbf{C}$, and so constant. This is a contradiction. □

EXERCISE 1. Let f be a continuous function on $\overline{\Delta(R)}$ which is holomorphic in $\Delta(R)$. Show that

$$f(z) = \frac{1}{2\pi i} \int_{C(0;R)} \frac{f(\zeta)}{\zeta - z} d\zeta, \qquad z \in \Delta(R).$$

EXERCISE 2. Let $\{f_n\}_{n=0}^{\infty}$ be a sequence of continuous functions on $\overline{\Delta(R)}$ which are holomorphic on $\Delta(R)$. Assume that $\{f_n\}_{n=0}^{\infty}$ converges uniformly on the circle $C(0; R)$. Show that $\{f_n\}_{n=0}^{\infty}$ converges uniformly on $\overline{\Delta(R)}$.

EXERCISE 3. Let $R > 0$, $R \neq 1$, and let C denote a curve from R to $-R$ on $C(0; R)$ with anti-clockwise orientation. Show that

$$\int_C \frac{1}{1+z^2} dz = \begin{cases} -2 \arctan R & (0 < R < 1), \\ \pi - 2 \arctan R & (R > 1). \end{cases}$$

EXERCISE 4. Let $f(z)$ be a holomorphic function on $\mathbf{C}$ satisfying

$$|f(z)| \leqq M|z|^n, \qquad |z| \geqq R,$$

where $M > 0$ and $R > 0$ are constants. Show that $f(z)$ is a polynomial of degree at most n.

EXERCISE 5. Let f be a continuous function on $\overline{\Delta(1)}$ which is holomorphic in $\Delta(1)$. Assume that $f(e^{i\theta})$ is constant on an arc $\{e^{i\theta}; 0 \leqq \theta_1 < \theta < \theta_2 \leqq 2\pi\}$. Then, show that $f(z)$ is constant on $\overline{\Delta(1)}$.

EXERCISE 6. Let f be a holomorphic function on D. For an arbitrary point $w \in f(D)$, show that the inverse image $f^{-1}(w)$ is either the whole of D, or a discrete subset of D.

3.6. Mean Value Theorem and Harmonic Functions

Let $D \subset \mathbf{C}$ be a domain, let f be a holomorphic function on D, and let $a \in D$. By Theorem (3.5.3) we have

$$f(a) = \frac{1}{2\pi}\int_0^{2\pi} f(a + re^{i\theta})d\theta, \qquad 0 < r < d(a; \partial D). \tag{3.6.1}$$

This is called the *mean value theorem* for holomorphic functions. We set $a = 0$ and $r = 1$ for the sake of simplicity in what follows. The value $f(0)$ of f at the center 0 of the disk $\Delta(1)$ is expressed by the integration of f over the boundary circle $C(0; 1)$ of $\Delta(1)$. We consider a similar expression for a point of $\Delta(1)$ other than the center.

Let $z \in \Delta(1)$ and take

$$\phi_z(\zeta) = \frac{\zeta - z}{-\bar{z}\zeta + 1}, \qquad \zeta \in \Delta(1).$$

As seen in Theorem (2.8.11), ϕ_z and ϕ_z^{-1} are injective and surjective mappings between $\Delta(1)$, and holomorphic in $\Delta(1/|z|)$, a neighborhood of $\overline{\Delta(1)}$. Therefore $f \circ \phi_z^{-1}(\zeta)$ is holomorphic in a neighborhood of $\overline{\Delta(1)}$, and

$$f \circ \phi_z^{-1}(0) = \frac{1}{2\pi i}\int_{C(0;1)} \frac{f \circ \phi_z^{-1}(w)}{w} dw.$$

Changing the parameter by $w = \phi_z(\zeta)$, $\zeta = e^{i\theta}$, $0 \leqq \theta \leqq 2\pi$, one gets from (3.2.18)

$$\begin{aligned} f(z) &= \frac{1}{2\pi i}\int_0^{2\pi} f(e^{i\theta})\frac{\phi_z(e^{i\theta})}{\phi_z'(e^{i\theta})} ie^{i\theta} d\theta \\ &= \frac{1}{2\pi}\int_0^{2\pi} f(e^{i\theta})\frac{1 - |z|^2}{|e^{i\theta} - z|^2} d\theta. \end{aligned} \tag{3.6.2}$$

(3.6.3) THEOREM. *Let f be a continuous function on $\overline{\Delta(a; R)}$ which is holomorphic in $\Delta(a; R)$. Then*

$$f(z) = \frac{1}{2\pi}\int_0^{2\pi} f(a + Re^{i\theta})\frac{R^2 - |z - a|^2}{|Re^{i\theta} - (z - a)|^2} d\theta, \qquad z \in \Delta(a; R).$$

PROOF. By the change of variable $w = (z-a)/R$, it is reduced to the case where $R = 1$ and $a = 0$. Let $z \in \Delta(1)$ be an arbitrary point, and take $0 < r < 1$. Applying (3.6.2) to $g(\zeta) = f(r\zeta)$, one obtains

$$f(rz) = \frac{1}{2\pi}\int_0^{2\pi} f(re^{i\theta})\frac{1-|z|^2}{|e^{i\theta}-z|^2}d\theta.$$

Letting $r \to 1$, we have the desired formula. □

The function of ζ and z

$$K(\zeta, z) = \frac{|\zeta|^2 - |z|^2}{|\zeta - z|^2} > 0, \qquad \zeta \neq z,$$

is called the *Poisson kernel.* The formula of Theorem (3.6.3) is written as

(3.6.4) $$f(z) = \frac{1}{2\pi}\int_{C(a;R)} f(\zeta)K(\zeta - a, z - a)d\theta.$$

This is called the *Poisson integral.*

Let $a = 0$ and set $f(z) = u(z) + iv(z)$. Then it follows from (3.6.4) that

(3.6.5) $$u(z) = \frac{1}{2\pi}\int_{C(a;R)} u(\zeta)K(\zeta, z)d\theta.$$

The same holds for $v(z)$. Since

(3.6.6) $$K(\zeta, z) = \mathrm{Re}\,\frac{\zeta + z}{\zeta - z},$$

the function

$$\hat{u}(z) = \frac{1}{2\pi}\int_{C(0;R)} u(\zeta)\frac{\zeta + z}{\zeta - z}d\theta$$

is holomorphic on $\Delta(R)$, and $\mathrm{Re}\,\hat{u}(z) = u(z)$. The above integral is called the *complex Poisson integral.* Thus we have

(3.6.7) $$f(z) = \hat{u}(z) + i\,\mathrm{Im} f(0).$$

The second order differential operator

(3.6.8) $$\Delta = \left(\frac{\partial^2}{\partial x^2} + \frac{\partial^2}{\partial y^2}\right) = 4\partial_z\overline{\partial}_z = 4\overline{\partial}_z\partial_z$$

is called the *Laplacian.* Since $\overline{\partial}_z f = 0$, we have the so-called *Laplace equation*

(3.6.9) $$\Delta u = 0.$$

A real-valued C^2-function $u(z) = u(x, y)$ which satisfies the partial defferential equation (3.6.9) is called a *harmonic function.* The real and imaginary parts of a holomorphic function are harmonic. Conversely, for a given harmonic function $u(x, y)$ we will construct a holomorphic function f with real part u.

Let D be a simply connected domain. Let $u(x, y)$ be a harmonic function on D. Then $\partial_z u$ is holomorphic by (3.6.8) and (3.6.9). Fixing a point $a \in D$, by Theorem (3.4.15) we have a holomorphic function

$$f(z) = \int_a^z 2\partial_z u dz + u(a),$$

which is a primitive function of $\partial_z u$. Let $0 < r < d(a; \partial D)$. Then

$$\begin{aligned} f(a + re^{i\theta}) &= \int_0^r 2\partial_z u(a + te^{i\theta})e^{i\theta} dt + u(a) \\ &= \int_0^r \left(\frac{\partial u}{\partial x} \cos\theta + \frac{\partial u}{\partial y} \sin\theta \right) dt \\ &\quad + \int_0^r \left(\frac{\partial u}{\partial x} \sin\theta - \frac{\partial u}{\partial y} \cos\theta \right) dt + u(a). \end{aligned}$$

Setting $a = a_1 + ia_2$, one gets

$$\begin{aligned} \mathrm{Re} f(a + re^{i\theta}) &= \int_0^r \frac{d}{dt} u(a_1 + t\cos\theta, a_2 + t\sin\theta) dt + u(a) \\ &= u(a_1 + r\cos\theta, a_2 + r\sin\theta) - u(a_1, a_2) + u(a) \\ &= u(a + re^{i\theta}). \end{aligned}$$

Set $D' = \{z \in D; \mathrm{Re} f(z) = u(z)\}$. Since $D' \ni a$, $D' \neq \emptyset$. Clearly, D' is closed. To conclude that $D' = D$, it is sufficient to show that D' is open. Let $b \in D'$ be an arbitrary point. For $z \in \Delta(b; d(b; \partial D))$

$$\begin{aligned} f(z) &= \int_a^z 2\partial_z u dz + u(a) \\ &= \int_a^b 2\partial_z u dz + \int_b^z 2\partial_z u(z) dz + u(a). \end{aligned}$$

The above computation and the assumption $b \in D'$ yield

$$\begin{aligned} \mathrm{Re} f(z) &= \mathrm{Re} \int_a^b 2\partial_z u dz + u(a) + \mathrm{Re} \int_b^z 2\partial_z u(z) dz \\ &= u(b) + u(z) - u(b) = u(z). \end{aligned}$$

Hence $\Delta(b; \partial D) \subset D'$, and so D' is open.

(3.6.10) THEOREM. i) *For a harmonic function u on a simply connected domain D, there exists a holomorphic function $f = u + iu^*$ on D with real part u, which is unique up to a purely imaginary constant.*

ii) *Let u be a harmonic function on a domain D. If $u|U \equiv 0$ on a non-empty open subset $U \subset D$, then $u \equiv 0$ on D.*

iii) *Let u and D be as in* ii). *Let $a \in D$, and $0 < r < d(a; \partial D)$. Then for $z \in \Delta(a; r)$*

$$u(z) = \frac{1}{2\pi} \int_{C(a;r)} u(\zeta) K(\zeta - a, z - a) d\theta, \qquad \zeta = a + re^{i\theta}.$$

In particular,

$$u(a) = \frac{1}{2\pi} \int_{C(a;r)} u(a + re^{i\theta}) d\theta \qquad (\textit{the mean value theorem}).$$

iv) *Let u be as in* ii). *If u is not constant, then u does not attain the maximum nor the minimum in D.*

PROOF. i) It remains to show the uniqueness; this follows from Theorem (3.1.8).

ii) Let D' denote the set of points $z \in D$ such that there is a neighborhood $V \subset D$ of z with $u|V \equiv 0$. By definition, D' is open. Since $D' \supset U$, $D' \neq \emptyset$. Let $a \in \overline{D'} \cap D$. There is a point $b \in D'$ such that $a \in \Delta(b; d(b; \partial D))$.

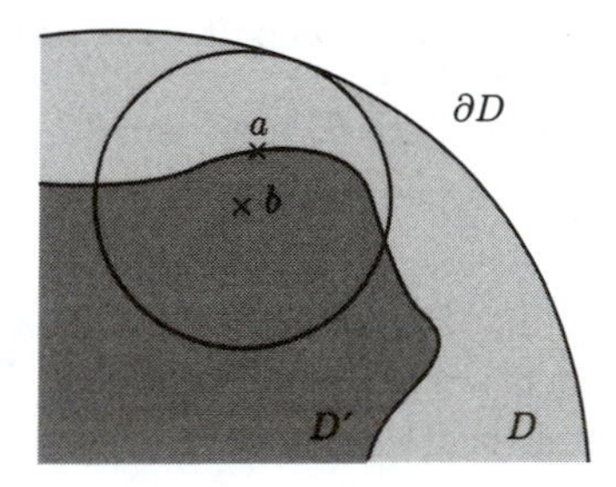

FIGURE 37

Let f be a holomorphic function on $\Delta(b; d(b; \partial D))$ such that $\mathrm{Re} f = u$. Since $b \in D'$, f is constant, so that $u|\Delta(b; d(b; \partial D)) \equiv 0$. Thus $a \in D'$. Since D is connected, $D' = D$.

iii) This follows from i) and (3.6.5).

iv) Assume that u attains the maximum value at $a \in D$. (For the minimum, it suffices to take $-u$.) For $0 < r < d(a; \partial D)$

$$\begin{aligned} &\frac{1}{2\pi} \int_0^{2\pi} (u(a) - u(a + re^{i\theta})) d\theta \\ &= u(a) - \frac{1}{2\pi} \int_0^{2\pi} u(a + re^{i\theta}) d\theta = 0. \end{aligned}$$

Since $u(a) - u(a + re^{i\theta})$ is continuous in θ and non-negative,

$$u(a + re^{i\theta}) = u(a), \qquad 0 \leqq \theta \leqq 2\pi.$$

Therefore $u|\Delta(a; d(a; \partial D)) \equiv u(a)$, and it follows from ii) that $u \equiv u(a)$ on D. This is a contradiction. $\square$

The function u^* in i) is called the *adjoint harmonic function* of u, and it is unique up to a constant term.

EXAMPLE. a) The adjoint harmonic function of x is y.
b) The adjoint harmonic function of $\log|z|$ is $\arg z$.

EXERCISE 1. Show that $x^2 - y^2$ is harmonic. What is its adjoint harmonic function?

The importance of harmonic functions for the study of holomorphic functions is made clear by the existence of adjoint harmonic functions. Generally speaking, it is easier to construct a harmonic function than to construct a holomorphic function. The problem of finding a harmonic function on a domain D with prescribed boundary values on ∂D is called the *Dirichlet problem.* We solve it for a disk:

(3.6.11) THEOREM. *Let h be a continuous real valued function on $C(0;1)$. Then the real valued function defined by*

$$u(z) = \frac{1}{2\pi}\int_0^{2\pi} h(e^{i\theta})K(e^{i\theta}, z)d\theta$$

is harmonic on $\Delta(1)$, and for every point $e^{it} \in C(0;1)$

$$\lim_{z \to e^{it}} u(z) = h(e^{it}).$$

PROOF. It follows from (3.6.6) that $u(z)$ is the real part of the holomorphic function

$$f(z) = \frac{1}{2\pi}\int_{C(0;1)} h(e^{i\theta})\frac{e^{i\theta}+z}{e^{i\theta}-z}d\theta,$$

and hence harmonic in $\Delta(1)$. Applying (3.6.5) to $u \equiv 1$, one gets

$$1 = \frac{1}{2\pi}\int_0^{2\pi} h(e^{i\theta})K(e^{i\theta}, z)d\theta.$$

Noting that $K(e^{i\theta}, z) > 0$, we have for a small $\delta > 0$

$$\begin{aligned} |u(z) - h(e^{it})| &= \frac{1}{2\pi}\left|\int_0^{2\pi} \{h(e^{i\theta}) - h(e^{it})\}K(e^{i\theta}, z)d\theta\right| \\ &\leqq \frac{1}{2\pi}\int_{t-\delta}^{t+\delta} |h(e^{i\theta}) - h(e^{it})|K(e^{i\theta}, z)d\theta \\ &\quad + \frac{1}{2\pi}\left(\int_{t-\pi}^{t-\delta} + \int_{t+\delta}^{t+\pi}\right) |h(e^{i\theta}) - h(e^{it})|K(e^{i\theta}, z)d\theta. \end{aligned}$$

For an arbitrary $\epsilon > 0$ there is a small $\delta > 0$ such that

$$|h(e^{i\theta}) - h(e^{it})| < \epsilon, \qquad t - \delta \leqq \theta \leqq t + \delta.$$

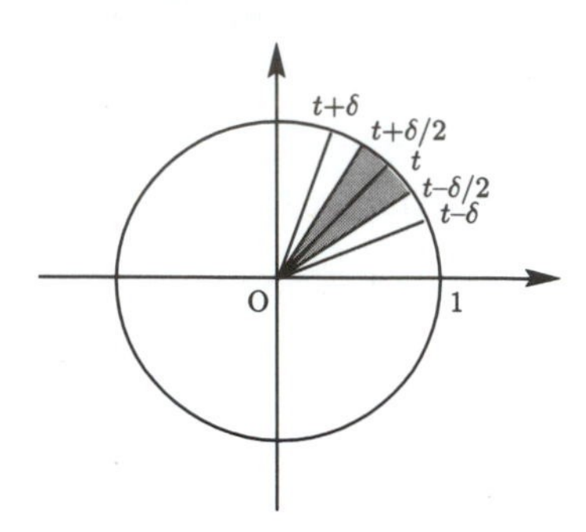

FIGURE 38

On the other hand, for $|\theta - t| \geqq \delta$ and $|\arg z - t| < \delta/2$

$$K(e^{i\theta}, z) = \frac{1 - |z|^2}{(1-|z|)^2 + 4|z|\sin^2 \frac{\theta - \arg z}{2}} < \frac{1-|z|^2}{4|z|\sin^2\frac{\delta}{4}}.$$

Thus it follows that for $|\arg z - t| < \delta/2$

$$\begin{aligned} |u(z) - h(e^{it})| &\leqq \frac{\epsilon}{2\pi}\int_{t-\delta}^{t+\delta} K(e^{i\theta}, z)d\theta + \frac{\max|h|}{\pi}\frac{1-|z|^2}{4|z|\sin^2\frac{\delta}{4}}(2\pi - 2\delta) \\ &< \epsilon + \frac{1}{2}(\max|h|)\frac{1-|z|^2}{|z|\sin^2\frac{\delta}{4}}. \end{aligned}$$

The above second term is less than ϵ for $|z|$ sufficiently close to 1. Therefore there is a $\delta' > 0$ such that

$$|u(z) - h(e^{it})| < 2\epsilon, \qquad z \in \Delta(1) \cap \Delta(e^{it}; \delta'),$$

so that $\lim_{z \to e^{it}} u(z) = h(e^{it})$. □

REMARK. By the above proof one sees that if h is bounded, $\lim_{z\to e^{it}} u(z) = h(e^{it})$ holds at a point e^{it} where h is continuous.

EXERCISE 2. Let D and D' be domains of $\mathbf{C}$, and let $f : D \to D'$ be a holomorphic function with values in D' (which is called a holomorphic mapping from D to D'). Show that if u' is a harmonic function on D', so is the composite $u \circ f$ on D.

EXERCISE 3 (*Harnack inequality*). Let $u(z)$ be a harmonic function on $\Delta(R)$ such that $u(z) \geqq 0$. Show that

$$\frac{R-|z|}{R+|z|}u(0) \leqq u(z) \leqq \frac{R+|z|}{R-|z|}u(0).$$

EXERCISE 4. Show that if u is a harmonic function on $\mathbf{C}$, and $u \geqq 0$, then u is constant.

EXERCISE 5. Express a harmonic function u on $\Delta(1)$ by the Poisson integral which has boundary values 1 on $C(0;1)\cap\{\operatorname{Im} z>0\}$ and -1 on $C(0;1)\cap\{\operatorname{Im} z<0\}$.

3.7. Holomorphic Functions on the Riemann Sphere

Let D be a connected open subset of the Riemann sphere $\widehat{\mathbf{C}}$. Then D is called a domain of $\widehat{\mathbf{C}}$, as in the case of $\mathbf{C}$. On the Riemann sphere, there are complex coordinates z on $\widehat{\mathbf{C}}\setminus\{\infty\}$ and $\tilde{z}$ on $\widehat{\mathbf{C}}\setminus\{0\}$. They are related on $\mathbf{C}^*=\widehat{\mathbf{C}}\setminus\{0,\infty\}$ by

$$z\tilde{z}=1.$$

On $\mathbf{C}^*$, $z=1/\tilde{z}$ is a holomorphic function of $\tilde{z}$, and the converse holds, too. Let f be a function on D. On $\mathbf{C}^*$, there are two coordinates, z and $\tilde{z}$. If f is holomorphic as a function of z on $D\cap\mathbf{C}^*$, then by Theorem (3.1.9) it is holomorphic with respect to $\tilde{z}$; the converse holds, too. We define f to be complex differentiable at ∞, provided that $\infty\in D$, if f is complex differentiable as a function of $\tilde{z}$ at $\tilde{z}=0$. We say that f is holomorphic on D if f is complex differentiable at every point of D. We know that the complex coordinates z and $\tilde{z}$ are related by $z\tilde{z}=1$, and represent the same point of the Riemann sphere $\widehat{\mathbf{C}}$. When f is considered as a function of z (resp., $\tilde{z}$), we write $f(z)$ (resp., $f\circ\tilde{z}$). For example, we have

$$f\circ\tilde{z}=f\left(\frac{1}{z}\right),\qquad f\circ\tilde{z}(0)=f(\infty).$$

The various properties of holomorphic functions on domains of $\mathbf{C}$ remain valid for those on domains of $\widehat{\mathbf{C}}$. We need, however, the following notion of differential to deal with curvilinear integrals. We set

$$U_0=D\cap\widehat{\mathbf{C}}\setminus\{\infty\},\qquad U_1=D\cap\widehat{\mathbf{C}}\setminus\{0\}.$$

A pair $\omega=\{\eta_0(z),\eta_1(\tilde{z})\}$ of holomorphic functions, $\eta_0(z)$ on U_0 and $\eta_1(\tilde{z})$ on U_1 is called a *differential* on D if on $U_0\cap U_1$ the following holds:

$$\eta_0(z)=\eta_1\left(\frac{1}{z}\right)\left(-\frac{1}{z^2}\right)=\eta_1(\tilde{z})(-\tilde{z}^2). \tag{3.7.1}$$

To represent this change of variables, we write

$$\omega=\eta_0(z)dz. \tag{3.7.2}$$

Here dz is considered to be transformed to $dz=(-1/\tilde{z}^2)d\tilde{z}$. Then, on $U_0\cap U_1$,

$$\omega=\eta_0(z)dz=\eta_1(\tilde{z})d\tilde{z}.$$

Let $C(\phi:I\to D)$ be a curve in D. Assume that $C\subset D\cap\mathbf{C}\setminus\{0\}$ and it has a finite length. Then

$$\int_C\eta_0(z)dz=\int_C\eta_1(\tilde{z})d\tilde{z},$$

where C in the right side is a curve given by $\tilde{z} \circ \phi(t) = 1/\phi(t)$. Thus we may define the curvilinear integral of ω along C by

$$\int_C \omega = \int_C \eta_0(z)dz = \int_C \eta_1(\tilde{z})d\tilde{z}. \tag{3.7.3}$$

For a general curve $C(\phi : [T_0, T_1] \to D)$, if there is a partition $(d) : T_0 = t_0 < t_1 < \cdots < t_n = T_1$ such that $\phi([t_{j-1}, t_j]) \subset U_0$ or $\phi([t_{j-1}, t_j]) \subset U_1$, and such that

$$\int_{C|[t_{j-1},t_j]} |dz| < +\infty, \quad \text{or} \quad \int_{C|[t_{j-1},t_j]} |d\tilde{z}| < +\infty, \tag{3.7.4}$$

we define the curvilinear integral of ω along C by

$$\int_C \omega = \sum_{j=1}^{n} \int_{C|[t_{j-1},t_j]} \omega. \tag{3.7.5}$$

If it exists, it is independent of the choice of the partition (d). The curvilinear integral $\int_C f(z)dz$ with $C \subset \mathbf{C}$ is considered as a curvilinear integral of the differential $f(z)dz$ along C.

A differential $\omega = \{\eta_0(z), \eta_1(\tilde{z})\}$ on D is called a *holomorphic differential* if $\eta_0(z)$ (resp., $\eta_1(\tilde{z})$) is holomorphic in z (resp., $\tilde{z}$). A holomorphic function F on D is called a *primitive function* of ω if

$$\frac{d}{dz}F(z) = \eta_0(z), \qquad z \in U_0,$$
$$\frac{d}{d\tilde{z}}F \circ \tilde{z} = \eta_1(\tilde{z}), \qquad \tilde{z} \in U_1.$$

We write $dF = \omega$ for this fact, and call it the *differential* of F. If G is another primitive function of ω, then the difference $F - G$ is constant.

All properties of holomorphic functions and curvilinear integrals of holomorphic functions on domains of $\mathbf{C}$ which have been proved up to the last section also hold for holomorphic functions and for holomorphic differentials on domains of $\widehat{\mathbf{C}}$ which may contain the infinite point ∞. For example, let ω be a holomorphic differential on a domain $D \subset \widehat{\mathbf{C}}$ and let C be a curve in D. Take a piecewise linear curve $\tilde{C}$ for C as in Lemma (3.3.4). Since a piecewise linear curve always satisfies (3.7.4), one may define the curvilinear integral of ω along C by

$$\int_C \omega = \int_{\tilde{C}} \omega.$$

As in the proof of Lemma (3.4.11), the above right side is independent of the choice of $\tilde{C}$ so long as $\tilde{C}$ satisfies (3.3.5), and Theorem (3.2.19) extends to

(3.7.6) THEOREM. *Let ω and D be as above. Let F be a primitive function of ω, and let C be a curve in D with initial point z_0 and terminal point z_1. Then*

$$\int_C \omega = F(z_1) - F(z_0).$$

Cauchy's integral Theorem (3.4.14) extends to

(3.7.7) THEOREM. *Let ω and D be as above. Let C be a closed curve in D which is homotopic to a point. Then*

$$\int_C \omega = 0.$$

(3.7.8) THEOREM. *Any function holomorphic on the whole $\widehat{\mathbf{C}}$ is constant.*

PROOF. Let f be a holomorphic function on $\widehat{\mathbf{C}}$. Since $\widehat{\mathbf{C}}$ is compact, $|f|$ attains the maximum at some point $a \in \widehat{\mathbf{C}}$. If $a \neq \infty$, Theorem (3.5.21) implies that f is constant; if $a = \infty$, then the function $f \circ \tilde{z}$, holomorphic in $\tilde{z}$, gives rise to a constant. Thus f is constant. □

If we deal with non-constant holomorphic functions on D, by this theorem we may exclude the case of $D = \widehat{\mathbf{C}}$. If $D \neq \widehat{\mathbf{C}}$, we take $b \in \widehat{\mathbf{C}} \setminus D$. If $b = \infty$, then $D \subset \mathbf{C}$; if $b \neq \infty$, we may reduce the case to $D \subset \mathbf{C}$ through the transformation $z \to 1/(z-b)$.

Problems

1. Show that $f(z) = x^2 + iy^2$ satisfies the Cauchy-Riemann equations at the origin 0, but is not holomorphic in any neighborhood of 0.

2. Let $f(z)$ be a holomorphic function on a domain D. Prove that $\overline{f(\bar{z})}$ is a holomorphic function on $D' = \{\bar{z}; z \in D\}$.

3. Let $a \in \mathbf{C}^*$ and define $f(z) = (1+z)^\alpha = e^{\alpha \log(1+z)}$. Take a branch of $f(z)$ on $\Delta(1)$, and expand it to a Taylor series about 0.

4. Let $C(\phi : [a,b] \to \mathbf{C})$ be a curve. Show that

$$L(C) = \lim_{|(d)| \to 0} L(C; (d)) \quad (\leqq +\infty),$$

where (d) is a partition of $[a,b]$.

5. Let C be the perimeter of the square $\{(x,y); -r \leqq x \leqq 2r, -r \leqq y \leqq 2r\}$ $(r > 0)$ with anti-clockwise orientation. Compute the curvilinear integral of $f(z) = x^2 + iy^2$ along C.

6. i) Let $f(z) = (z-\alpha)/(z-\beta)$ with $\alpha \neq \beta$ $(\in \mathbf{C})$. Show that f is holomorphic in a neighborhood of ∞.
ii) Let L denote the line segment from α to β. Show that $D = \widehat{\mathbf{C}} \setminus L$ is simply connected.

7. Let f be a holomorphic function on $\Delta(R)$ such that $|f(z)| \leqq M$ for all $z \in \Delta(R)$. Prove that

$$|f^{(n)}(z)| \leqq \frac{MRn!}{(R-|z|)^{n+1}}, \qquad z \in \Delta(R), \quad n = 1, 2, \ldots .$$

8. Let f be a holomorphic function on a bounded domain $D \subset \mathbf{C}$ which is continuously extended over the closure $\overline{D}$. Show that if f does not take the value 0 on D, and $|f|$ is constant on ∂D, then f is constant. What happens if f may take the value 0 on D?

9. Let f be a holomorphic function on $\mathbf{C}$. Show that if $\mathrm{Re} f \geq 0$, then f is constant.

10. (*Hadamard's three circles theorem*) Let f be a holomorphic function on a domain $\{z \in \mathbf{C}; R_1 < |z| < R_2\}$ with $0 < R_1 < R_2$, which is called an annulus. Set $M(r) = \max\{|f(z)|; |z| = r\}$ for $r \in (R_1, R_2)$. Show that $\log M(r)$ is a convex function with respect to $\log r$: that is, for $R_1 < r_1 < r_2 < r_3 < R_2$

$$\log M(r_2) \leqq \frac{\log r_3 - \log r_2}{\log r_3 - \log r_1} \log M(r_1) + \frac{\log r_2 - \log r_1}{\log r_3 - \log r_1} \log M(r_3).$$

11. Let f be a holomorphic function of a neighborhood of $\overline{\Delta(1)}$. Show that

$$\frac{1}{2\pi i} \int_{C(0;1)} \frac{\overline{f(z)}}{z-a} dz = \begin{cases} \overline{f(0)} & (|a| < 1), \\ \overline{f(0)} - \overline{f\left(\dfrac{1}{\bar{a}}\right)} & (|a| > 1). \end{cases}$$

12. Show that $\displaystyle \frac{1}{2\pi} \int_0^{2\pi} \log|a - e^{i\theta}| d\theta = \begin{cases} 0 & (|a| \leqq 1), \\ \log|a| & (|a| > 1). \end{cases}$

13. Find the adjoint harmonic function of the harmonic function $x + y + x^2 - y^2$.

14. Show that $u(x,y) = e^x(x\cos y - y \sin y)$ is harmonic, and find a holomorphic function $f(z)$ with real part $u(x,y)$.

15. Let $\{u_n(x,y)\}_{n=0}^{\infty}$ be a sequence of harmonic functions on a domain D, converging to $u(x,y)$ uniformly on compact subsets. Prove that $u(x,y)$ is harmonic.

16. Let $f(z)$ be a bounded function, holomorphic on $\mathbf{H}$, and continuous up to $\mathbf{R}$. Prove that

$$f(z) = \frac{1}{\pi i} \int_{-\infty}^{\infty} \frac{tz+1}{(t-z)} \cdot \frac{1}{t^2+1} \mathrm{Re} f(t) dt + \mathrm{Im} f(i).$$

17. Let $h(t)$ be a bounded continuous function on $\mathbf{R}$. Show that the function

$$u(x,y) = \frac{1}{\pi} \int_{-\infty}^{\infty} \frac{y}{(t-x)^2 + y^2} h(t) dt$$

is harmonic on $\mathbf{H}$. Moreover, as $x \to t_0 \in \mathbf{R}$ and $y \to +0$, $u(x,y) \to h(t_0)$.

CHAPTER 4

Residue Theorem

We begin in this chapter with Laurent series, and then explain meromorphic functions and the residue theorem. By making use of it, we will prove the Prinzip von der Gebietstreue (open mapping theorem), and the inverse function theorem. In the last section we explain methods for applying the residue theorem to calculate various kinds of definite integrals. These are direct consequences derived from Cauchy's integral theorem proved in Chapter 3; nevertheless, the reader may sense the depth of content of the theorem.

4.1. Laurent Series

For $a \in \mathbf{C}$ and $0 \leqq r_1 < r_2 \leqq +\infty$ we set

$$R(a; r_1, r_2) = \{z \in \mathbf{C}; r_1 < |z - a| < r_2\},$$

which is called a *ring domain* or an *annulus*. In the case of $a = 0$ we write

$$R(r_1, r_2) = R(0; r_1, r_2).$$

For the sake of simplicity we assume $a = 0$, and let $f(z)$ be a holomorphic function on $R(r_1, r_2)$. Let $z \in R(r_1, r_2)$ be an arbitrary point, and take $r_1' < r_2'$ so that

$$r_1 < r_1' < |z| < r_2' < r_2.$$

Take r with $0 < r < d(z; \partial R(r_1', r_2'))$. As in Figure 39 on the next page, let C_1 denote a curve which starts from $r_1' e^{i \arg z}$ and anti-clockwisely rounds once on $C(0; r_1')$, let C_2 be the line segment from $r_1' e^{i \arg z}$ to $(|z| - r)e^{i \arg z}$, let C_3 denote the half-circle of $C(z; r)$ from $(|z| - r)e^{i \arg z}$ to $(|z| + r)e^{i \arg z}$ with clockwise orientation, let C_4 be the line segment from $(|z| + r)e^{i \arg z}$ to $r_2' e^{i \arg z}$, let C_5 be the circle $C(0; r_2')$ with initial point $r_2' e^{i \arg z}$ and with anti-clockwise orientation, and finally let C_6 be the half circle of $C(z; r)$ from $(|z| + r)e^{i \arg z}$ to $(|z| - r)e^{i \arg z}$ with clockwise orientation. Then, in the domain $R(r_1, r_2) \setminus \{z\}$ the curve

$$C = -C_1 + C_2 + C_3 + C_4 + C_5 - C_4 + C_6 - C_2 \tag{4.1.1}$$

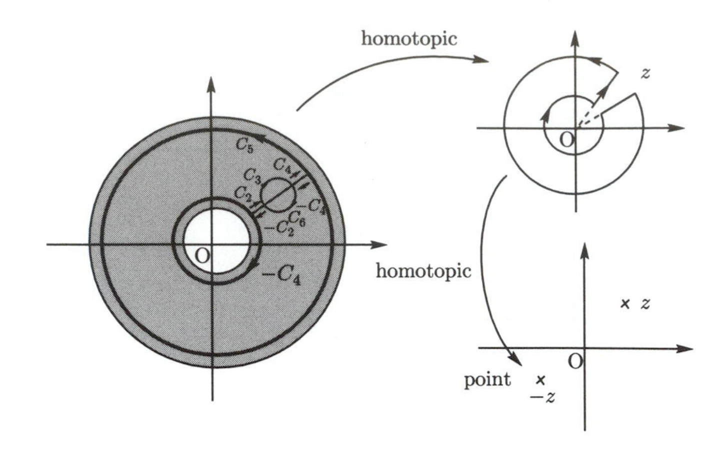

FIGURE 39

is homotopic to a point, so that $\int_C \frac{f(\zeta)}{\zeta - z} d\zeta = 0$. It follows that

$$\begin{aligned} f(z) &= \frac{-1}{2\pi i} \int_{C_3+C_4} \frac{f(\zeta)}{\zeta - z} d\zeta \\ &= \frac{-1}{2\pi i} \int_{C_1} \frac{f(\zeta)}{\zeta - z} d\zeta + \frac{1}{2\pi i} \int_{C_5} \frac{f(\zeta)}{\zeta - z} d\zeta. \end{aligned}$$

Since $|\zeta/z| < 1$ on C_1, we have

$$\begin{aligned} &\frac{-1}{2\pi i} \int_{C_1} \frac{f(\zeta)}{\zeta - z} d\zeta = \frac{1}{2\pi i} \int_{C_1} \frac{f(\zeta)}{1 - \frac{\zeta}{z}} d\zeta \cdot \frac{1}{z} \\ &= \frac{1}{2\pi i} \int_{C_1} f(\zeta) \sum_{n=0}^{\infty} \left(\frac{\zeta}{z}\right)^n d\zeta \cdot \frac{1}{z} = \sum_{n=1}^{\infty} \frac{1}{2\pi i} \int_{C_1} f(\zeta)\zeta^{n-1} d\zeta \cdot z^{-n}. \end{aligned}$$

The last power series of the above expression is absolutely convergent, and is written as

$$\sum_{n=-1}^{-\infty} \frac{1}{2\pi i} \int_{C_1} f(\zeta)\zeta^{-n-1} d\zeta \cdot z^n.$$

Similarly we have

$$\frac{1}{2\pi i} \int_{C_5} \frac{f(\zeta)}{\zeta - z} d\zeta = \sum_{n=0}^{\infty} \frac{1}{2\pi i} \int_{C_5} f(\zeta)\zeta^{-n-1} d\zeta \cdot z^n.$$

Since $f(\zeta)\zeta^{-n-1}$ is holomorphic in $R(r_1, r_2)$, one infers in the same way as above that

$$\frac{1}{2\pi i} \int_{C_1} f(\zeta)\zeta^{-n-1} d\zeta = \frac{1}{2\pi i} \int_{C_5} f(\zeta)\zeta^{-n-1} d\zeta.$$

This shows that for $n \in \mathbf{Z}$

$$a_n = \frac{1}{2\pi i} \int_{C(0;r)} \frac{f(\zeta)}{\zeta^{n+1}} d\zeta, \qquad r_1 < r < r_2, \tag{4.1.2}$$

is independent of the choice of r. Therefore we have the expression

$$f(z) = \sum_{n=-1}^{-\infty} a_n z^n + \sum_{n=0}^{\infty} a_n z^n = \sum_{n=-\infty}^{\infty} a_n z^n. \tag{4.1.3}$$

The power series $\sum_{n=-1}^{-\infty} a_n z^n$ in $1/z$ converges absolutely and uniformly on compact subsets of $R(r_1, \infty)$, and so does $\sum_{n=0}^{\infty} a_n z^n$ in $\Delta(r_2)$. Hence, $\sum_{n=-\infty}^{\infty} a_n z^n$ converges absolutely and uniformly on compact subsets of $R(r_1, r_2)$, and is called the *Laurent series* of f.

As in the case of the Taylor series, the derived function $f'(z)$ is written as

$$f'(z) = \sum_{n=-\infty}^{\infty} n a_n z^{n-1}. \tag{4.1.4}$$

Thus, if $a_{-1} = 0$, the function f has the primitive function F, which is written as

$$F(z) = \sum_{\substack{n=-\infty \\ n\neq -1}}^{\infty} \frac{a_n}{n+1} z^{n+1} + \text{const.} \tag{4.1.5}$$

As in Theorem (3.5.20), ii), for $r_1 < r < r_2$ we have

$$\sum_{n=-\infty}^{\infty} |a_n|^2 r^{2n} = \frac{1}{2\pi} \int_0^{2\pi} |f(re^{i\theta})|^2 d\theta. \tag{4.1.6}$$

In the annulus $R(a; r_1, r_2)$ we have the following, called the Laurent series of f about a:

$$f(z) = \sum_{n=-\infty}^{\infty} a_n (z-a)^n, \qquad r_1 < |z| < r_2. \tag{4.1.7}$$

We consider the case of $r_1 = 0$ ($R(a; 0, r_2) = \Delta(a; r_2) \setminus \{a\}$), and assume that f is not constant. If there are infinitely many coefficients $a_n \neq 0$ with $n < 0$, then a is called an *isolated essential singularity* of f. If there are only finitely many $a_n \neq 0$ with $n < 0$, f is said to have *at most a pole* at a, and f is expressed as

$$f(z) = \frac{a_{-m}}{(z-a)^m} + \frac{a_{-m+1}}{(z-a)^{m-1}} + \cdots + a_0 + \cdots, \qquad a_{-m} \neq 0.$$

If $m > 0$, a is called a *pole* of *order* m of f, and one writes $f(a) = \infty$. If $a_n = 0$ for all $n < 0$, then f is holomorphic in $\Delta(a; r_2)$, and

$$f(z) = a_m (z-a)^m + a_{m+1}(z-a)^{m+1} + \cdots, \qquad a_m \neq 0, \quad m \geqq 0.$$

If $m > 0$, a is called a *zero* of *order* m of f.

For $a = \infty$, we define the notion of zeros and poles in the same way as above by making use of the complex variable $\tilde{z}$. Note that

$$\{\tilde{z}; 0 < |\tilde{z}| < R\} = \left\{z; |z| > \frac{1}{R}\right\}.$$

The relation between the Laurent series of f with respect to $\tilde{z}$ and that with respect to z is given by

(4.1.8) $$f(z) = \sum_{n=-\infty}^{\infty} a_n z^n = \sum_{n=-\infty}^{\infty} a_{-n}\tilde{z}^n = f \circ \tilde{z}.$$

Therefore we have the following:

(4.1.9) $f(z) = \sum_{n=-\infty}^{\infty} a_n z^n$ is holomorphic at ∞ if and only if $a_n = 0$ for all $n \geqq 1$;

(4.1.10) the differential

$$\omega = \left(\sum_{n=-\infty}^{\infty} a_n z^n\right) dz = \left(\sum_{n=-\infty}^{\infty} a_{-n}\tilde{z}^n\right)\left(-\frac{1}{\tilde{z}^2}\right) d\tilde{z}$$

is holomorphic at ∞ if and only if $a_n = 0$ for all $n \geqq -1$.

EXERCISE 1. Obtain the Laurent series of $f(z) = e^{1/z}$ about $z = 0$.

EXERCISE 2. Obtain the Laurent series of $f(z) = \dfrac{1}{z^5(1+z)}$ about $z = 0$.

4.2. Meromorphic Functions and Residue Theorem

Let D be a domain in $\widehat{\mathbf{C}}$. A function on D which has at most a pole at every point of D is called a *meromorphic function*. Let f be a meromorphic function on D. If $a \in D$ is a pole of f, the function

$$\tilde{w} \circ f(z) = \frac{1}{f(z)}$$

is holomorphic in a neighborhood of a. In this sense f uniquely defines a continuous mapping

(4.2.1) $$f : D \to \widehat{\mathbf{C}}.$$

For example, a *rational function* $P(z)/Q(z)$, defined as a ratio of two polynomials $P(z)$ and $Q(z)$ with $Q(z) \not\equiv 0$, is meromorphic on $\widehat{\mathbf{C}}$. By definition, the set E of all poles of f is a discrete subset of D. If $E = \emptyset$, then, of course, f is holomorphic in D. Two meromorphic functions f and g on D are said to be equal if there is a discrete subset $F \subset D$ such that f and g are holomorphic on $D \setminus F$, and if $f = g$ on $D \setminus F$. This property is independent of the choice of F by the identity Theorem (2.4.14). If $f \neq 0$, $1/f$ is meromorphic, and the sum and product of two meromorphic functions are meromorphic. Thus the set of all meromorphic functions on D forms a field. For meromorphic functions we have the following identity theorem.

(4.2.2) THEOREM. *Let f_1 and f_2 be meromorphic functions on D. Let $A \subset D$ be a subset with an accumulation point in D. If $f_1 = f_2$ on A, then $f_1 = f_2$ on D.*

PROOF. We may assume that $f_2 = 0$. Let $z_0 \in D$ be an accumulation point of A, and E be the set of all poles of f. If $z_0 \notin E$, Theorem (2.4.14) implies that $f_1 \equiv 0$ on $D \setminus E$, so that $f_1 = 0$. Assume that $z_0 \in E$. If $z_0 \neq \infty$, we take $0 < r < d(z_0; \partial D)$ with $\Delta(z_0; r) \cap E \setminus \{z_0\} = \emptyset$. Set

$$g_1(z) = \frac{1}{f_1(z)}, \qquad z \in \Delta(z_0; r).$$

Then g_1 is a holomorphic function on $\Delta(z_0; r)$. Therefore f_1 does not take the value 0 on $\Delta(z_0; r)$. This contradicts the fact that z_0 is an accumulation point of A. When $z_0 = \infty$, by making use of the complex coordinate $\tilde{z}$ we make the same argument as above about $\tilde{z} = 0$. □

Let $a \in \mathbf{C}$ and let f be a holomorphic function on $\Delta(a; r_2) \setminus \{a\}$ expanded as (4.1.7). We define the *residue* of f at a by

$$\text{(4.2.3)} \qquad \operatorname{Res}(a; f) = a_{-1} = \frac{1}{2\pi i} \int_{C(a;r)} f(\zeta) d\zeta \qquad (0 < r < r_2).$$

When $a = \infty$, we consider a holomorphic function f on $\{\tilde{z} \in \mathbf{C}; 0 < |\tilde{z}| < R\}$ $= \{z \in \mathbf{C}; |z| > 1/R\}$. Then we have the Laurent series

$$f(z) = \sum_{n=-\infty}^{\infty} a_n z^n = \sum_{n=-\infty}^{\infty} a_{-n} \tilde{z}^n = f \circ \tilde{z},$$

and we define the *residue* of f at ∞ by

$$\text{(4.2.4)} \quad \operatorname{Res}(\infty; f) = -a_{-1} = \frac{1}{2\pi i} \int_{\{|\tilde{z}|=r\}} f \circ \tilde{z} \cdot \frac{-1}{\tilde{z}^2} d\tilde{z}, \qquad 0 < r < R.$$

By this definition we see that the residue is a notion not for the holomorphic function f, but for the differential $\omega = f(z)dz$. That is, no matter whether a is ∞ or not, taking a small circle C around a with anti-clockwise orientation, we set

$$\text{(4.2.5)} \qquad \operatorname{Res}(a; \omega) = \frac{1}{2\pi i} \int_C \omega.$$

It follows that $\operatorname{Res}(a; \omega) = \operatorname{Res}(a; f)$ for all $a \in \widehat{\mathbf{C}}$.

In general, let $D \subset \widehat{\mathbf{C}}$ be a domain, and let $\omega = f(z)dz = g(\tilde{z})d\tilde{z}$ be a differential on D. If the coefficient $f(z)$ (resp., $g(\tilde{z})$) is a meromorphic function of z (resp., $\tilde{z}$), ω is called a *meromorphic differential* on D. We say that ω has a zero (resp., pole) at $a \in D$ if $a \in \mathbf{C}$ and $f(a) = 0$ (resp., ∞), or if $a = \infty$ and $g(0) = 0$ (resp., ∞). The set of zeros and poles of ω is discrete in D.

Let C be a closed curve in D which is homotopic to a point in D. Let

$$\Phi : [0,1] \times [0,1] \to D$$

be the homotopy connecting C to a point. Assume that there is a point $b \in \widehat{\mathbf{C}} \setminus \Phi([0,1] \times [0,1])$. For $a \in D \setminus C$ we define the *rotation number* $n_b(a; C)$ with respect to b by

(4.2.6) i) The case of $a \neq \infty$: If $b \neq \infty$,

$$n_b(a; C) = \frac{1}{2\pi i} \int_C \left(\frac{1}{\zeta - a} - \frac{1}{\zeta - b} \right) d\zeta.$$

If $b = \infty$,

$$n_\infty(a; C) = \frac{1}{2\pi i} \int_C \frac{1}{\zeta - a} d\zeta.$$

ii) The case of $a = \infty$:

$$n_b(\infty; C) = \frac{1}{2\pi i} \int_C \frac{-1}{\zeta - b} d\zeta.$$

In the case of $b = \infty$ we set

$$n(a; C) = n_\infty(a; C).$$

Note that if $a \neq \infty$ and $b \neq \infty$, the differential $(1/(\zeta - a) - 1/(\zeta - b))d\zeta$ is holomorphic at ∞.

EXERCISE 1. Let $a, b \in \mathbf{C} \setminus C(0;1)$. Obtain $n_b(a; C(0;1))$. Here $C(0;1)$ is endowed with the anti-clockwise orientation.

(4.2.7) THEOREM. *Let D, C, and b be as above. Let $E = \{a_\nu; \nu = 1, 2, \ldots\}$ be a discrete subset of D such that $E \cap C = \emptyset$, and let ω be a holomorphic differential on $D \setminus E$. Then*

$$\frac{1}{2\pi i} \int_C \omega = \sum_{\nu=1}^{\infty} n_b(a_\nu; C) \mathrm{Res}(a_\nu; \omega).$$

Here the number of a_ν with $n_b(a_\nu; C) \neq 0$ is finite.

PROOF. If $a_\nu \notin \Phi([0,1] \times [0,1])$, then the differential to define $n_b(a_\nu; C)$ in (4.2.6) is holomorphic in a neighborhood of $\Phi([0,1] \times [0,1])$. Thus, by Theorem (3.7.7), $n_b(a_\nu; C) = 0$. Changing the order of numbering of the a_ν, we may assume that

$$E \cap \Phi([0,1] \times [0,1]) = \{a_\nu; 1 \leqq \nu \leqq \nu_0\}, \qquad 0 \leqq \nu_0 < +\infty.$$

Fix a domain $D' \subset D$ such that $b \notin D'$ and $D' \cap E = \{a_\nu; 1 \leqq \nu \leqq \nu_0\}$. About each a_ν we write $\omega = f(z)dz$ (or $g(\tilde{z})d\tilde{z}$), and expand $f(z)$ (or $g(\tilde{z})$) to a Laurent series. Suppose first that $a_\nu = \infty$ and $b \neq \infty$. We set

$$f(z) = \sum_{n=-\infty}^{-1} \alpha_n (z - a_\nu)^n + \sum_{n=0}^{\infty} \alpha_n (z - a_\nu)^n,$$

$$\eta_\nu = \left\{ \sum_{n=-\infty}^{-2} \alpha_n \left((z - a_\nu)^n - (z - b)^n \right) \right\} dz + \alpha_{-1} \left(\frac{1}{z - a_\nu} - \frac{1}{z - b} \right) dz.$$

If $b = \infty$, we set $(z - b)^n = 0$ for $n < 0$ in the above equations. Note that η_ν converges absolutely and uniformly on compact subsets of $\widehat{\mathbf{C}} \setminus \{a_\nu, b\} (\supset D')$, and hence is holomorphic there. The first term of η_ν has a primitive function

$$\sum_{n=-\infty}^{-2} \frac{\alpha_n}{n+1} \left((z - a_\nu)^{n+1} - (z - b)^{n+1} \right).$$

By Theorem (3.7.6) the curvilinear integral of the first term of η_ν along C is zero. It follows from this and the definition that

$$\text{(4.2.8)} \qquad \frac{1}{2\pi i} \int_C \eta_\nu = n_b(a_\nu; C) \operatorname{Res}(a_\nu; \omega).$$

In the case of $a_\nu = \infty$, we similarly define η_ν by making use of $g(\tilde{z})$, and deduce that

$$\text{(4.2.9)} \qquad \frac{1}{2\pi i} \int_C \eta_\nu = n_b(\infty; C) \operatorname{Res}(\infty; \omega).$$

The differential $\omega - \sum_{\nu=1}^{\nu_0} \eta_\nu$ is holomorphic in D'. Since C is homotopic to a point in D', it follows from Theorem (3.7.7) that

$$\int_C \omega - \sum_{\nu=1}^{\nu_0} \int_C \eta_\nu = 0.$$

Thus, (4.2.8) and (4.2.9) imply the required formula. $\square$

We explain a practical method to calculate residues. If $f(z)$ has a pole of order 1 at $a \in \mathbf{C}$, then

$$\text{(4.2.10)} \qquad \operatorname{Res}(a; f) = \lim_{z \to a} (z - a) f(z).$$

If $a \in \mathbf{C}$ is a pole of order m of f,

$$\text{(4.2.11)} \qquad \operatorname{Res}(a; f) = \frac{1}{(m-1)!} \left. \frac{d^{m-1}}{dz^{m-1}} \right|_{z=a} (z - a)^m f(z).$$

EXERCISE 2. Prove (4.2.10) and (4.2.11).

If $a = \infty$, the equations similar to (4.2.10) and (4.2.11) hold in terms of $\tilde{z}$. That is, if f is holomorphic in a neighborhood of ∞ and has a zero at ∞, then

$$\text{Res}(\infty; f) = \lim_{|z|\to\infty} -zf(z). \tag{4.2.12}$$

If f has a pole of order m at ∞,

$$\text{Res}(\infty; f) = \lim_{|z|\to\infty} \frac{(-1)^m}{(m+1)!} z^{m+2} f^{(m+1)}(z). \tag{4.2.13}$$

In the case where f is holomorphic at ∞, this formula is valid with $m = 0$.

EXERCISE 3. Prove (4.2.12) and (4.2.13).

Let $D \subset \mathbf{C}$ be a bounded domain surrounded by finitely many Jordan closed curves, C_i, $1 \leqq i \leqq k$, as in Figure 40 below. When a description depending on geometric intuition is used, every C_i is assumed to be endowed with orientation so that D is lying on the left.

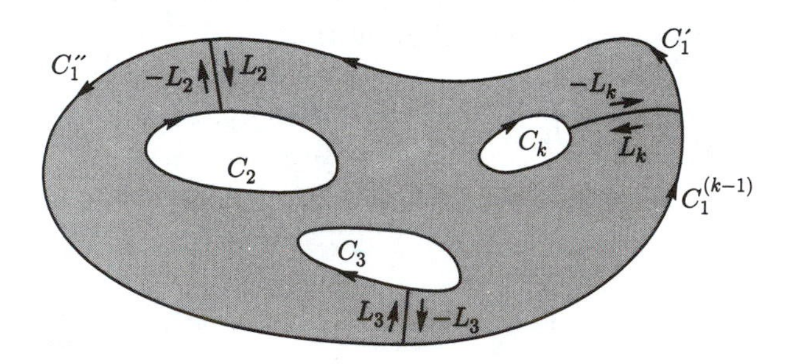

FIGURE 40

Let the numbering of the C_i be as in Figure 40. Then C_1 is called the outer boundary of D, and $C_2, \dots, C_k$ are called the inner boundaries of D. Let D' be a domain containing the closure $\overline{D} = D \cup \left(\bigcup_{i=1}^k C_i\right)$, and let f be a meromorphic function on D'. Connect C_i $(2 \leqq i \leqq k)$ and C_1 by suitable curves L_i. Then, as in (4.1.1) we obtain a curve $C_1' + L_2 + C_2 - L_2 + C_1'' + \cdots + C_1^{k-1}$ which is homotopic to a point in D', where $C_1 = C_1' + \cdots + C_i^{(k-1)}$. Assume that f has no pole on $\bigcup_{i=1}^k C_i$, and take the L_i so that f has no pole on them, either. Then Theorem (4.2.7) takes the following form:

$$\sum_{i=1}^k \frac{1}{2\pi i} \int_{C_i} f(z)dz = \sum_{a\in D} \text{Res}(a; f). \tag{4.2.14}$$

In the above, the description was based on geometric intuition. We will, however, rigorously check the conditions when we use it. We will summarize in the last section of this chapter how to apply the residue theorem to calculate various definite integrals.

EXERCISE 4. Let E be a finite set of $\widehat{\mathbf{C}}$, and let ω be a holomorphic differential on $\widehat{\mathbf{C}} \setminus E$. Show that

$$\sum_{a \in E} \mathrm{Res}(a; \omega) = 0.$$

EXERCISE 5. Show that a meromorphic function on $\widehat{\mathbf{C}}$ is a rational function.

EXERCISE 6 (Fractional expansion of a rational function). Let $f(z) = \dfrac{P(z)}{Q(z)}$ be a rational function, where $P(z)$ and $Q(z)$ are coprime. Let $a_i \in \mathbf{C}$, $1 \leqq i \leqq k$, be zeros of $Q(z)$, and expand $f(z)$ to a Laurent series about a_i:

$$f(z) = \sum_{j=1}^{p_i} \frac{\alpha_{ij}}{(z-a_i)^j} + \sum_{j=0}^{\infty} \beta_{ij}(z-a_i)^j, \qquad 1 \leqq i \leqq k.$$

Let the Laurent series of $f(z)$ about ∞ be

$$f(z) = \sum_{j=0}^{p_0} \alpha_{0j} z^j + \sum_{j=-1}^{-\infty} \beta_{0j} z^j.$$

Then, show that

$$f(z) = \sum_{j=0}^{p_0} \alpha_{0j} z^j + \sum_{i=1}^{k} \sum_{j=1}^{p_i} \frac{\alpha_{ij}}{(z-a_i)^j}.$$

EXERCISE 7. What are the residues of $f(z) = z/(1+z^2)$ at poles?

EXERCISE 8. What are the residues of $f(z) = (1+z)/(a^2+z^2)^2$ at poles?

EXERCISE 9. What are the residues of $f(z) = e^{z+1/z}$ at $z = 0, \infty$?

4.3. Argument Principle

Let D be a domain of $\widehat{\mathbf{C}}$, and assume that $D \neq \widehat{\mathbf{C}}$. Let f be a meromorphic function in D. We call

$$\omega = \frac{1}{f(z)} \frac{df(z)}{dz} dz = \frac{1}{f \circ \tilde{z}} \frac{df \circ \tilde{z}}{d\tilde{z}} d\tilde{z} \tag{4.3.1}$$

the *logarithmic differential* of f. If $a \in D$ is not a zero nor a pole of f, then ω is holomorphic in its neighborhood. Let $a \in D$ be a zero of order m of f. If $a \neq \infty$, we have

$$\begin{aligned}
f(z) &= a_m(z-a)^m + a_{m+1}(z-a)^{m+1} + \cdots, \qquad a_m \neq 0, m \geqq 1, \\
\frac{df}{dz}(z) &= a_m m(z-a)^{m-1} + a_{m+1}(m+1)(z-a)^m + \cdots, \\
\omega &= \frac{a_m m \left(1 + \frac{a_{m+1}(m+1)}{a_m m}(z-a) + \cdots\right)}{a_m(z-a)\left(1 + \frac{a_{m+1}}{a_m}(z-a) + \cdots\right)} dz \\
&= \left(\frac{m}{z-a} + a_0' + a_1'(z-a) + \cdots\right) dz.
\end{aligned}$$

Thus

$$\text{(4.3.2)} \qquad \operatorname{Res}(a;\omega) = m.$$

If $a = \infty$, by making use of $\tilde{z}$, we similarly deduce that

$$f \circ \tilde{z} = a_m \tilde{z}^m + \cdots, \qquad a_m \neq 0, m \geqq 1,$$
$$\frac{df \circ \tilde{z}}{d\tilde{z}} = a_m m \tilde{z}^{m-1} + \cdots,$$

so that $\operatorname{Res}(\infty;\omega) = m$. Suppose that a is a pole of order p of f. If $a \neq \infty$, we have

$$f(z) = \frac{a_{-p}}{(z-a)^p} + \frac{a_{-p+1}}{(z-a)^{p-1}} + \cdots, \qquad a_{-p} \neq 0, p \geqq 1,$$
$$f'(z) = \frac{-pa_{-p}}{(z-a)^{p+1}} + \frac{-(p-1)a_{-p+1}}{(z-a)^p} + \cdots,$$
$$\omega = \left(\frac{-p}{z-a} + a'_0 + a'_1(z-a) + \cdots\right) dz.$$

Thus

$$\text{(4.3.3)} \qquad \operatorname{Res}(a;\omega) = -p.$$

This holds also for $a = \infty$. The following theorem is called the *argument principle.*

(4.3.4) THEOREM. *Let $D \neq \widehat{\mathbf{C}}$ be a domain, and take $c \in \widehat{\mathbf{C}} \setminus D$. Let f be a meromorphic function in D, and let ω be its logarithmic differential. Let C be a closed curve in D such that f has no zero nor pole on C and it is homotopic to a point in D. Let $a_\nu \in D, \nu = 1, 2, \ldots$ (resp., $b_\mu \in D, \mu = 1, 2, \ldots$), be zeros (resp., poles) of order m_ν (resp., p_μ) of f. Then*

$$\frac{1}{2\pi i}\int_C \omega = \sum_\nu n_c(a_\nu; C) m_\nu - \sum_\mu n_c(b_\mu; C) p_\mu.$$

PROOF. This easily follows from (4.3.2), (4.3.3), and Theorem (4.2.7). □

Let $D \subset \mathbf{C}$ be a bounded domain as in (4.2.14), and let $\partial D = \bigcup_{i=1}^k C_i$ and $c = \infty$. Let f be a meromorphic function in a neighborhood of $\overline{D}$ without zeros or poles on ∂D. Then Theorem (4.3.4) takes the form

$$\text{(4.3.5)} \qquad \sum_{i=1}^k \frac{1}{2\pi i}\int_{C_i} \frac{f'(z)}{f(z)} dz = \sum m_\nu - \sum p_\mu.$$

Here the first term on the right represents the total number of zeros of f counting multiplicities, and the second that of poles. Note that the curvilinear integral

$$\frac{1}{2\pi i}\int_{C_i} \frac{f'(z)}{f(z)} dz$$

is the rotation number of the curve $f(C_i)$ around 0. Therefore, if C is a closed curve in D such that f has no zero nor pole on C and the image $f(C)$ is contained in a simply connected domain D' inside $\mathbf{C} \setminus \{0\}$, then the logarithmic differential ω of f has a primitive function $\log f$, once we choose a branch of log on D'. Then, it follows from Theorem (3.7.6) that

$$\int_C \frac{1}{f(z)} \frac{df}{dz}(z) dz = 0. \tag{4.3.6}$$

If in Theorem (4.3.4) D is a bounded domain of $\mathbf{C}$, $c = \infty$, and f is holomorphic, then the sum $\sum_\nu n(a_\nu; C) m_\nu$ is called the *number of zeros of f surrounded by C*. The next theorem is called *Rouché's theorem.*

(4.3.7) THEOREM. *Let $D \subset \mathbf{C}$ be a bounded domain, let f be a holomorphic function on D, and let C be a curve in D. Assume that C is homotopic to a point, and that there is no zero of f on C. Let g be a holomorphic function on D such that $|g(z)| < |f(z)|$ for $z \in C$. Then the number of zeros of f surrounded by C is equal to that of $f + g$.*

PROOF. Let M be the number of zeros of f surrounded by C. Then, that of $f + g$ is

$$\begin{aligned} M' &= \frac{1}{2\pi i} \int_C \frac{f' + g'}{f + g} dz = \frac{1}{2\pi i} \int_C \left(\frac{f'}{f} + \frac{(g/f)'}{1 + g/f} \right) dz \\ &= M + \frac{1}{2\pi i} \int_C \frac{(g/f)'}{1 + g/f} dz. \end{aligned}$$

Since $|g/f| < 1$ on C, $1 + g(z)/f(z) \in \Delta(1; 1)$ for $z \in C$. Taking a branch of log on the simply connected domain $\Delta(1; 1)$ $(\not\ni 0)$, we have

$$\frac{(g/f)'}{1 + g/f} = \frac{d}{dz} \log \left(1 + \frac{g}{f} \right).$$

Hence, it follows from (4.3.6) that

$$\frac{1}{2\pi i} \int_C \frac{(g/f)'}{1 + g/f} dz = 0,$$

and that $M = M'$. □

Let D and C_j be as in (4.2.14), and let f and g be holomorphic functions on D with $|g(z)| < |f(z)|$ on C. Then

(4.3.8) the number of zeros of $f + g$ in D equals that of f in D.

EXERCISE 1. Set $f_1(z) = z^n + a_1 z^{n-1} + \cdots + a_n$, and $f_2(z) = z^n$. Show by making use of Rouché's theorem that for a sufficient large $R > 0$ the number of zeros of f_1 surrounded by $C(0; R)$ equals that of f_2, i.e., n.

We give some applications of the above theorem. The first is *Hurwitz's theorem.*

(4.3.9) THEOREM. *Let $D \subset \widehat{\mathbf{C}}$ be a domain, and let $\{f_n\}_{n=0}^{\infty}$ be a sequence of holomorphic functions in D converging uniformly on compact subsets of D to f. If f_n, $n = 0, 1, \ldots$, do not take the value zero, then either f has no zero, or $f \equiv 0$.*

PROOF. First of all, f is holomorphic (Theorem (3.5.19)). Assume that $f \not\equiv 0$. Take $a \in D$. Suppose $a \in \mathbf{C}$; otherwise, use the coordinate $\tilde{z}$ about ∞. Since the set of zeros of f is discrete, there is a $\delta > 0$ such that $\overline{\Delta(a;\delta)} \subset D$, and

$$(\overline{\Delta(a;\delta)} \setminus \{a\}) \cap \{f = 0\} = \emptyset.$$

Here, $\{f = 0\}$ stands for $\{z \in D; f(z) = 0\}$. For $C(a;\delta)$ with anti-clockwise orientation

$$\frac{1}{2\pi i} \int_{C(a;\delta)} \frac{f'(z)}{f(z)} dz \geqq 1.$$

Since $\{f_n\}$ converges uniformly on $\overline{\Delta(a;\delta)}$, there is an n_0 such that

$$|f_n(z) - f(z)| < \min\{|f(w)|; w \in C(a;\delta)\}, \quad z \in C(a;\delta), n \geqq n_0.$$

It follows from Theorem (4.3.7) that the number of zeros of f in $\Delta(a;\delta)$ equals that of $f + (f_n - f) = f_n$ $(n \geqq n_0)$. Thus f_n has a zero in $\Delta(a;\delta)$; this is a contradiction. □

Let $D \subset \widehat{\mathbf{C}}$ be a domain. A meromorphic function f in D is said to be *univalent* if f is injective (one-to-one) as a mapping from D into $\widehat{\mathbf{C}}$.

(4.3.10) THEOREM. *Let $\{f_n\}_{n=0}^{\infty}$ be a sequence of holomorphic functions in a domain $D \subset \widehat{\mathbf{C}}$ such that it converges uniformly on compact subsets of D to f. If all f_n are univalent, then f is either univalent or constant.*

PROOF. Assume that f is not constant, nor univalent. Then there are two points $z_1, z_2 \in D$ with $f(z_1) = f(z_2) = w_0$. Take neighborhoods U_i of z_i, $i = 1, 2$, with $U_1 \cap U_2 = \emptyset$. Since $\{f_n - w_0\}_{n=0}^{\infty}$ converges to $f - w_0$ uniformly on compact subsets of D, it follows from Theorem (4.3.9) that there are some n and $z_1' \in U_i, i = 1, 2$, such that $f_n(z_i') = w_0$. That is, $f_n(z_1') = f_n(z_2')$, and this is a contradiction. □

For a meromorphic function f in D and for a point $w \in \widehat{\mathbf{C}}$, a point $z \in D$ such that $f(z) = w$ is called a *w-point* of f; an ∞-point of f is nothing but a pole of f. When $w \neq \infty$ and the order of zero of $f - w$ at z is m, m is called the *order* of the w-point z; the order of the ∞-point z of f is defined as that of the pole of f at z.

(4.3.11) THEOREM. *Let f be a meromorphic function in a domain $D \subset \widehat{\mathbf{C}}$. Let $z_0 \in D$ be a w_0-point of order m of f with $w_0 = f(z_0)$. Then there are neighborhoods V of z_0 and W of w_0 such that for an arbitrary $w \in W \setminus \{w_0\}$ there are exactly m distinct w-points, $z_1, \ldots, z_m$, of order 1 of f in V.*

PROOF. We assume that z_0 and w are not ∞; otherwise, we argue similarly by using $\tilde{z}$ or $\tilde{w}$ about ∞. By assumption we may expand f to a Taylor series about z_0:

$$f(z) = w_0 + a_m(z - z_0)^m + \cdots, \qquad a_m \neq 0.$$

There is $0 < \delta < d(z_0; \partial D)$ such that

$$\overline{\Delta(z_0;\delta)} \cap \{f = w_0\} = \{z_0\},$$

$$\overline{\Delta(z_0;\delta)} \cap \{f' = 0\} \setminus \{z_0\} = \emptyset, \tag{4.3.12}$$

$$\frac{1}{2\pi i} \int_{C(z_0;\delta)} \frac{f'(z)}{f(z) - w_0} dz = m. \tag{4.3.13}$$

There is $0 < \epsilon < 1$ such that

$$|f(z) - w_0| > \epsilon, \qquad z \in C(z_0; \delta).$$

Applying Theorem (4.3.7) to $f(z) - w = (f(z) - w_0) + (w_0 - w)$ with $w \in \Delta(w_0; \epsilon) \setminus \{w_0\}$, we infer that the number of zeros of $f(z) - w_0$ in $\Delta(z_0; \delta)$ is equal to that of $f(z) - w$ for $w \in \Delta(w_0; \epsilon)$. If $z \in \Delta(z_0; \delta)$ and $f(z) = w \in \Delta(w_0; \epsilon)$, then z is a w-point of order 1 of f by (4.3.12). Therefore there must be exactly m such points z. □

The following theorem is called the *Prinzip von der Gebietstreue (open mapping theorem).*

(4.3.14) COROLLARY. *Let f be a non-constant meromorphic function in a domain $D \subset \widehat{\mathbf{C}}$. Then, as f is considered to be a continuous mapping from D into $\widehat{\mathbf{C}}$, the image $f(D)$ is a domain of $\widehat{\mathbf{C}}$.*

PROOF. By Theorem (4.3.11) $f(D)$ is an open subset. Take two arbitrary points $f(z_1), f(z_2) \in f(D)$. There is a curve $C(\phi)$ connecting z_1 and z_2 in D. Then $f \circ \phi$ is a curve connecting $f(z_1)$ and $f(z_2)$. Thus, $f(D)$ is connected. □

We give one more application of Theorem (4.3.11), which is called the *inverse function theorem.*

(4.3.15) THEOREM. *Let f be a holomorphic function in a neighborhood of $z_0 \in \mathbf{C}$, and set $w_0 = f(z_0)$. If $f'(z_0) \neq 0$, there are neighborhoods U of z_0 and W of w_0 such that $f(U) = W$ and the restriction $f|U : U \to W$ of f to U has a holomorphic inverse function $g : W \to U$ of $f|U$ with $g(w_0) = z_0$.*

PROOF. Take neighborhoods U of z_0 and W of w_0 as in Theorem (4.3.11). Since $f'(z_0) \neq 0$, $m = 1$, so that for an arbitrary $w \in W$ there is a unique point

$z \in U$ with $f(z) = w$. Write $g(w) = z$. We first check the continuity of g. Take an arbitrary $w' \in W$, and set $z' = g(w')$. In a neighborhood V' of z', we have

(4.3.16)
$$\begin{aligned} f(z) - f(z') &= a_1(z - z') + a_2(z - z')^2 + \cdots \qquad (a_1 \neq 0) \\ &= (z - z')(a_1 + a_2(z - z') + \cdots) \\ &= (z - z')h(z). \end{aligned}$$

Since $h(z') = a_1 \neq 0$, after taking V' smaller if necessary, there is a positive number δ such that

(4.3.17)
$$|h(z)| \geqq \delta, \qquad z \in V'.$$

Then $f(V') = W'$ is an open neighborhood of w'. It follows from (4.3.16) and (4.3.17) that

$$\delta |g(w) - g(w')| \leqq |w - w'|.$$

Hence $g(w) - g(w') \to 0$ as $w \to w'$; that is, g is continuous. Again by (4.3.16) we have for $w \in W'$ with $w \neq w'$

$$\frac{g(w) - g(w')}{w - w'} = \frac{1}{h(g(w))}.$$

The continuity of g implies that $\lim_{w \to w'} h(g(w)) = h(g(w')) = h(z') = f'(z)$, and so

$$\lim_{w \to w'} \frac{g(w) - g(w')}{w - w'} = g'(w') = \frac{1}{f'(z')}.$$

Thus, g is complex differentiable in W'; i.e., g is holomorphic. Then we may set $U = g(W) = f^{-1}(W)$. □

(4.3.18) THEOREM. *Let f be a univalent meromorphic function in a domain $D \subset \widehat{\mathbf{C}}$. Then there is a meromorphic function g in the domain $W = f(D) \subset \widehat{\mathbf{C}}$ such that the mapping $g : W \to \widehat{\mathbf{C}}$ is the inverse mapping of $f : D \to W$.*

PROOF. By Theorem (4.3.14), the image $W = f(D) \subset \widehat{\mathbf{C}}$ is a domain, and we have the inverse mapping $g : W \to D$ of f since f is univalent. Take an arbitrary point $z_0 \in D$, and set $w_0 = f(z_0)$. It follows from Theorem (4.3.11) and the univalence of f that z_0 is always a w_0-point of order 1. Therefore, in the case of $z_0 \neq \infty$ and $w_0 \neq \infty$, $f'(z_0) \neq 0$. By Theorem (4.3.15) g is holomorphic in a neighborhood of w_0. In the case of $z_0 = \infty$ and $w_0 \neq \infty$, we have that $f \circ \tilde{z}/d\tilde{z}(0) \neq 0$. Thus $\tilde{z} \circ g(w) = 1/g(w)$ is meromorphic in a neighborhood of w_0, and hence $g(w)$ is meromorphic there. In the case of $z_0 = \infty$ and $w_0 = \infty$, we see that $\tilde{w} \circ f = 1/f$ is holomorphic in a neighborhood of $\tilde{z} = 0$. Thus, for the same reason as above, g is meromorphic in a neighborhood of ∞. □

It should be noted that Theorem (4.3.18) does not hold for real differentiable functions, nor for real analytic functions. For example, take $f(t) = t^3$ for $t \in \mathbf{R}$. Then $f : \mathbf{R} \to \mathbf{R}$ is injective, but the inverse function $\sqrt[3]{t} \in \mathbf{R}$ is not differentiable at $t = 0$.

Let f be a holomorphic function in a neighborhood of $z_0 \in \mathbf{C}$. If $f'(z_0) \neq 0$, then by Theorem (4.3.15) f has an inverse function g defined in a neighborhood of $w_0 = f(z_0)$. In what follows, we consider the case of $f'(z_0) = 0$. For the sake of simplicity, we assume that $z_0 = w_0 = 0$. Let m be the order of zero of f at 0. Then $m > 1$, and we have

$$\begin{aligned} f(z) &= a_m z^m + a_{m+1} z^{m+1} + \cdots \\ &= a_m z^m \left(1 + \frac{a_{m+1}}{a_m} + \cdots\right) \\ &= a_m z^m h_1(z) \qquad (a_m \neq 0, h_1(0) = 1). \end{aligned}$$

Taking $\delta > 0$ small, we have

$$h_1(z) \in \Delta(1;1), \qquad z \in \Delta(\delta).$$

Fixing one branch of log in $\Delta(1;1)$, we get $h_1(z) = e^{\log h_1(z)}$, and set

$$h_2(z) = \exp\left(\frac{1}{m}\log h_1(z)\right).$$

Taking one value of $\sqrt[m]{a_m}$, we set

$$h_3(z) = \sqrt{a_m} z h_2(z).$$

Then $f(z) = (h_3(z))^m$, and

$$\begin{aligned} h_3(z) &= \sqrt[m]{a_m} z + \cdots, \\ h_3'(0) &= \sqrt[m]{a_m} \neq 0. \end{aligned}$$

It follows that $v(z) = h_3(z)$ has the inverse function $z = g_1(v)$ with $g_1(0) = 0$. Let the Taylor series expansion of g_1 be

$$g_1(v) = c_1 v + c_2 v^2 + \cdots.$$

Since $w = f(z) = v^m$, we write $v = \sqrt[m]{w}$, and set

$$g_2(w) = c_1 \sqrt[m]{w} + c_2 \left(\sqrt[m]{w}\right)^2 + \cdots. \tag{4.3.19}$$

Then $f(g_2(w)) = w$, and so one may consider $g_2(w) = f^{-1}(w)$. But, $g_2(w)$ is a holomorphic function of $v = \sqrt[m]{w}$, and for the variable w, $g_2(w) = g_1(\sqrt[m]{w})$ is a multi-valued function with m values if $w \neq 0$. In general a series such as (4.3.19) in the powers of $\sqrt[m]{w}$ is called a *Puiseux series*, and in the present case, in particular, (4.3.19) is called the *Puiseux series expansion* of the inverse function $f^{-1}(w)$.

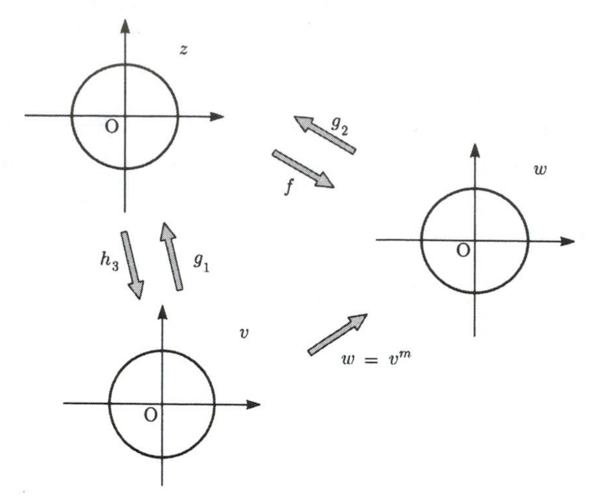

FIGURE 41

EXERCISE 2. Show that the numbers of zeros and poles of a rational function counting multiplicities over the Riemann sphere $\widehat{\mathbf{C}}$ are equal.

EXERCISE 3. Set $f(z) = z^2/(1+z)$. Obtain the Puiseux series expansion of $f^{-1}(w)$ about $w = 0$.

4.4. Residue Calculus

In this section we explain how to calculate the values of some definite integrals as applications of residue Theorem (4.2.7) or (4.2.14), classifying the definite integrals into several types.

i) Let $R(x, y)$ be a rational function of two variables x and y such that it has no pole on the circle $C(0;1)$. We want to calculate the value of

$$I = \int_0^{2\pi} R(\cos t, \sin t)dt.$$

For $z = e^{it}$ we have

$$\cos t = \frac{1}{2}\left(z + \frac{1}{z}\right), \qquad \sin t = \frac{1}{2i}\left(z - \frac{1}{z}\right),$$
$$dt = \frac{1}{iz}dz.$$

It follows from Theorem (4.2.7) that

$$\begin{aligned} I &= \int_{C(0;1)} R\left(\frac{1}{2}\left(z + \frac{1}{z}\right), \frac{1}{2i}\left(z - \frac{1}{z}\right)\right)\frac{1}{iz}dz \\ &= 2\pi \sum_{a\in\Delta(1)} \operatorname{Res}\left(a; \frac{1}{z}R\left(\frac{1}{2}\left(z + \frac{1}{z}\right), \frac{1}{2i}\left(z - \frac{1}{z}\right)\right)\right). \end{aligned}$$

EXAMPLE. We calculate $I = \displaystyle\int_0^{2\pi} \frac{dt}{a+\cos t}$, $a > 1$. Then

$$\begin{aligned}
I &= \int_{C(0;1)} \frac{1}{a+\frac{1}{2}\left(z+\frac{1}{z}\right)} \frac{1}{iz} dz = \frac{2}{i}\int_{C(0;1)} \frac{1}{z^2+2az+1} dz \\
&= \frac{2}{i}\int_{C(0;1)} \frac{1}{\left\{z-(-a-\sqrt{a^2-1})\right\}\left\{z-(-a+\sqrt{a^2-1})\right\}} dz \\
&= \frac{2}{i}\cdot 2\pi i \operatorname{Res}\left(-a+\sqrt{a^2-1}; \frac{1}{\left\{z-(-a-\sqrt{a^2-1})\right\}\left\{z-(-a+\sqrt{a^2-1})\right\}}\right) \\
&= 4\pi \lim_{z\to -a+\sqrt{a^2-1}} \frac{\left\{z-(-a+\sqrt{a^2-1})\right\}}{\left\{z-(-a-\sqrt{a^2-1})\right\}\left\{z-(-a+\sqrt{a^2-1})\right\}} \\
&= \frac{2\pi}{\sqrt{a^2-1}}.
\end{aligned}$$

In the latter half we used (4.2.10) to obtain the value of the residue.

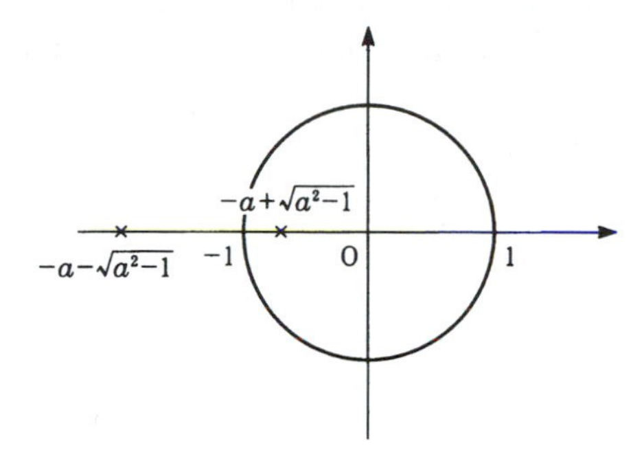

FIGURE 42

EXERCISE 1. Calculate the following definite integrals.

i) $\displaystyle\int_0^{\pi} \frac{d\theta}{1-2a\cos\theta+a^2}$ $\quad (0 < a < 1)$.

ii) $\displaystyle\int_0^{\frac{\pi}{2}} \frac{d\theta}{a+\sin^2\theta}$ $\quad (a > 0)$.

iii) $\displaystyle\int_0^{2\pi} \frac{\cos 2\theta}{a+\cos\theta} d\theta$ $\quad (a > 1)$.

ii) Let $R(z)$ be a rational function of z. We want to get

$$I = \int_{-\infty}^{\infty} R(x)dx = \int_{-\infty}^{\infty} R(z)dz. \tag{4.4.1}$$

Express $R(z) = P(z)/Q(z)$ with coprime polynomials $P(z)$ and $Q(z)$. For the convergence of the integral (4.4.1) at $\pm\infty$ the following condition is necessary and sufficient:

$$\deg P - \deg Q \leqq -2. \tag{4.4.2}$$

If $Q(z) = 0$ has a real root, the integral (4.4.1) does not exist. We add the following condition:

(4.4.3) The orders of poles of $R(z)$ on the real axis are at most 1.

Nonetheless, if $R(z)$ has a pole $b \in \mathbf{R}$, the integral (4.4.1) does not exist in the ordinary sense; we have to take the *principal value* of integration defined by

$$\text{p.v.} \int_{b-\delta}^{b+\delta} R(x)dx = \lim_{\epsilon \to +0} \left(\int_{b-\delta}^{b-\epsilon} + \int_{b+\epsilon}^{b+\delta} R(x)dx \right).$$

Here $\delta > 0$ is chosen so that b is the only pole of $R(z)$ contained in the interval $[b-\delta, b+\delta]$. The existence of the above limit will be proved below. We take the integral (4.4.1) in this way, call it the *principal value* of integration, and write

$$I = \text{p.v.} \int_{-\infty}^{\infty} R(x)dx.$$

Let $a_i \in \mathbf{C}$, $\operatorname{Im} a_i > 0$, $1 \leqq i \leqq k$, be all the roots of $Q(z) = 0$ in the upper half-plane. Let $b_j \in \mathbf{R}$, $1 \leqq j \leqq l$, be all the real roots of $Q(z) = 0$. Then we are going to show that

$$\text{(4.4.4)} \quad \text{p.v.} \int_{-\infty}^{\infty} R(x)dx = 2\pi i \sum_{i=1}^{k} \operatorname{Res}(a_i; R(z)) + \pi i \sum_{j=1}^{l} \operatorname{Res}(b_j; R(z)).$$

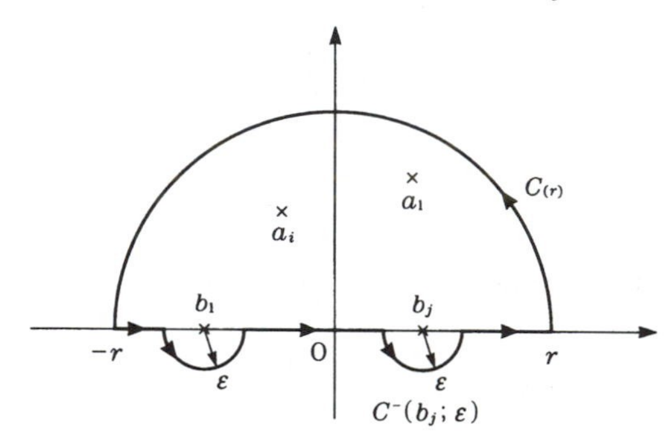

FIGURE 43

Take $r \geqq r_0 > \max_{i,j}\{|a_i|, |b_j|\}$, and small $\epsilon > 0$ so that $\overline{\Delta(a_i;\epsilon)}$ do not intersect $\mathbf{R}$, and $\overline{\Delta(a_i;\epsilon)}$ and $\overline{\Delta(b_j;\epsilon)}$ are mutually disjoint. Let $C_{(r)}$ denote the half circle from r to $-r$ in the upper half plane, and let $C_{(r,\epsilon)}$ denote the curve from $-r$ to r on the real axis where about every b_j the curve goes around the half circle $C^{-}(b_j;\epsilon)$ of $C(b_j;\epsilon)$ in the lower half plane (see Figure 43). Then

$$\text{(4.4.5)} \quad \int_{C_{(r)}+C_{(r,\epsilon)}} R(z)dz = 2\pi i \left\{ \sum \operatorname{Res}(a_i; R(z)) + \sum \operatorname{Res}(b_j; R(z)) \right\}.$$

By condition (4.4.2) there are an $M > 0$ and a $r_0 > 0$ such that for all $r \geqq r_0$

$$|R(re^{i\theta})| \leqq \frac{M}{r^2},$$

and hence

$$\left|\int_{C_{(r)}} R(z)dz\right| \leqq \int_{C_{(r)}} \frac{M}{r^2}dz = \frac{\pi M}{r} \to 0 \qquad (r \to \infty).$$

Expand $R(z)$ about every b_j to a Laurent series:

$$\begin{aligned} R(z) &= \frac{c_{-1}}{z-b_j} + c_0 + c_1(z-b_j) + \cdots \\ &= \frac{c_{-1}}{z-b_j} + h(z). \end{aligned}$$

We get

$$\int_{C^-(b_j;\epsilon)} \left\{\frac{c_{-1}}{z-b_j} + h(z)\right\} dz \to \pi i c_{-1} \qquad (\epsilon \to 0).$$

Since $c_{-1} = \operatorname{Res}(b_j; R(z))$, we obtain (4.4.4) by letting $r \to \infty$ and $\epsilon \to 0$ in (4.4.5).

EXAMPLE. We calculate $I = \text{p.v.}\int_0^\infty \frac{1}{x^4-1}dx = \frac{1}{2}\text{p.v.}\int_{-\infty}^\infty \frac{1}{x^4-1}dx$. Since

$$Q(z) = z^4 - 1 = (z-1)(z+1)(z-i)(z+i),$$

conditions (4.4.2) and (4.4.3) are satisfied. We have

$$\operatorname{Res}\left(1; \frac{1}{Q(z)}\right) = \frac{1}{4}, \quad \operatorname{Res}\left(-1; \frac{1}{Q(z)}\right) = -\frac{1}{4}, \quad \operatorname{Res}\left(i; \frac{1}{Q(z)}\right) = \frac{i}{4}.$$

Therefore

$$\begin{aligned} I &= \frac{1}{2}\left[2\pi i \operatorname{Res}\left(i; \frac{1}{Q(z)}\right) + \pi i\left\{\operatorname{Res}\left(1; \frac{1}{Q(z)}\right) + \operatorname{Res}\left(-1; \frac{1}{Q(z)}\right)\right\}\right] \\ &= \frac{1}{2}2\pi i\left(\frac{i}{4}\right) = -\frac{\pi}{4}. \end{aligned}$$

EXERCISE 2. Calculate the following integrals.

i) $\text{p.v.}\int_{-\infty}^\infty \frac{1}{x^2-1}dx.$ ii) $\int_0^\infty \frac{x^2}{x^4+1}dx.$

iii) $\text{p.v.}\int_{-\infty}^\infty \frac{x^2+1}{x^4+x^2-2}dx.$ iv) $\int_0^\infty \frac{1}{(x^2+1)^n}dx \quad (n = 1, 2, \ldots).$

iii) We let $R(z) = P(z)/Q(z)$ be as in ii), but relax condition (4.4.2) to

$$\deg P - \deg Q \leqq -1. \tag{4.4.6}$$

We keep condition (4.4.3). Then the following holds:

$$\text{p.v.}\int_{-\infty}^\infty R(x)e^{ix}dx = 2\pi i \sum_{\operatorname{Im} a > 0} \operatorname{Res}(a; R(z)e^{iz}) + \pi i \sum_{\operatorname{Im} b = 0} \operatorname{Res}(b; R(z)e^{iz}). \tag{4.4.7}$$

The proof is the same as in ii); the only difference is to show that

$$\int_{C_{(r)}} R(z)e^{iz}dz \to 0 \qquad (r \to \infty)$$

under condition (4.4.6). Set $z = re^{i\theta}$ $(0 \leqq \theta \leqq \pi)$. Then $|e^{ire^{i\theta}}| = e^{-r\sin\theta}$, and there is some $M > 0$ such that $|R(re^{i\theta})| \leqq Mr$. Therefore

$$\left| \int_{C_{(r)}} R(z)e^{iz} \right| \leqq M \int_0^\pi e^{-r\sin\theta} d\theta.$$

If one knows Lebesgue integrals, it is clear that the above right side converges to 0 as $r \to \infty$. Relying only on Riemann integrals, one deduces the same from the following estimate:

$$\begin{aligned}\int_0^\pi e^{-r\sin\theta}d\theta = 2\int_0^{\pi/2} e^{-r\sin\theta}d\theta \leqq 2\left[-\frac{\pi}{2r}e^{-2r\theta/\pi}\right]_0^{\pi/2} \\ = \frac{\pi}{r} - \frac{\pi}{r}e^{-r} \to 0 \qquad (r \to \infty).\end{aligned}$$

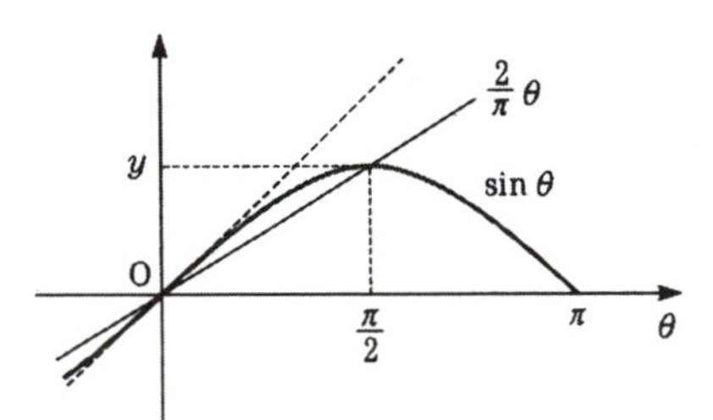

FIGURE 44

EXAMPLE. Set $I = \displaystyle\int_0^\infty \frac{\sin x}{x}dx = \frac{1}{2}\text{p.v.}\int_{-\infty}^\infty \frac{\sin x}{x}dx.$

Note that $(\sin x)/x$ is continuous at 0 if the value there is defined to be 1. Since $(\cos x)/x$ is an odd function, we have the principal value of integral,

$$\text{p.v.}\int_{-\infty}^\infty \frac{\cos x}{x}dx = 0.$$

Thus

$$2I = \text{p.v.}\int_{-\infty}^\infty \frac{e^{iz}}{iz}dz.$$

Since $R(z) = 1/iz$ satisfies conditions (4.4.6) and (4.4.3), it follows from (4.4.7) that

$$2I = \pi\text{Res}\left(0; \frac{e^{iz}}{iz}\right) = \pi.$$

Hence, $I = \pi/2$.

REMARK 1. The integral $I = \displaystyle\int_0^\infty \frac{\sin x}{x} dx$ is not absolutely convergent; that is,

$$\int_0^\infty \left| \frac{\sin x}{x} \right| dx = \infty.$$

REMARK 2. To calculate $I' = \text{p.v.} \int_{-\infty}^{\infty} R(x) e^{-ix} dx$, one has to take a curve in the lower half plane as in Figure 45.

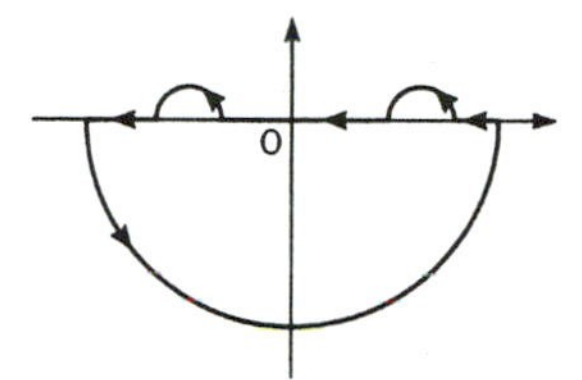

FIGURE 45

Then

$$I' = -2\pi i \sum_{\operatorname{Im} a<0} \operatorname{Res}(a; R(z)e^{iz}) - \pi i \sum_{\operatorname{Im} b=0} \operatorname{Res}(b; R(z)e^{iz}).$$

EXERCISE 3. Calculate the following integrals.

i) $\text{p.v.} \displaystyle\int_{-\infty}^{\infty} \frac{x^2}{x^3-1} e^{ix} dx.$ ii) $\text{p.v.} \displaystyle\int_{\infty}^{\infty} \frac{\sin x}{a+x} dx \quad (a \in \mathbf{R}).$

iii) $\displaystyle\int_0^\infty \frac{\cos x}{x^2+1} dx.$ iv) $\displaystyle\int_0^\infty \frac{\sin^3 x}{x} dx.$

iv) Let $R(z) = P(z)/Q(z)$ be a rational function which has no pole at 0 nor on the positive real axis, and satisfies (4.4.6). Let $0 < \alpha < 1$. We calculate

$$I = \int_0^\infty \frac{R(x)}{x^\alpha} dx.$$

It is clear that the integral converges at 0 and ∞. Set $D = \mathbf{C} \backslash \{z \in \mathbf{R}; z \geqq 0\}$, and take a branch of $\log z$ on D so that $\lim_{z \to 1} \log z = 0$. Consider a meromorphic function on D,

$$f(z) = \frac{R(z)}{z^\alpha} = R(z) e^{-\alpha \log z}.$$

Let $r > 0$ be sufficiently large so that $\Delta(r)$ contains all the poles of $R(z)$. Take sufficiently small $\epsilon > 0$ and $\epsilon' > 0$ to form a curve $C(\epsilon, \epsilon', r)$ as in Figure 46:

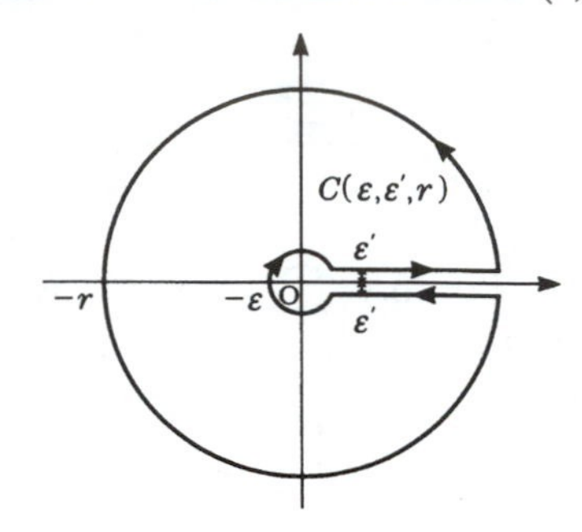

FIGURE 46

Then we have

$$\int_{C(\epsilon,\epsilon',r)} f(z)dz = 2\pi i \sum_{a \in D} \operatorname{Res}(a; f). \tag{4.4.8}$$

Letting $\epsilon' \to 0$, (4.4.8) is decomposed to

(4.4.9)

$$\begin{aligned} &\int_{-C(0;\epsilon)} f(z)dz + \int_{-\epsilon}^{r} \lim_{\epsilon' \to +0} f(x+i\epsilon')dx + \int_{C(0;r)} f(z)dz \\ &+ \int_{r}^{-\epsilon} \lim_{\epsilon' \to +0} f(x-i\epsilon')dx = 2\pi i \sum_{a \in D} \operatorname{Res}(a; f(z)). \end{aligned}$$

For $x \in \mathbf{R}, x > 0$,

$$\lim_{\epsilon' \to +0} f(x+i\epsilon')dx - \lim_{\epsilon' \to +0} f(x-i\epsilon') = \frac{R(x)}{x^\alpha}\left(1 - e^{-2\alpha\pi i}\right).$$

There are $M > 0$ and $M' > 0$ such that

$$\begin{aligned} \left|\int_{-C(0;\epsilon)} f(z)dz\right| &\leqq \frac{M}{\epsilon^\alpha} 2\pi\epsilon = 2\pi M \epsilon^{1-\alpha} \to 0 \qquad (\epsilon \to 0), \\ \left|\int_{C(0;r)} f(z)dz\right| &\leqq \frac{M'}{r^{1+\alpha}} 2\pi r = \frac{2\pi M'}{r^\alpha} \to 0 \qquad (r \to \infty). \end{aligned}$$

It follows from (4.4.9) that

$$I = \int_0^\infty \frac{R(x)}{x^\alpha} dx = \frac{2\pi i}{1 - e^{-2\alpha\pi i}} \sum_{a \in D} \operatorname{Res}(a; R(z)e^{-\alpha \log z}).$$

EXAMPLE. Set $I = \int_0^\infty \frac{1}{(1+x)x^\alpha}dx$, $0 < \alpha < 1$. The only pole of $f(z) = \frac{e^{-\alpha \log z}}{1+z}$ is -1 and its order is 1. The residue there is

$$\operatorname{Res}(-1; f(z)) = e^{-\alpha \log(-1)} = e^{-\alpha\pi i}.$$

Therefore,

$$I = \frac{2\pi i e^{-\alpha\pi i}}{1 - e^{-2\alpha\pi i}} = \frac{2\pi i}{e^{\alpha\pi i} - e^{-\alpha\pi i}} = \frac{\pi}{\sin \alpha\pi}.$$

EXERCISE 4. Calculate the following integrals.

i) $\int_0^\infty \frac{x}{(1+x^2)x^\alpha}dx \qquad (0 < \alpha < 1).$

ii) $\int_0^\infty \frac{1}{(1+x)^n x^\alpha}dx \qquad (0 < \alpha < 1, n = 1, 2, \dots).$

iii) $\int_0^\infty \frac{1}{(x^2+a^2)x^\alpha}dx \qquad (a > 0, 0 < \alpha < 1).$

v) Let $R(z) = P(z)/Q(z)$ be a rational function without pole at 0 nor on the positive real axis, satisfying (4.4.2). Assume moreover that all the coefficients of $P(z)$ and $Q(z)$ are real. We calculate

$$I = \int_0^\infty R(x) \log x \, dx.$$

The convergence of the integral at 0 and ∞ may be clear. Let D and $C(\epsilon, \epsilon', r)$ be as in iv), and take a branch of log as in iv). Set $f(z) = R(z)(\log z)^2$. Then

$$\int_{C(\epsilon,\epsilon',r)} f(z) = 2\pi i \sum_{a\in D} \operatorname{Res}(a; f(z)).$$

Letting $\epsilon' \to 0$, we get

$$\int_{-C(0;\epsilon)} f(z)dz + \int_\epsilon^r R(x)\{(\log x)^2 - (\log x + 2\pi i)^2\}dx + \int_{C(0;r)} f(z)dz = 2\pi i \sum_{a\in D} \operatorname{Res}(a; f(z)).$$

There are $M, M' > 0$ such that

$$\left|\int_{-C(0;\epsilon)} R(z)(\log z)^2 dz\right| \leqq M((\log \epsilon)^2 + 4\pi^2)2\pi\epsilon \quad (\epsilon \to 0),$$

$$\left|\int_{C(0;r)} R(z)(\log z)^2 dz\right| \leqq \frac{M'}{r^2}((\log r)^2 + 4\pi^2)2\pi r \quad (r \to \infty).$$

Therefore,

$$\begin{aligned}&-4\pi i\int_0^\infty R(x)\log x dx+4\pi^2\int_0^\infty R(x)dx\\&=2\pi\sum_{a\in D}\operatorname{Res}(a;R(z)(\log z)^2).\end{aligned}$$

EXAMPLE. Calculate $I=\displaystyle\int_0^\infty\frac{\log x}{a+x^2}dx$, $a>0$. Note that $R(z)=1/(a+z^2)$ has no pole at 0 nor on the positive real axis, and satisfies (4.4.2). Its poles are $z=\pm\sqrt{a}i$, of order 1. One has

$$\begin{aligned}\operatorname{Res}\left(+\sqrt{a}i;\frac{(\log z)^2}{a+z^2}\right)&=\frac{(\log\sqrt{a}+\pi i/2)^2}{2\sqrt{a}i},\\\operatorname{Res}\left(-\sqrt{a}i;\frac{(\log z)^2}{a+z^2}\right)&=\frac{(\log\sqrt{a}+\pi i/2)^2}{-2\sqrt{a}i}.\end{aligned}$$

Hence,

$$I=-\frac{1}{2}\operatorname{Re}\left\{-\frac{\pi}{\sqrt{a}}(\log\sqrt{a}+\pi i)\right\}=\frac{\log a}{4\sqrt{a}}\pi.$$

EXERCISE 5. Calculate the following integrals.

i) $\displaystyle\int_0^\infty\frac{x\log x}{(x+a)^3}dx\qquad(a>0).$

ii) $\displaystyle\int_0^\infty\frac{\log x}{(a+x)(b+x)}dx\qquad(a,b>0).$

iii) $\displaystyle\int_0^\infty\frac{\log x}{(1+x^2)^2}dx.$

Problems

1. Obtain the Laurent series of the following functions.

i) $\dfrac{e^{1/z}}{1-z}$ (about $z=0$). ii) $\sin\dfrac{z}{z-1}$ (about $z=1$).

iii) $\dfrac{z^3}{1+z^2}$ (about $z=\infty$).

2. What are the poles and residues of the following functions on $\mathbf{C}$ or $\widehat{\mathbf{C}}$?

i) z^n $(n=\pm1,\pm2,\dots)$. ii) $\tan z$ (on $\mathbf{C}$).

iii) $e^{1/z}$. iv) $\dfrac{z}{z^2+5z+6}$. v) $\left(z+\dfrac{1}{z+1}\right)^3$.

3. Calculate the following curvilinear integrals.

i) $\displaystyle\int_{C(0;1)}\left(\frac{1}{z(z^2+2)}+\frac{1}{z^2}\right)dz.$

ii) $\displaystyle\int_{C(0;1)}\frac{e^z}{z^n}dz\qquad(n=1,2,\dots).$

iii) $\displaystyle\int_{C(0;2)} z^n(z-1)^m dz \qquad (n,m \in \mathbf{Z}).$

iv) $\displaystyle\int_{C(0;1)} \frac{1}{(z-a)^n(z-b)^m} dz \qquad (|a|<1<|b|, \quad m,n=1,2,\ldots).$

v) $\displaystyle\int_C \frac{e^{az}}{z^2} dz \quad (C=\{z=a+iy; -\infty<y<\infty\}, \alpha \in \mathbf{R}, \operatorname{Re} a \neq 0).$

4. i) How many roots of the equation, $z^6-2z^5+7z^4+z^3-z+1=0$, are there in $\Delta(1)$?

ii) How many roots of the equation, $z^4-6z+3=0$, are there in the annulus $R(1,2)$?

5. Let $f(z)$ be a holomorphic function on $\Delta(R)$ such that $|f(z)|<\theta R$ $(0<\theta<1)$. Show that there is a $z \in \Delta(R)$ with $z=f(z)$.

6. Let f,g be holomorphic functions in a neighborhood of $\overline{\Delta(1)}$ such that $|f(z)|>|g(z)|$ on $C(0;1)$. Then, show that $zf(z)+g(z)$ has a zero in $\Delta(1)$.

7. Let $\{f_n\}_{n=1}^{\infty}$ be a sequence of holomorphic functions in a domain $D \subset \widehat{\mathbf{C}}$ which converges uniformly on compact subsets to f with $f \not\equiv 0$. Then, show that a zero of f is a zero of $f_n, n=1,2,\ldots$, or an accumulation point of zeros of $f_n, n=1,2,\ldots$.

8. Calculate the following integrals.

i) $\displaystyle\int_0^{\frac{\pi}{2}} \frac{\cos^2 x}{a+\sin^2 x} dx \quad (a>0).$ ii) $\displaystyle\int_0^{\infty} \frac{\cos x}{x^2+a^2} dx \quad (a \in \mathbf{R}).$

iii) p.v. $\displaystyle\int_{-\infty}^{\infty} \frac{x^2}{x^3-1} e^{-2\pi i x} dx.$ iv) $\displaystyle\int_0^{2\pi} \frac{\cos x}{a+\cos x} dx \quad (a>1).$

v) $\displaystyle\int_0^{\infty} \frac{x \sin x}{x^2+a^2} dx \qquad (a \in \mathbf{R}).$

vi) $\displaystyle\int_0^{\infty} \frac{x+b}{(x^2+1)x^a} dx \qquad (0<a<1, b \in \mathbf{R}).$

vii) p.v. $\displaystyle\int_0^{\infty} \frac{1}{(1-x^2)x^{\alpha}} dx \qquad (0<\alpha<1).$

viii) $\displaystyle\int_0^{\infty} \frac{(\log x)^2}{x^2+1} dx.$ ix) $\displaystyle\int_0^{\infty} \frac{\log x}{x^2-1} dx.$

CHAPTER 5

Analytic Continuation

The analytic continuation is defined, and the fundamental extension theorem of Riemann is proved. After the reflection principle and the monodromy theorem are proved, the construction of universal coverings is described, and then the notion of Riemann surface is introduced. We will not go into function theory on Riemann surfaces, but readers will comprehend the necessity of this notion.

5.1. Analytic Continuation

Let $D \subset \widehat{\mathbf{C}}$ be a domain, and let f be a meromorphic function in D. Let $D' \subset \widehat{\mathbf{C}}$ be another domain, and assume that $D' \cap D \neq \emptyset$. If there is a meromorphic function g in D' such that $f \equiv g$ on a connected component U of $D' \cap D$, we define a function $\tilde{f}$ on $D \cup D'$ so that $\tilde{f} = f$ on D and $\tilde{f} = g$ on D'. The function $\tilde{f}$ thus defined is not necessarily one valued, for the values of f and g on other components of $D' \cap D$ may be different. We call $\tilde{f}$ the *analytic continuation* of f from D to $D \cup D'$ (or sometimes simply D').

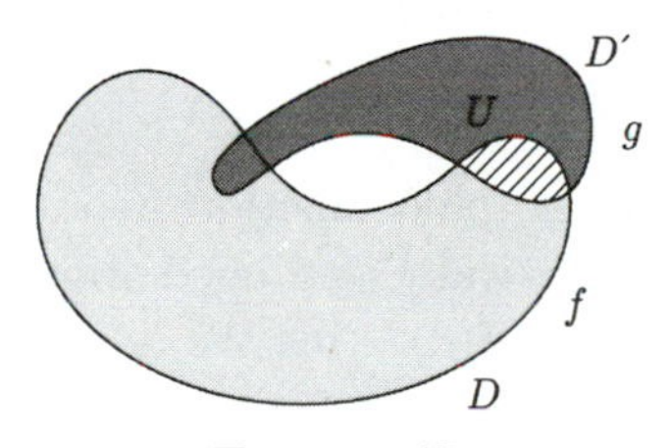

FIGURE 47

In general, we call it the *analytic continuation* extending a meromorphic function to one defined on a larger domain, allowing the resulting function to be multi-valued. The above meromorphic function g is unique if it exists and if it coincides with f on U (Theorem (4.2.2)). This property is called the *uniqueness of analytic continuation.* The following theorem is called *Riemann's extension theorem.*

(5.1.1) THEOREM. *Let f be a holomorphic function in $\Delta(r) \setminus \{0\}$. If there is a*

$0 < r' < r$ such that f is bounded on $\Delta(r') \setminus \{0\}$, then f is uniquely extended to a holomorphic function on $\Delta(r)$.

PROOF. Expand f to a Laurent series:

$$f(z) = \sum_{n=-\infty}^{\infty} a_n z^n.$$

By the assumption, there is an $M > 0$ such that $|f(r'' e^{i\theta})| \leqq M$ for all $0 < r'' \leqq r'$ and $0 \leqq \theta \leqq 2\pi$. It follows from (4.1.6) that

$$\sum_{n=-\infty}^{\infty} |a_n|^2 r''^{2n} = \frac{1}{2\pi} \int_0^{2\pi} |f(r'' e^{i\theta})|^2 d\theta \leqq M^2.$$

Letting $r'' \searrow 0$, we have that $a_n = 0$ for $n < 0$, and that

$$f(z) = \sum_{n=0}^{\infty} a_n z^n.$$

The right hand side defines a holomorphic function in $\Delta(r)$. The uniqueness is clear. $\square$

As in Theorem (5.1.1), let f be a holomorphic function defined in a neighborhood of a point a except at a itself. If f has an analytic continuation so that it is holomorphic about a, a is called a *removable singularity*. As a corollary of Theorem (5.1.1) we have *Casorati-Weierstrass' theorem*:

(5.1.2) THEOREM. *Let f be a holomorphic function in $\Delta(r) \setminus \{0\}$ such that it has 0 as an isolated essential singularity. Then the image $f(\Delta(r) \setminus \{0\})$ is dense in $\widehat{\mathbf{C}}$; that is, $\overline{f(\Delta(r) \setminus \{0\})} = \widehat{\mathbf{C}}$.*

PROOF. Assume that $\overline{f(\Delta(r) \setminus \{0\})} \neq \widehat{\mathbf{C}}$. Then there are a point $p \in \widehat{\mathbf{C}}$ and a neighborhood U of p such that $\overline{f(\Delta(r) \setminus \{0\})} \cap U = \emptyset$. Suppose that $p = \infty$. Take $\delta > 0$ such that $\{\tilde{w}; |\tilde{w}| < \delta\} \subset U$. Since $w\tilde{w} = 1$,

$$f(\Delta(r) \setminus \{0\}) \subset \overline{\Delta(1/\delta)}.$$

Hence, f is a bounded holomorphic function in $\Delta(r) \setminus \{0\}$. It follows from Theorem (5.1.1) that f is analytically continued about 0; this is a contradiction. If $p \in \mathbf{C}$, we consider $g(z) = 1/(f(z) - p)$. We deduce from the above that g is analytically continued about 0, so that $f(z) = p + 1/g(z)$ has at most a pole at 0; this is again a contradiction. $\square$

In the above theorem, it can actually be proved that the image $f(\Delta(r) \setminus \{0\})$ contains all points of $\widehat{\mathbf{C}}$ except for at most two points. This is called the big Picard theorem, which will be proved in Chapter 6.

A holomorphic function on all of $\mathbf{C}$ is called an *entire function*. Let f be an entire function. Then one of the following three cases holds:

i) f is a constant function.

ii) f has a pole at ∞.

iii) f has an isolated essential singularity at ∞.

These correspond respectively to the following cases, if f is expanded to a Taylor series

$$f(z) = \sum_{n=0}^{\infty} a_n z^n :$$

i) $a_n = 0,\ n \geqq 1$.

ii) There is some number n_0 such that $a_n = 0,\ n \geqq n_0$; i.e., $f(z)$ is a polynomial.

iii) There are infinitely many n with $a_n \neq 0$.

In the last case iii), f is said to be *transcendental.*

(5.1.3) COROLLARY. *The image $f(\mathbf{C})$ of a non-constant entire function f is dense in $\widehat{\mathbf{C}}$.*

PROOF. If f is a polynomial, the equation $f(z) - w = 0$ has a root for every $w \in \mathbf{C}$ (Theorem (3.5.23)). Thus $f(\mathbf{C}) = \mathbf{C}$, and $\overline{f(\mathbf{C})} = \widehat{\mathbf{C}}$. If f is transcendental, it suffices to apply Theorem (5.1.2) for the function $f \circ \tilde{z}$ in $\tilde{z} \neq 0$. $\square$

In the above corollary, if f is transcendental, then in fact it will be proved in Chapter 6 that f takes every point of $\widehat{\mathbf{C}}$ infinitely many times except for at most two points; this is called the little Picard theorem.

The next theorem is called *Schwarz' reflection principle.*

(5.1.4) THEOREM. *Let $D \subset \mathbf{C}$ be a domain such that the boundary ∂D contains an open interval L on the real axis, and $\partial D \setminus L$ does not accumulate at any point of L. Let f be a continuous function on $D \cup L$ which is holomorphic in D. Assume that f takes real values on L. Then f is analytically continued to $\tilde{D} = D \cup \{z \in \mathbf{C}; \overline{z} \in D \cup L\}$.*

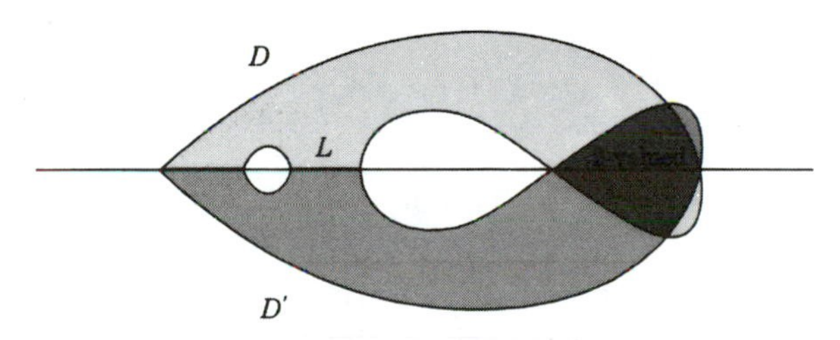

FIGURE 48

PROOF. Set $D' = \{z' \in \mathbf{C}; \overline{z}' \in D\}$, and

$$g(z') = \overline{f(\overline{z}')}, \qquad z' \in D' \cup L.$$

On L, $g = f$. We expand f about an arbitrary point $\overline{a}'$ with $a' \in D'$ to a Taylor series:

$$f(z) = \sum_{n=0}^{\infty} a_n (z - \overline{a}')^n, \qquad z \in \Delta(\overline{a}'; r) \subset D.$$

Then

$$g(z') = \overline{\sum a_n (\overline{z}' - \overline{a}')^n} = \sum \overline{a}_n (z' - a')^n, \qquad z' \in \Delta(a'; r).$$

Hence g is analytic in D', i.e., holomorphic there. Define a function $\tilde{f}$ on $\tilde{D} = D \cup L \cup D'$ so that $\tilde{f} = f$ on $D \cup L$, and $\tilde{f} = g$ on $D' \cup L$. The function $\tilde{f}$ is holomorphic in $D \cup D'$, and continuous on $\tilde{D}$. It remains to prove the analyticity of $\tilde{f}$ about an arbitrary point $a \in L$. Take a small $\delta > 0$ with $\Delta(a; \delta) \subset \tilde{D}$. To prove that $\tilde{f}$ is holomorphic in $\Delta(a; \delta)$ it suffices by Morera's Theorem (3.5.18) to show that for an arbitrary triangle contour C (with the anti-clockwise orientation) inside $\Delta(a; \delta)$

$$\int_C \tilde{f}(z)dz = 0. \tag{5.1.5}$$

If $C \subset D$ or $C \subset D'$, the interior E of C is also contained in D or D', and so Cauchy's integral Theorem (3.4.14) implies (5.1.5). Assume that $C \cap L \neq \emptyset$. Taking a sufficiently small $\epsilon > 0$, we set

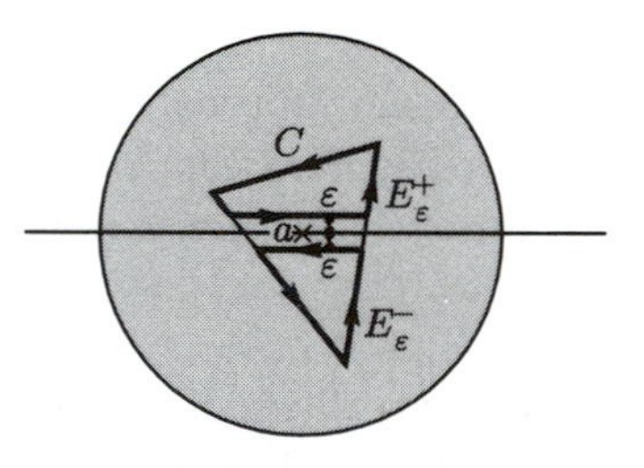

FIGURE 49

$$E^+_\epsilon = \{z \in E; \operatorname{Im} z > \epsilon\}, \qquad E^-_\epsilon = \{z \in E; \operatorname{Im} z > -\epsilon\}.$$

Suppose that $E^+_\epsilon \neq \emptyset$ and $E^-_\epsilon \neq \emptyset$. Let $C^\pm_\epsilon$ be the closed piecewise linear curves of the contours of $E^\pm_\epsilon$ with anti-clockwise orientation. Then

$$\int_{C^\pm_\epsilon} \tilde{f}(z)dz = 0.$$

Moreover,

$$\int_{C^+_\epsilon} \tilde{f}(z)dz \to \int_{C \cap \{\operatorname{Im} z \geqq 0\}} \tilde{f}(z)dz + \int_{L \cap \overline{E}} f(z)dz \qquad (\epsilon \to 0),$$

$$\int_{C^-_\epsilon} \tilde{f}(z)dz \to \int_{C \cap \{\operatorname{Im} z \leqq 0\}} \tilde{f}(z)dz - \int_{L \cap \overline{E}} f(z)dz \qquad (\epsilon \to 0).$$

Here one notes that in the second terms of the above right sides the orientation of the closed interval $L \cap \overline{E} \subset \mathbf{R}$ is the natural one, compatible with the order of the real number $\mathbf{R}$. Thus we get (5.1.5). If one of $E^\pm_\epsilon$ is empty, say $E^-_\epsilon = \emptyset$,

$$\int_{C^+_\epsilon} \tilde{f}(z)dz \to \int_C \tilde{f}(z)dz.$$

Therefore, in either case, (5.1.5) holds. □

A curve $\phi : [T_0, T_1] \to \widehat{\mathbf{C}}$ is called an *analytic curve* if for an arbitrary point $t_0 \in (T_0, T_1)$ with $z_0 = \phi(t_0) \neq \infty$, there are a $\delta > 0$ with $(t_0 - \delta, t_0 + \delta) \subset (T_0, T_1)$ and real analytic functions ϕ_1, ϕ_2 on $(t_0 - \delta, t_0 + \delta)$ such that

$$\phi(t) = \phi_1(t) + i\phi(t), \qquad t_0 - \delta < t < t_0 + \delta;$$

when $z_0 = \infty$, the same property as above with respect to the coordinate $\tilde{z}$ is to be satisfied. In particular, if for any $t \in (T_0, T_1)$ $\phi'(t) \neq 0$, or $\frac{d}{dt}\tilde{z} \circ \phi(t) \neq 0$, then ϕ is called a *non-singular* real analytic curve.

Let $D \subset \widehat{\mathbf{C}}$ be a domain. Assume that a part of the boundary ∂D is represented by a non-singular analytic curve $C(\phi : [T_0, T_1] \to \partial D)$. Set

$$\overset{\circ}{C} = C \setminus \{\phi(T_0), \phi(T_1)\}.$$

Assume that the set $\partial D \setminus \overset{\circ}{C}$ does not accumulate at any point of $\overset{\circ}{C}$.

Let $t_0 \in (T_0, T_1)$ be an arbitrary point, and set $z_0 = \phi(t_0)$. Assume for a moment that $z_0 \neq \infty$. Taking a sufficiently small $\delta > 0$, we have a Taylor expansion of ϕ about t_0: For $t \in (t_0 - \delta, t_0 + \delta)$

$$\phi(t) = z_0 + \sum_{n=1}^{\infty} c_n (t - t_0)^n, \qquad c_n \in \mathbf{C}. \tag{5.1.6}$$

Introduce a new complex coordinate

$$\xi = t + is, \qquad t, s \in \mathbf{R}.$$

It follows from (5.1.6) that

$$\phi(\xi) = z_0 + \sum_{n=1}^{\infty} c_n (\xi - t_0)^n, \qquad |\xi - t_0| < \delta,$$

is a convergent power series, and defines a holomorphic function. Since $\phi'(t_0) = c_1 \neq 0$, by the inverse function Theorem (4.3.15) $\phi(\xi)$ may be assumed to be injective on $\Delta(t_0; \delta)$. Set

$$\begin{array}{rclrcl} U^+ & = & \Delta(t_0;\delta) \cap \{\operatorname{Im}\xi > 0\}, & U^- & = & \Delta(t_0;\delta) \cap \{\operatorname{Im}\xi < 0\}, \\ L & = & (t_0 - \delta, t_0 + \delta), & C_0 & = & \phi(L), \\ V & = & \phi(\Delta(t_0;\delta)), & V^{\pm} & = & \phi(U^{\pm}). \end{array}$$

By the assumption we may suppose, after taking a smaller $\delta > 0$ if necessary, that $\partial D \cap V = C_0$. It does not lose generality to assume that $V^+ \subset D$ (if $V^- \subset D$, one may replace s by $-s$). Then $\xi \in U^+$ and $\bar{\xi} \in U^-$ are the reflection points of each other with respect to L. We say that $\phi(\xi)$ and $\phi(\bar{\xi})$ are the *reflection points* of each other with respect to C_0.

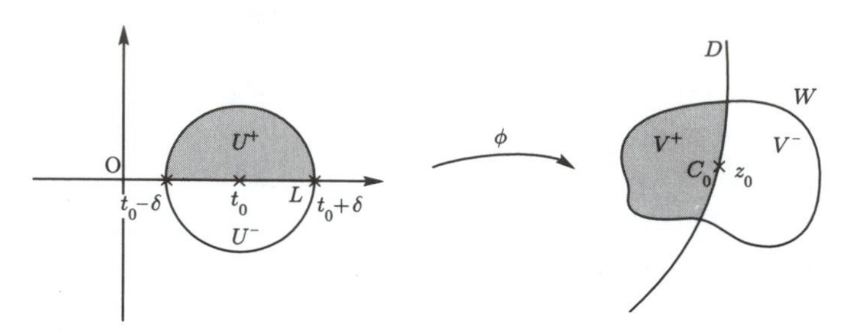

FIGURE 50

We show that this concept of reflection does not depend on the choice of parameters of ϕ. Let $\hat{\phi} : [\hat{T}_0, \hat{T}_1] \to \partial D$ be a change of parameter of ϕ, which gives C and is non-singular and analytic. Take $\hat{t}_0$ with $\hat{\phi}(\hat{t}_0) = \phi(t_0)$, and define $\Delta(\hat{t}_0; \hat{\delta})$, $\hat{U}^{\pm}$, $\hat{L}$, $\hat{V}$, ..., etc., as above. Let $\hat{\delta} > 0$ be sufficiently small so that the values of the holomorphic function $\phi^{-1} \circ \hat{\phi}$ on $\Delta(\hat{t}_0; \hat{\delta})$ are contained in $\Delta(t_0; \delta)$. Set

$$h = \phi^{-1} \circ \hat{\phi} | \Delta(\hat{t}_0; \hat{\delta}) : \Delta(\hat{t}_0; \hat{\delta}) \to \Delta(t_0; \delta).$$

The holomorphic function h takes real values on $\hat{L}$: $h(\hat{L}) \subset L$. Therefore it follows from Theorem (5.1.4) and the uniqueness of the analytic continuation that for $\hat{\xi} \in \Delta(\hat{t}_0; \hat{\delta})$, $h(\bar{\hat{\xi}}) = \overline{h(\hat{\xi})}$. Setting $\xi = h(\hat{\xi})$, we have

$$\phi(\xi) = \hat{\phi}(\hat{\xi}), \qquad \phi(\bar{\xi}) = \hat{\phi}(\bar{\hat{\xi}}).$$

Hence, two points which are reflection points with respect to $\hat{C}_0$ are also reflection points with respect to C_0.

In the case of $z_0 = \infty$, we use the coordinate $\tilde{z}$ to define the notion of reflection points as well.

EXERCISE 1. Show that if C is a circle (the case of a line is included), the notion of reflection points defined in Chapter 2, §8 coincides with one defined here.

(5.1.7) THEOREM. *Let $D_i \subset \widehat{\mathbf{C}}$, $i = 1, 2$, be domains such that a part C_i of ∂D_i is a simple non-singular analytic curve and $\partial D_i \setminus \overset{\circ}{C}_i$ does not accumulate at any point of $\overset{\circ}{C}_i$. Let f be a meromorphic function on D_1 with values in D_2 as a mapping into $\widehat{\mathbf{C}}$. Assume that f is extended to a continuous mapping over $D_1 \cup \overset{\circ}{C}_1$ so that $f(\overset{\circ}{C}_1) \subset \overset{\circ}{C}_2$. Then, if $z \in D_1$ is sufficiently close to $\overset{\circ}{C}_1$, to its reflection point z^* with respect to $\overset{\circ}{C}_1$ there corresponds the reflection point $f(z)^*$ of $f(z)$ with respect to $\overset{\circ}{C}_2$, so that f is analytically continued, crossing $\overset{\circ}{C}_1$, to a domain containing $\overset{\circ}{C}_1$ as a meromorphic function.*

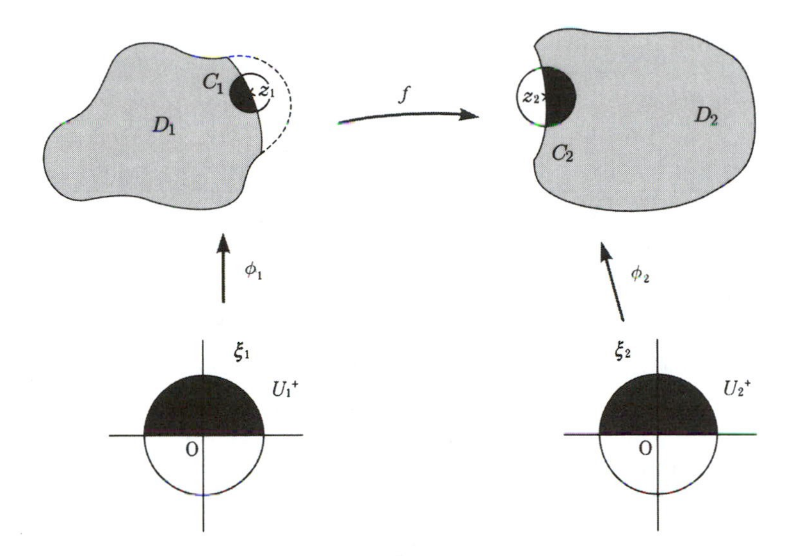

FIGURE 51

PROOF. Take arbitrarily a point $z_1 \in \overset{\circ}{C}_1$, and set $z_2 = f(z_1) \in \overset{\circ}{C}_2$. We suppose that in a neighborhood of z_i, C_i is represented by a simple non-singular analytic curve $\phi_i : (-\delta_i, \delta_i) \to C_i$ with $\delta_i > 0$. As described above, we extend ϕ_i over a complex variable $\xi_i \in \Delta(\delta_i)$. We may assume that $\phi_i : \Delta(\delta_i) \to \phi_i(\Delta(\delta_i))$ is injective. Let U_i^+ denote the part of $\Delta(\delta_i)$ in the upper half plane. Take $\delta_1 > 0$ sufficiently small. Then the composite function $\phi_2^{-1} \circ f \circ \phi_1 : U_1^+ \to U_2^+$ is holomorphic, extends continuously over the real axis $(-\delta_1, \delta_1)$, and takes real values on it. By Schwarz' reflection principle, Theorem (5.1.4), $\phi_2^{-1} \circ f \circ \phi_1$ extends to a holomorphic function g on $\Delta(\delta_1)$ with values in $\Delta(\delta_2)$. Set $h_1 = \phi_2 \circ g \circ \phi_1^{-1}$ on $V_1 = \phi_1(\Delta(\delta_1))$. Then f is analytically continued to h_1 on V_1. We write $V_1(z_1)$ for this V_1, and set $W = \bigcup_{z_1 \in \overset{\circ}{C}_1} V_1(z_1)$. Then W is a domain, $W \supset \overset{\circ}{C}_1$, and there is a meromorphic function h such that $h = f$ on $W \cap D_1$. Thus f is analytically continued to $D_1 \cup W$. $\square$

REMARK. When $C_1(\phi : [T_0, T_1] \to \partial D)$ is a closed curve in Theorem (5.1.7), there are two possible cases:

i) There is a change $\hat{\phi} : [\hat{T}_0, \hat{T}_1] \to \partial D$ of parameter of ϕ such that $\hat{\phi}$ is non-singular analytic and for some $\hat{t}_0 \in (\hat{T}_0, \hat{T}_1)$, $\hat{\phi}(t_0) = \phi(T_0)$.

ii) There is no such change of parameter of ϕ.

If C_2 is not of case i) at $f \circ \phi(T_0)$, f cannot be analytically continued to a domain containing $\phi(T_0)$ no matter whether C_1 is of case ii) or i) (see Figure 52, below). If C_1 (resp., C_2) is of case i) at $\phi(T_0)$ (resp., $f \circ \phi(T_0)$), then f is analytically continued to a domain containing C_1.

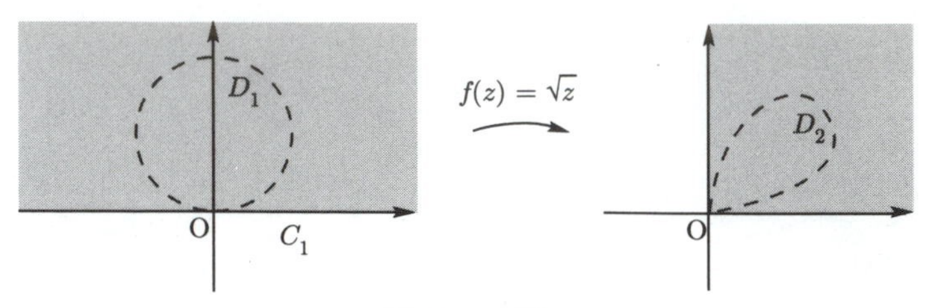

FIGURE 52

EXERCISE 2. Let $f(z)$ be a continuous function on $\overline{\Delta(R)}$ that is holomorphic in $\Delta(R)$, and let $|f(z)| \equiv K$ (constant) on the boundary $C(0; R)$. Show that $f(z)$ is a polynomial.

EXERCISE 3. Show that if an entire function $f(z)$ takes real values on the real axis, and purely imaginary values on the imaginary axis, then $f(z)$ is an odd function.

5.2. Monodromy Theorem

A meromorphic (resp., holomorphic) *function element* about a point $a \in \widehat{\mathbf{C}}$ is defined as a pair (F_a, U_a) of a meromorphic (resp., holomorphic) function in a disk neighborhood U_a of a, and U_a is called the *disk of definition.* We sometimes abbreviate (F_a, U_a) by F_a, calling it a meromorphic or holomorphic function element about a. Let $C(\phi : [0,1] \to \widehat{\mathbf{C}})$ be a curve. Assume that a meromorphic function element $(F_{\phi(t)}, U_t)$ is given for every $t \in [0,1]$. We write $F_t = F_{\phi(t)}$. We say that $\{(F_t, U_t)\}_{t\in[0,1]}$ is an *analytic continuation* (of (F_0, U_0)) along C if for an arbitrary $t_0 \in [0,1]$

$$F_{t_0} \equiv F_t \qquad \text{on } U_{t_0} \cap U_t, \quad t \in (t_1, t_2),$$

where

$$t_1 = \inf\{t \in [0, t_0] : \phi([t, t_0]) \subset U_{t_0}\},$$
$$t_2 = \sup\{t \in [t_0, 1] : \phi([t_0, t]) \subset U_{t_0}\}.$$

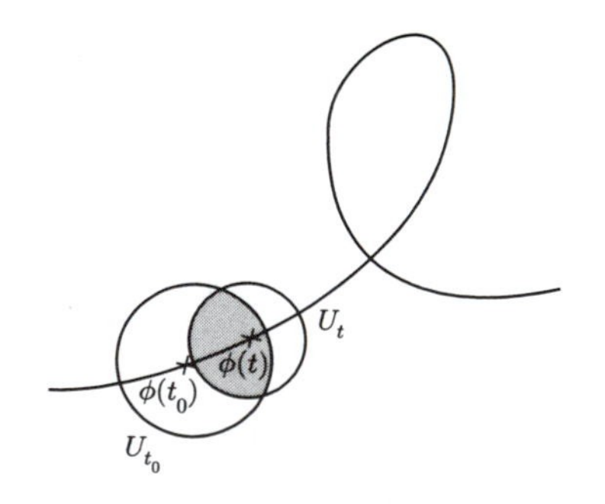

FIGURE 53

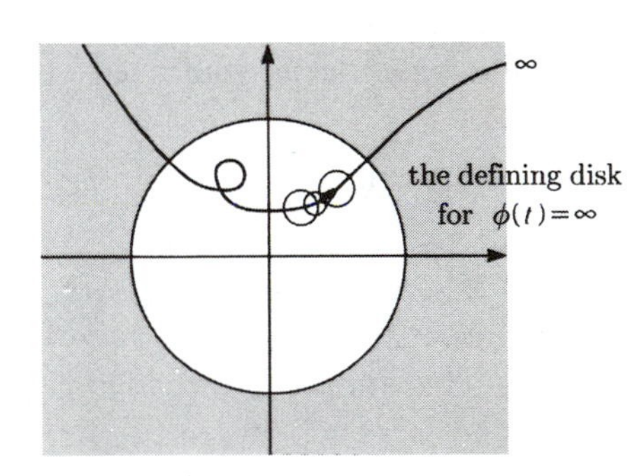

FIGURE 54

In this case, the disk U_t of definition of every F_t is written as

$$U_t = \begin{cases} \Delta(\phi(t); r_t), & \phi(t) \neq \infty, \\ \{\tilde{z} \in \mathbf{C}; |\tilde{z}| < r_t\}, & \phi(t) = \infty, \end{cases}$$

and by increasing r_t if necessary we may assume (cf. Figures 53 and 54) that

$$\inf\{r_t; t \in [0,1]\} > 0. \tag{5.2.1}$$

(5.2.2) THEOREM. *Let $\{(F_t, U_t)\}_{t\in[0,1]}$ and $\{(G_t, V_t)\}_{t\in[0,1]}$ be two analytic continuations along a curve $C(\phi : [0,1] \to \widehat{\mathbf{C}})$. If $F_0 = G_0$ in a neighborhood of $\phi(0)$, then $F_1 = G_1$ in a neighborhood of $\phi(1)$.*

PROOF. Set $E = \{t \in [0,1]; F_t = G_t \text{ in a neighborhood of } \phi(t)\}$. By the assumption, $E \neq \emptyset$, and E is clearly open in $[0,1]$. Take an arbitrary $t_0 \in \overline{E}$. Take a connected neighborhood U_{t_0} of $\phi(t_0)$ so that it is contained in the disks of definition of F_{t_0} and G_{t_0}. If $t' \in E$ is sufficiently close to t_0, $\phi(t') \in U_{t_0}$. Since F_{t_0} and G_{t_0} coincide on a neighborhood of $\phi(t')$, $F_{t_0} = G_{t_0}$ on U_{t_0} by Theorem (4.2.2). Thus $t_0 \in E$, and so E is closed in $[0,1]$. We have that $E = [0,1]$. □

Let $C_j(\phi_j : [0,1] \times [0,1] \to \widehat{\mathbf{C}})$, $j = 0, 1$, be two curves which are homotopic to each other. Let the homotopy connecting them be

$$\Phi : [0,1] \times [0,1] \to \widehat{\mathbf{C}},$$
$$\Phi(t,0) = \phi_0(t), \quad \Phi(t,1) = \phi_1(t).$$

Then we have the following *monodromy theorem*:

(5.2.3) THEOREM. *Let $C_j, j = 1, 2$, and Φ be as above. Suppose that for every $(t,s) \in [0,1] \times [0,1]$ a meromorphic function element $F_{(t,s)}$ about $\Phi(t,s)$ is given so that for each fixed $s \in [0,1]$ $\{F_{(t,s)}\}_t$ gives rise to an analytic continuation along $C_s(\phi_s = \Phi(\cdot, s))$. If $F_{(0,0)} = F_{(0,s)}$ in a neighborhood of the initial point $\mathrm{P} = \phi_0(0)$ $(0 \leqq s \leqq 1)$, then $F_{(1,0)} = F_{(1,1)}$ in a neighborhood of the terminal point $\mathrm{Q} = \phi_0(1)$.*

PROOF. Set $S = \{s \in [0,1]; F_{(1,s)} = F_{(1,0)} \text{ in a neighborhood of Q}\}$. Since $0 \in S$, $S \neq \emptyset$. We write $U_{(t,s)}$ for the disk of definition of $F_{(t,s)}$. Note that $U_{(t,s)}$ is connected. Fix arbitrarily $s_0 \in S$. We may assume (5.2.1) for $\{F_{(t,s_0)}\}_t$. By the uniform continuity of Φ there is $\delta > 0$ such that for every $s \in (s_0 - \delta, s_0 + \delta) \cap [0,1]$

$$\Phi(t,s) \in U_{t,s_0}, \qquad 0 \leqq t \leqq 1. \tag{5.2.4}$$

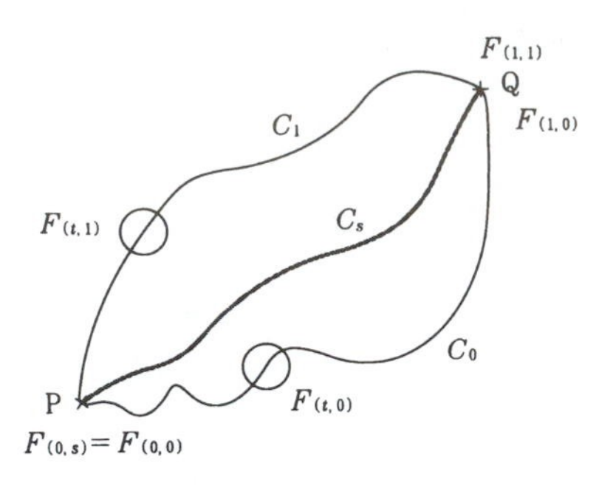

FIGURE 55

FIGURE 56

We fix $s \in (s_0 - \delta, s_0 + \delta) \cap [0, 1]$, and set

$$T = \{t \in [0,1]; F_{(t,s)} = F_{(t,s_0)} \text{ on } U_{(t,s)} \cap U_{(t,s_0)}\}.$$

Since $0 \in T$, $T \neq \emptyset$. In the same way as in the proof of Theorem (5.2.2) one sees that T is open and closed in $[0, 1]$. Hence $T = [0, 1]$, and so in a neighborhood of Q, $F_{(1,s)} = F_{(1,s_0)} = F_{(1,0)}$. It follows that S is open. In the above arguments, without assuming $s_0 \in S$, we infer that if there is an s which satisfies (5.2.4) and belongs to S, then $s_0 \in S$. Hence S is also closed, and then $S = [0, 1]$. □

Let f be a meromorphic function on a domain $D \subset \widehat{\mathbf{C}}$. Let $\mathrm{P} \in D$, and consider the meromorphic function element obtained by the restriction of f to a disk neighborhood of P in D. We consider meromorphic function elements $(F_{\mathrm{Q}}, U_{\mathrm{Q}})$ produced by all possible analytic continuations of f along curves in $\widehat{\mathbf{C}}$ with initial point P, and set $\tilde{D} = \bigcup U_{\mathrm{Q}}$. Then $\tilde{D}$ is a domain containing D, and we have a multi-valued meromorphic function $\tilde{f}(\mathrm{Q}) = F_{\mathrm{Q}}(\mathrm{Q})$. It follows from Theorems (3.3.6) and (5.2.3) that

(5.2.5) THEOREM. *Let $\tilde{f}$ and $\tilde{D}$ be as above.*

i) *Let $\mathrm{Q} \in \tilde{D}$. Identify the meromorphic function elements $(F_{\mathrm{Q}}, U_{\mathrm{Q}})$ and $(G_{\mathrm{Q}}, V_{\mathrm{Q}})$ if F_{Q} and G_{Q} are identical in a neighborhood of* Q *contained in $U_{\mathrm{Q}} \cap V_{\mathrm{Q}}$. Then the number of meromorphic function elements of $\tilde{f}$ defined about* Q *is at most countable.*

ii) *Let $D_0 \subset \tilde{D}$ be a simply connected domain such that for a point $\mathrm{Q} \in D_0$ a meromorphic function element $(F_{\mathrm{Q}}, U_{\mathrm{Q}})$ about* Q *can be analytically continued along any curve in D_0 with initial point* Q. *Then $\tilde{f}$ defines a one-valued meromorphic function f_0 on D_0.*

The above f_0 in ii) is called a *branch* of $\tilde{f}$. The identification of meromorphic function elements used in the above i) induces an equivalence relation in $\{(F_{\mathrm{Q}}, U_{\mathrm{Q}})\}$. Let $[F_{\mathrm{Q}}, U_{\mathrm{Q}}]$ denote the equivalence class of $(F_{\mathrm{Q}}, U_{\mathrm{Q}})$. We denote by X the set of all $[F_{\mathrm{Q}}, U_{\mathrm{Q}}]$. We introduce a topology on X as follows. It is sufficient to define neighborhoods of $[F_{\mathrm{Q}}, U_{\mathrm{Q}}] \in X$. For every $\mathrm{P} \in U_{\mathrm{Q}}$, we take a disk neighborhood $U_P \subset U_Q$ and the restriction $F_P = F_Q|U_P$ of F_Q to U_P. We define the system of neighborhoods of $[F_Q, U_Q]$ by

$$\{[F_P, U_P]; P \in V\},$$

where V runs over all neighborhoods of Q in U_Q. Thus X is a topological space, which is called a *Riemann surface* defined by $\tilde{f}$. Defining $\tilde{f}([F_Q, U_Q]) = F_Q(Q)$, $\tilde{f}$ gives rise to a one-valued function on X.

(5.2.6) EXAMPLE. The Riemann surface defined by the multi-valued meromorphic function

$$\tilde{f}(z) = \log z, \quad z \in \tilde{D} = \mathbf{C} \setminus \{0\},$$

is obtained as follows. Take countably many complex planes $\mathbf{C}_j$, $-\infty < j < \infty$. Let L_j^+ (resp., L_j^-) denote the real positive axis in $\mathbf{C}_j \setminus \{0\}$ as accumulation points of the upper (resp., lower) half-plane, and distinguish between L_j^+ and L_j^-. Then we identify L_j^+ with L_{j-1}^- of $\mathbf{C}_{j-1} \setminus \{0\}$ for all $-\infty < j < \infty$ (see Figure 57). Let X denote the resulted topological space. Defining $\tilde{f}(1) = 0$ at $1 \in \mathbf{C}_0 \setminus \{0\}$, we obtain a one-valued function $\tilde{f}$ on X.

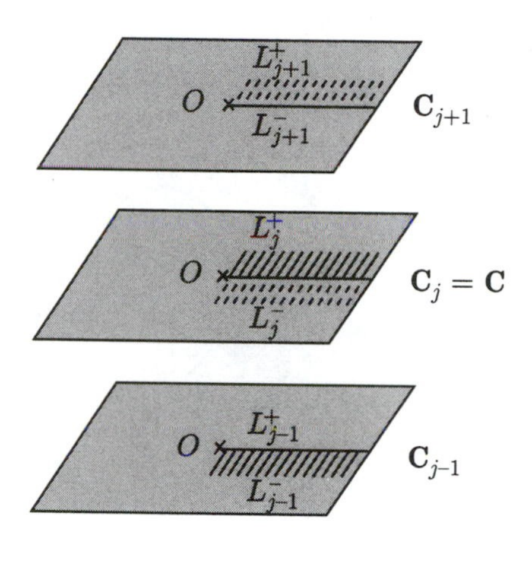

FIGURE 57

Making use of $g(w) = e^w$, we identify injectively $G_j = \{z \in \mathbf{C}; 2j\pi \leqq \operatorname{Im} z < 2(j+1)\pi\}$ with $(\mathbf{C}_j \setminus \{0\}) \cup L_j^+$:

$$g|G_j : G_j \to (\mathbf{C}_j \setminus \{0\}) \cup L_j^+.$$

Since $\bigcup_{j=-\infty}^{\infty} G_j = \mathbf{C}$, X is identified with $\mathbf{C}$. The logarithmic function $w = \log z$ is the inverse function of the holomorphic function $z = z(w) = e^w$, and so the above Riemann surface X is also called the *Riemann surface of the inverse function* of $z(w) = e^w$.

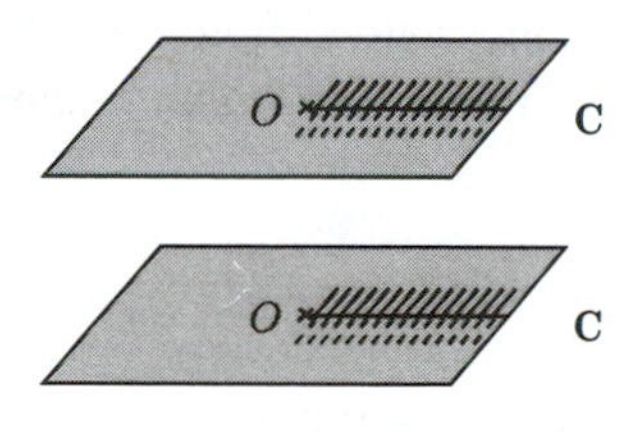

FIGURE 58

(5.2.7) EXAMPLE. The Riemann surface X of the inverse function $w = \sqrt{z}$ of $z(w) = w^2$ is obtained by gluing (identifying) two copies of $\mathbf{C}$ along the non-negative real axis (see Figure 58). Let D_0 be a simply connected domain in $\mathbf{C} \setminus \{0\}$, and take a branch $f_0(z)$ of $\sqrt{z}$ on D_0. The other branch of $\sqrt{z}$ is $f_1(z) = -f_0(z)$. Set $G_j = f_j(D_0)$, $j = 0, 1$. Then G_j is a domain of $\mathbf{C} \setminus \{0\}$, the mapping

$$f_j : D_0 \to G_j$$

is injective, and $G_0 \cap G_1 = \emptyset$.

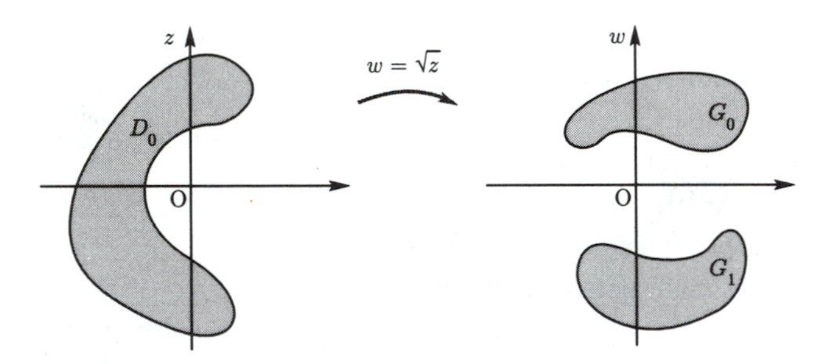

FIGURE 59

In general, if the function $\tilde{f}$ in Theorem (5.2.5) is expanded to a Puiseux series with an integer $p \geqq 2$,

$$\begin{aligned} \tilde{f}(z) &= \sum_{n \geqq -N}^{\infty} a_n(\sqrt[p]{z-a})^n & & (a \neq \infty, 0 \leqq N < \infty), \\ \tilde{f} \circ \tilde{z} &= \sum_{n \geqq -N}^{\infty} a_n(\sqrt[p]{\tilde{z}})^n & & (a = \infty, 0 \leqq N < \infty), \end{aligned}$$

and if p cannot be reduced to a smaller one by another Puiseux series expansion, then a is called a *branch point*, and $p - 1$ is called the *order of the branch*. If in a neighborhood of a $\tilde{f}$ is expressed as a meromorphic function of $\log(z - a)$ or $\log \tilde{z}$, and is an infinitely-many-valued function, then a is called a *logarithmic branch point*. For instance, if $\alpha \in \mathbf{R}$ is irrational, then

$$\tilde{f}(z) = z^{\alpha}$$

has logarithmic branch points at $z = 0$ and ∞.

(5.2.8) EXAMPLE. We consider the Riemann surface X of the inverse cosine function $w = \cos^{-1} z$.

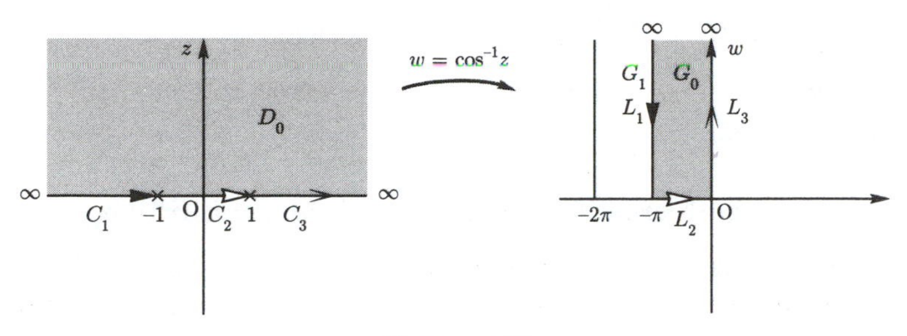

FIGURE 60

Let L_1 denote the half line $\operatorname{Re} w = -\pi$, $\operatorname{Im} w \geqq 0$ in the w-plane from ∞ to $-\pi$. Let L_2 denote the line segment of the real axis from $-\pi$ to 0, and let L_3 denote the half line $\operatorname{Re} w = 0$, $\operatorname{Im} w \geqq 0$ from 0 to ∞. Let G_0 be the domain surrounded by them. Let C_j be the image of L_j mapped by $z = \cos w$ with orientation for $j = 1, 2, 3$. Then C_1 runs from ∞ to -1 on the negative real axis, C_2 from -1 to 1 on the real axis, and C_3 from 1 to ∞ on the positive real axis. Since a holomorphic function preserves the orientation except at points where the derivative vanishes, G_0 is injectively mapped to the upper half plane D_0. This mapping is analytically continued by the reflection principle, Theorem (5.1.7) or Theorem (5.1.4). The reflection of D_0 with respect to C_1 is the lower half plane D_1, and its image by $\cos^{-1} z$ is the reflection G_1 of G_0 with respect to L_1. We carry out this continuation process with respect to C_j, $j = 1, 2, 3$, as far as possible. The branch points of $\cos^{-1} z$ are -1 and 1, and their values are $2n\pi$ and $(2n+1)\pi$ $(n \in \mathbf{C})$, respectively. The orders of branch are 1. Thus the Riemann surface X is identified with $\mathbf{C}$.

EXERCISE 1. Construct the Riemann surface of $\sin^{-1} z$.

EXERCISE 2. What is the Riemann surface of $\sqrt{1 - \cos z}$? Show that in fact it defines two 1-valued entire function on $\mathbf{C}$.

Let $F(z_1, \ldots, z_n)$ be a complex function of n variables, $z_1, \ldots, z_n$. Set $z_j = x_j + iy_j$, $1 \leqq j \leqq n$. If F is continuously differentiable in the variables x_j and y_j, $1 \leqq j \leqq n$, and if for every j the Cauchy-Riemann equations

$$\frac{1}{2}\left(\frac{\partial}{\partial x_j} - \frac{1}{i}\frac{\partial}{\partial y_j}\right) F = 0, \qquad 1 \leqq j \leqq n, \tag{5.2.9}$$

hold, then F is said to be holomorphic in $(z_1, \ldots, z_n)$.

Let $f_j(z), 1 \leqq j \leqq n$, be n holomorphic functions of z, and assume that the composite $F(f_1(z), \ldots, f_n(z))$ is defined. Then, if F is holomorphic, so is $F(f_1(z), \ldots, f_n(z))$. Let $C(\phi : [0,1] \to \widehat{\mathbf{C}})$ be a curve and let $\{f_{jt}\}_t$, $1 \leqq j \leqq n$, be analytic continuations of holomorphic functions along C. Let $F(w_1, \ldots, w_n)$

be a holomorphic function such that all composites $F(f_{1t}(z), \dots, f_{nt}(z))$ are defined. Then it follows from the identity Theorem (2.4.14) that

$$(5.2.10) \qquad F(f_{10}(z), \dots, f_{n0}(z)) \equiv 0 \Longrightarrow F(f_{11}(z), \dots, f_{n1}(z)) \equiv 0.$$

This is referred as the *principle of the permanence of the functional relation* of analytic continuation.

5.3. Universal Covering and Riemann Surface

Let $D \subset \widehat{\mathbf{C}}$ be a domain and fix a point $\mathrm{P} \in D$. Let $\pi_1(D)_{\mathrm{P}}$ denote the set of all homotopy classes $\{C\}$ of closed curves C in D with the initial and terminal point P. A multiplication in $\pi_1(D)_{\mathrm{P}}$ is defined by

$$\{C_1\} \cdot \{C_2\} = \{C_1 + C_2\}.$$

Here it is sometimes written as $\{C_1\} + \{C_2\}$, while the multiplication is not commutative in general. Let e denote the homotopy class $\{C\} \in \pi_1(D)_{\mathrm{P}}$ of C which are homotopic to the point P. We set $\{C\}^{-1} = \{-C\}$ for $\{C\} \in \pi_1(D)_{\mathrm{P}}$. Then $\pi_1(D)_{\mathrm{P}}$ forms a group with unit element e, and is called the *fundamental group* of D with base point P. For another point $\mathrm{Q} \in D$ we have a group isomorphism

$$(5.3.1) \qquad \Phi_{\mathrm{Q}}^{\mathrm{P}} : \pi_1(D)_{\mathrm{P}} \ni \{C\} \to \{C_{\mathrm{Q}}^{\mathrm{P}} + C - C_{\mathrm{Q}}^{\mathrm{P}}\} \in \pi_1(D)_{\mathrm{Q}},$$

where $C_{\mathrm{Q}}^{\mathrm{P}}$ is a curve from Q to P in D.

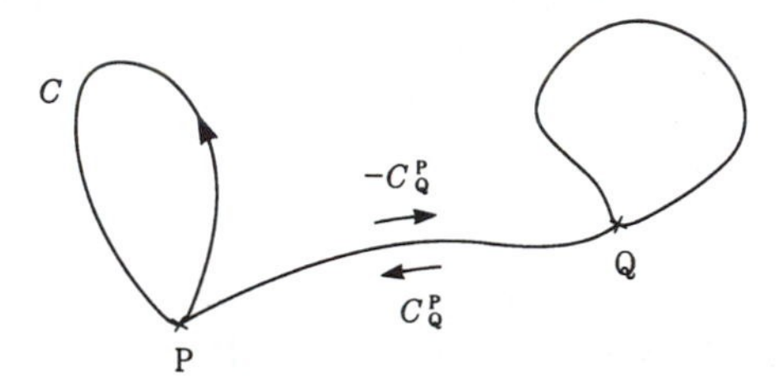

FIGURE 61

Thus the groups $\pi_1(D)_{\mathrm{P}}$ and $\pi_1(D)_{\mathrm{Q}}$ are mutually isomorphic as abstract groups. We simply denote it by $\pi_1(D)$, and call it the *fundamental group* of D. Let D' be another domain, and $f : D \to D'$ be a continuous mapping. Then we have a group homomorphism

$$(5.3.2) \qquad f_* : \pi_1(D)_{\mathrm{P}} \ni \{C\} \to \{f(C)\} \in \pi_1(D')_{f(\mathrm{Q})}.$$

Here $f(C)$ is defined by $f \circ \phi$ with $C = C(\phi)$.

The continuous mapping $f : D \to D'$ is called a *covering mapping*, and D is called the *covering* of D' if

(5.3.3) i) $f(D) = D'$;

ii) for every point $w \in D'$ there is a neighborhood $V \subset D'$ of w such that for any connected component U of $f^{-1}(V)$, the restriction of f to U,

$$f|U : U \to V,$$

is a *homeomorphism* (a surjective and injective continuous mapping with continuous inverse mapping).

EXAMPLE. 1) $f : \mathbf{C} \ni z \to e^z \in \mathbf{C}^*$ is a covering mapping.
2) $f : \mathbf{C}^* \ni z \to z^2 \in \mathbf{C}^*$ is a covering mapping.

EXERCISE 1. Let $f : D \to D'$ be a covering mapping. Show that $f^{-1}(w)$ has no accumulation point for every $w \in D'$.

EXERCISE 2. Show that the fundamental group of $\mathbf{C}^*$ is isomorphic to $\mathbf{Z}$.

Now, let D be as above, and fix a point $\mathrm{P}_0 \in D$. Let X be the set of all homotopy classes α_{P} of curves from P_0 to P in D, where P moves over D. Take an arbitrary point $\mathrm{P} \in D$ and a disk neighborhood U_{P}. For a point $\mathrm{Q} \in U_{\mathrm{P}}$ we take a curve $C_{\mathrm{P}}^{\mathrm{Q}}$ from P to Q in U_{P}. Then the homotopy class $\alpha_{\mathrm{P}} + \{C_{\mathrm{P}}^{\mathrm{Q}}\}$ is independent of the choice of $C_{\mathrm{P}}^{\mathrm{Q}}$, since U_{P} is simply connected.

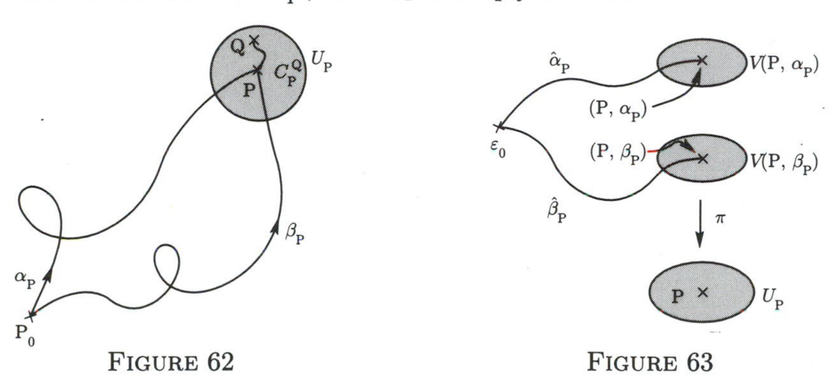

FIGURE 62 FIGURE 63

Set

$$V(\alpha_{\mathrm{P}}) = \{\alpha_{\mathrm{P}} + \{C_{\mathrm{P}}^{\mathrm{Q}}\}; \mathrm{Q} \in U_{\mathrm{P}}\}.$$

If $\alpha_{\mathrm{P}} \neq \beta_{\mathrm{P}}$, then

$$V(\alpha_{\mathrm{P}}) \cap V(\beta_{\mathrm{P}}) = \emptyset. \tag{5.3.4}$$

Let $\epsilon_0 \in X$ be the homotopy class of the constant curve P_0, and let

$$\pi : X \ni \alpha_{\mathrm{P}} \to \mathrm{P} \in D \tag{5.3.5}$$

be the natural projection. We define a topology of X by taking neighborhoods $V(\alpha_{\mathrm{P}})$ for each point $\alpha_{\mathrm{P}} \in X$; that is, a point sequence $\{\alpha_n\}_{n=0}^{\infty}$ of X converges to $\alpha_{\mathrm{P}} \in X$ if and only if for an arbitrary neighborhood $V(\alpha_{\mathrm{P}})$ of α_{P} there is a number n_0 such that $\alpha_n \in V(\alpha_{\mathrm{P}})$ for all $n \geqq n_0$. The restriction mapping

$$\pi|V(\alpha_{\mathrm{P}}) : V(\alpha_{\mathrm{P}}) \to U_{\mathrm{P}} \tag{5.3.6}$$

is a homeomorphism. Take two distinct points $\alpha_{\mathrm{P}} \neq \beta_{\mathrm{Q}} \in X$. If $\mathrm{P} \neq \mathrm{Q}$, then, taking U_{P} and U_{Q} with $U_{\mathrm{P}} \cap U_{\mathrm{Q}} = \emptyset$, we define $V(\alpha_{\mathrm{P}})$ and $V(\beta_{\mathrm{Q}})$ so that

$$V(\alpha_{\mathrm{P}}) \cap V(\beta_{\mathrm{Q}}) = \emptyset;$$

if $\mathrm{P} = \mathrm{Q}$, the same holds by (5.3.4). Therefore, X is a so-called Hausdorff topological space. For $\alpha_{\mathrm{P}} = \{C(\phi : [0,1] \to D)\}$ we have a continuous mapping

$$\hat{\phi} : [0,1] \ni t \to \alpha_{\phi(t)} = \{C(\phi|[0,t])\} \in X, \tag{5.3.7}$$

and $\hat{\phi}(0) = \epsilon_0$ and $\hat{\phi}(1) = \alpha_{\mathrm{P}}$. Thus X is arcwise connected. While we do not know if X is a domain of $\widehat{\mathbf{C}}$, the mapping $\pi : X \to D$ defined by (5.3.5) satisfies the properties of (5.3.3), i), ii). We also call this a covering. If we choose another $\mathrm{P}_0 \in D$, we have another X, which is homeomorphic to the original one, and so we may identify them as a topological space. We call this X the *universal covering* of D, and $\pi : X \to D$ the *universal covering mapping.*

EXERCISE 3. Let $\mathrm{P}_0 \in D$ and X be as above. We take another point $\mathrm{P}_0' \in D$, and then obtain another X' as X. Let $C_{\mathrm{P}_0}^{\mathrm{P}_0'}$ be a curve from P_0 to P_0'. Show that

$$\phi : X' \ni \alpha_{\mathrm{P}_0'} \to \{C_{\mathrm{P}_0}^{\mathrm{P}_0'}\} + \alpha_{\mathrm{P}_0'} \in X \tag{5.3.8}$$

is a homeomorphism.

An element $\gamma \in \pi_1(D) = \pi_1(D)_{\mathrm{P}_0}$ of the fundamental group defines a homeomorphism

$$\gamma : X \ni \alpha \to \gamma + \alpha \in X. \tag{5.3.9}$$

For two elements $\gamma_1, \gamma_2 \in \pi_1(D)$

$$\gamma_1 \circ \gamma_2 = \gamma_1 \cdot \gamma_2 \ (= \gamma_1 + \gamma_2), \tag{5.3.10}$$

where the left side is the composition of the mappings and the right side is the multiplication in the group $\pi_1(D)$. In this case, we say that the group $\pi_1(D)$ *acts* on X. The set $\{\gamma(\alpha); \gamma \in \pi_1(D)\}$ with an element $\alpha \in X$ is called the $\pi_1(D)$-*orbit* of α. It follows that

$$\{\gamma(\alpha_{\mathrm{P}}); \gamma \in \pi_1(D)\} = \{\gamma(\alpha_{\mathrm{Q}}); \gamma \in \pi_1(D)\} \Longleftrightarrow \mathrm{P} = \mathrm{Q}.$$

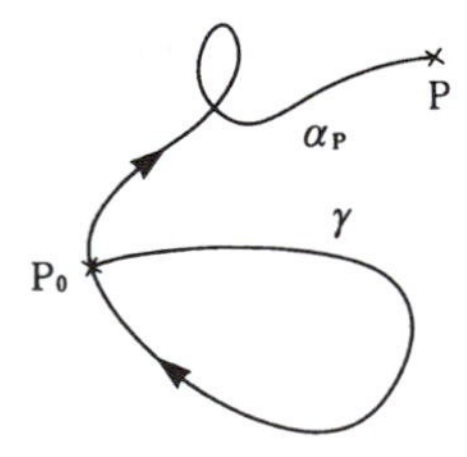

FIGURE 64

Considering a $\pi_1(D)$-orbit as one point, we get a quotient space $X/\pi_1(D)$ of X by the action of $\pi_1(D)$. We have

$$X/\pi_1(D) = D. \tag{5.3.11}$$

When an element γ of $\pi_1(D)$ is considered as a homeomorphism defined by (5.3.9), $\pi_1(D)$ is called the *deck transformation group* or the *covering transformation group*. For distinct $\alpha_{\mathrm{P}}, \beta_{\mathrm{P}} \in X$ with $\mathrm{P} \in D$, set

$$\gamma = \beta_{\mathrm{P}} - \alpha_{\mathrm{P}} \in \pi_1(D).$$

It follows from (5.3.9) that $\gamma : X \to X$ satisfies

$$\gamma(V(\alpha_{\mathrm{P}})) = V(\beta_{\mathrm{P}}). \tag{5.3.12}$$

In general, let Y be a closed interval of $\mathbf{R}$, an open interval of $\mathbf{R}$, a domain of $\widehat{\mathbf{C}}$, or some finite product of them. Let $f : Y \to D$ be a continuous mapping. A continuous mapping $\hat{f} : Y \to X$ is called a *lifting* of f if $f = \pi \circ \hat{f}$:

$$\begin{array}{ccc} & \hat{f} & X \\ & \nearrow & \downarrow \pi \\ Y & \rightarrow & D \\ & f & \end{array}$$

For a curve $C(\psi : [0,1] \to D)$ with the initial point $\mathrm{P} \in D$ and $\alpha_{\mathrm{P}} \in X$ we set

$$\hat{\psi} : [0,1] \ni t \to \alpha_{\mathrm{P}} + \{C|[0,t]\} \in X. \tag{5.3.13}$$

Then $\pi \circ \hat{\psi}(t) = \psi(t)$, so that $\hat{\psi}$ is a lifting of ψ and gives a curve $\hat{C}_{\alpha_{\mathrm{P}}}$, called a *lifting* of C. We see by (5.3.6) that a lifting of C is unique if the initial point α_{P} is given. Therefore, for a lifting $\hat{C}_{\beta_{\mathrm{P}}}$ of C with the initial point $\beta_{\mathrm{P}} \neq \alpha_{\mathrm{P}}$ we have

$$\hat{C}_{\alpha_{\mathrm{P}}} \cap \hat{C}_{\beta_{\mathrm{P}}} = \emptyset. \tag{5.3.14}$$

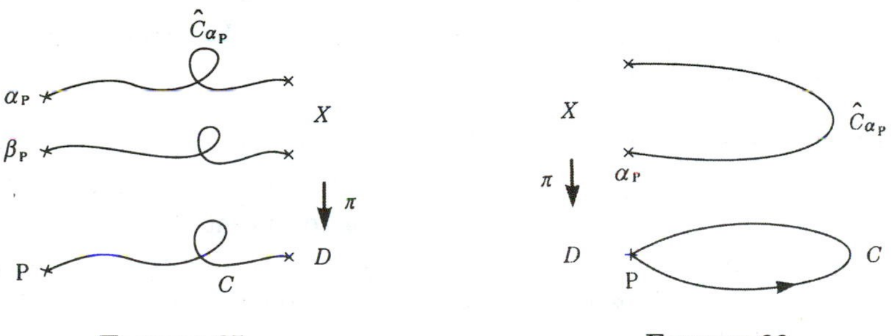

FIGURE 65 FIGURE 66

If C is closed, then $\gamma = \alpha_{\mathrm{P}} + \{C\} - \alpha_{\mathrm{P}} \in \pi_1(D)$ and it follows from (5.3.9) that

(5.3.15) i) $\gamma(\alpha_{\mathrm{P}}) =$ the terminal point of $\hat{C}_{\alpha_{\mathrm{P}}}$.

ii) $\{C\}$ is homotopic to a point if and only if $\gamma(\alpha_{\mathrm{P}}) = \alpha_{\mathrm{P}}$.

Since X is arcwise connected, the notion of curves, their homotopy, and simple connectedness is defined for X in the same way as in the case of domains of $\widehat{\mathbf{C}}$.

(5.3.16) LEMMA. *Let $\pi : X \to D$ be as above. Let $\Phi : [0,1] \times [0,1] \to D$ be a continuous mapping, and set* $\mathrm{P} = \Phi(0,0)$. *For an arbitrary $\alpha_{\mathrm{P}} \in X$ there exists a unique lifting $\hat{\Phi} : [0,1] \times [0,1] \to X$ with $\hat{\Phi}(0,0) = \alpha_{\mathrm{P}}$.*

PROOF. Take a lifting $\hat{\Phi}(\cdot, 0)$ of $\Phi(\cdot, 0) : [0,1] \ni t \to \Phi(t,0) \in D$ with $\hat{\Phi}(0,0) = \alpha_{\mathrm{P}}$. By making use of (5.3.6) we see that there exist a $\sigma > 0$ and a continuous mapping $\hat{\Phi} : [0,1] \times [0,\sigma] \to X$ such that $\hat{\Phi}(0,0) = \alpha_{\mathrm{P}}$ and $\pi \circ \hat{\Phi} = \Phi$. Let S be the set of all such $\sigma \in [0,1]$. In the same way as in the proof of the monodromy Theorem (5.2.3) we infer that S is open and closed, so that $S = [0,1]$.

To show the uniqueness, we take a continuous mapping $\hat{\Psi} : [0,1] \times [0,1] \to X$ such that $\Psi(0,0) = \alpha_{\mathrm{P}}$ and $\pi \circ \hat{\Psi} = \Phi$. Set

$$E = \{(t,s) \in [0,1] \times [0,1]; \hat{\Psi}(t,s) = \hat{\Phi}(t,s)\}.$$

Then E is clearly closed, and $E \ni (0,0)$. It follows from (5.3.6) that E is open in $[0,1] \times [0,1]$. Thus $E = [0,1] \times [0,1]$, and $\hat{\Psi} = \hat{\Phi}$. □

Let $\hat{\Phi}'$ be another lifting of Φ in Lemma (5.3.16). Then, by (5.3.12) there is a $\gamma \in \pi_1(D)$ such that

$$\hat{\Phi}' = \gamma \circ \hat{\Phi}. \tag{5.3.17}$$

(5.3.18) THEOREM. *Let D and X be as above.*

i) *X is simply connected.*

ii) *Let $D_0 \subset \widehat{\mathbf{C}}$ be an arbitrary simply connected domain, and let $f : D_0 \to D$ be a continuous mapping. Then, for an arbitrary $\alpha_{f(z_0)} \in X$ with $z_0 \in D_0$ there is a unique lifting $\hat{f} : D_0 \to X$ such that $\hat{f}(z_0) = \alpha_{f(z_0)}$ and $\pi \circ \hat{f} = f$. If $\hat{f}'$ is another lifting of f, then there is a $\gamma \in \pi_1(D)$ such that $\hat{f}' = \gamma \circ \hat{f}$.*

PROOF. i) Take a closed curve $\hat{C}$ in X with the initial point $\alpha_{\mathrm{P}} \in X$. Then $C = \pi(\hat{C})$ is a curve in D, and $\hat{C}$ is a lifting of C with the initial point α_{P}. It follows from (5.3.15) that C must be homotopic to a point. Let $\Phi : [0,1] \times [0,1] \to D$ be a homotopy connecting C and P. By Lemma (5.3.16) there is a lifting $\hat{\Phi} : [0,1] \times [0,1] \to X$ with $\hat{\Phi}(0,0) = \alpha_{\mathrm{P}}$. It follows that $\hat{\Phi}$ is a homotopy connecting $\hat{C}$ and α_{P}.

ii) Take an arbitrary point $z \in D_0$ and a curve C from z_0 to z. Let $\hat{C}$ be the lifting of $f(C)$ with the initial point $\alpha_{f(z_0)}$, and denote the terminal point by $\hat{f}(z)$. We see by Lemma (5.3.16) that $\hat{f}(z)$ is independent of the choice of C. For every point $z' \in D_0$ we take a sufficiently small neighborhood $W \subset D_0$ so that for neighborhoods $U_{f(z')}$ of $f(z')$ and $V(\hat{f}(z'))$ of $\hat{f}(z')$ satisfying (5.3.6), $f(W) \subset U_{f(z')}$, and

$$\hat{f}(z) \in V(\hat{f}(z')), \qquad z \in W.$$

Since $\pi \circ \hat{f}(z) = f(z)$ is continuous and $\pi|V(\hat{f}(z'))$ is a homeomorphism, $\hat{f}$ is continuous. The latter half follows from (5.3.17). □

EXERCISE 4. Let $\pi_j : X_j \to D_j$, $j = 1, 2$, be universal covering mappings, and let $f : D_1 \to D_2$ be a continuous mapping. Let $\hat{f} : X_1 \to X_2$ be a lifting of $f \circ \pi_1$, and let $f_* : \pi_1(D_1) \to \pi_1(D_2)$ be as in (5.3.2). Then, show that

$$\hat{f} \circ \gamma = f_*(\gamma) \circ \hat{f}, \qquad \gamma \in \pi_1(D_1).$$

EXERCISE 5. Set $D_j = \mathbf{C}^*, j = 1, 2$, and $\lambda : D_1 \ni z \to z^n \in D_2$.

i) Obtain all liftings of the curve C_1 given by $\phi_1 : [0,1] \ni t \to 1 + t \in D_2$.

ii) Obtain all liftings of the curve C_2 given by $\phi_2 : [0,1] \ni t \to e^{2\pi it} \in D_2$.

EXERCISE 6. The mapping $\pi : \mathbf{C}\ (= D_1) \ni z \to e^{2\pi iz} \in \mathbf{C}^*\ (= D_2)$ is a universal covering mapping. In this case, answer the questions similar to the above i) and ii).

Let X be as above. Then every point $\alpha_{\mathrm{P}} \in X$ has a neighborhood $V(\alpha_{\mathrm{P}})$ which is homeomorphic to a disk neighborhood U_{P} ($\subset D$) of P. We call $V(\alpha_{\mathrm{P}})$ a *disk neighborhood* of α_{P}. There are a complex coordinate z in U_{P} for P $\neq \infty$, and $\tilde{z}$ in U_∞. Through the restriction $\pi|V(\alpha_{\mathrm{P}})$, the points of $V(\alpha_{\mathrm{P}})$ are identified with the points of U_{P}, so that we introduce a complex coordinate $z_{V(\alpha_{\mathrm{P}})}$ in $V(\alpha_{\mathrm{P}})$ by making use of z or $\tilde{z}$. The mapping

$$z_{V(\alpha_{\mathrm{P}})} : V(\alpha_{\mathrm{P}}) \to z_{V(\alpha_{\mathrm{P}})}(V(\alpha_{\mathrm{P}}))(\subset \mathbf{C}) \tag{5.3.19}$$

is homeomorphic, and if $V(\alpha_{\mathrm{P}}) \cap V(\beta_{\mathrm{Q}}) \neq \emptyset$,

(5.3.20)
$$z_{V(\beta_{\mathrm{Q}})} \circ z_{V(\alpha_{\mathrm{P}})}^{-1} : z_{V(\alpha_{\mathrm{P}})}(V(\alpha_{\mathrm{P}}) \cap V(\beta_{\mathrm{Q}})) \to z_{V(\beta_{\mathrm{Q}})}(V(\alpha_{\mathrm{P}}) \cap V(\beta_{\mathrm{Q}}))$$

is a homeomorphism, and a holomorphic function as well. We call $z_{V(\alpha_{\mathrm{P}})}$ a *holomorphic local coordinate* in $V(\alpha_{\mathrm{P}})$ of X.

In general, if every point of a connected Hausdorff topological space has a neighborhood endowed with a complex function satisfying (5.3.19) and (5.3.20), then the space is called a *Riemann surface*, and the complex function is called a *holomorphic local coordinate*. The Riemann sphere $\widehat{\mathbf{C}}$ is a Riemann surface with holomorphic local coordinates z and $\tilde{z}$, which is compact and simply connected. The Riemann surfaces of a multi-valued meromorphic function and of the inverse function dealt with in the last section are Riemann surfaces in this sense. The above X gives rise to a simply connected Riemann surface.

EXAMPLE. The mapping $\pi : \mathbf{C}\ (= D_1) \ni z \to e^{2\pi iz} \in \mathbf{C}^*\ (= D_2)$ is a universal covering mapping, and $\mathbf{C}$ is the Riemann surface of the inverse function

$(1/2\pi i)\log w$ of the holomorphic function $w = e^{2\pi i z}$. The fundamental group of $\mathbf{C}^*$ is $\mathbf{Z}$, and acts on $\mathbf{C}$ by

$$\mathbf{C} \ni z \to z + n \in \mathbf{C}, \qquad n \in \mathbf{Z}.$$

REMARK. We have a "global" holomorphic coordinate z in $\mathbf{C}$ or in its domain D. For multi-valued meromorphic functions on D or for the universal covering of D, the coordinate z is no longer "global". Here it becomes necessary to introduce the notion of Riemann surfaces. For Riemann surfaces the notion of holomorphic functions and holomorphic mappings is defined in what follows, and all theory which has been described can be developed with suitable generalizations. It can be shown that a simply connected Riemann surface must be one of $\widehat{\mathbf{C}}$, $\mathbf{C}$, or $\Delta(1)$. This fact is known as the *uniformization of Riemann surfaces*, and the proof requires considerable preparations of function theory on Riemann surfaces. We will prove this fact for domains of $\widehat{\mathbf{C}}$ in Chapter 6.

In general, let X_1 and X_2 be Riemann surfaces. A continuous mapping $f : X_1 \to X_2$ is defined to be *holomorphic* if for every point $\mathrm{P} \in X_1$ the function $w \circ f(z)$ is holomorphic in z, where z (resp., w) is a holomorphic local coordinate in a disk neighborhood V (resp., W) of P (resp., $f(\mathrm{P})$) with $f(V) \subset W$. By (5.3.20) this definition is independent of the choice of holomorphic local coordinates z and w. Setting $X_2 = \mathbf{C}$, we call a holomorphic mapping $f : X_1 \to \mathbf{C}$ a *holomorphic function* on X_1. The mapping $f : D \to \widehat{\mathbf{C}}$ (cf. (4.2.1)) defined by a meromorphic function on D is a holomorphic mapping. Conversely, a holomorphic mapping $f : D \to \widehat{\mathbf{C}}$ with $f \not\equiv \infty$ is defined by a meromorphic function. Hence, a meromorphic function on D is a holomorphic mapping from D into $\widehat{\mathbf{C}}$ which is not constantly ∞.

EXERCISE 7. Let X be a Riemann surface. Show that the followings are defined independently by the choice of holomorphic local coordinates.

i) A function $u : X \to \mathbf{R}$ (or $\mathbf{C}$) is defined to be of class C^k ($k \leqq \infty$) if in a disk neighborhood V of an arbitrary point $\mathrm{P} \in X$ with a holomorphic local coordinate $z_V(\mathrm{Q}) = z = x + iy$ ($\mathrm{Q} \in V$), $u \circ z_V^{-1}$ is of class C^k as a function in (x, y).
ii) A continuous mapping $\phi : [T_0, T_1] \to X$ is said to be of class C^k if for every point $t_0 \in [T_0, T_1]$ the function $z_V \circ \phi(t)$ is of class C^k in a neighborhood of t_0, where z_V is a holomorphic local coordinate in a disk neighborhood V of $\phi(t_0)$. The notion of piecewise class C^k is defined in the same way as above.

Let $\pi : X \to D$ be the universal covering defined in (5.3.5), and consider X and D as Riemann surfaces. Of course, the mapping π is holomorphic. Let $D_0 \subset \widehat{\mathbf{C}}$ be a simply connected domain, and let $f : D_0 \to D$ be a holomorphic mapping. The following is clear by definition:

(5.3.21) A lifting $\hat{f}: D_0 \to X$ of f is holomorphic.

For a holomorphic mapping between Riemann surfaces, the Prinzip von der Gebietstreue holds as in Corollary (4.3.14). As in Theorem (4.3.18), the inverse mapping of a univalent (injective) holomorphic mapping $f: X_1 \to X_2$ between Riemann surfaces X_1 and X_2 is called a *biholomorphic mapping.* In this case, by Theorem (4.3.18) the inverse f^{-1} is holomorphic, too, and X_1 and X_2 are said to be mutually biholomorphic. In the special case of $X_1 = X_2$, a biholomorphic mapping $f: X_1 \to X_1$ is called a *holomorphic transformation*, and the set of all of them is denoted by $\mathrm{Aut}(X_1)$, which forms a group under the law of composition. We call $\mathrm{Aut}(X_1)$ the *holomorphic transformation group* of X_1.

Assume that for every point $\mathrm{P} \in X_2$ a disk neighborhood W_{P} with a holomorphic local coordinate w_{P} and a function $h_{\mathrm{P}}(w_{\mathrm{P}})$ in w_{P} are assigned. We call the family $h = \{h_{\mathrm{P}}(w_{\mathrm{P}})\}_{\mathrm{P}\in X_2}$ an *hermitian* (resp., *pseudo-*) *metric* or a *conformal* (resp., *pseudo-*) *metric* if the following conditions are satisfied:

(5.3.22) i) $h_{\mathrm{P}}(w_{\mathrm{P}}) > 0$ (resp., $\geqq 0$), and $h_{\mathrm{P}}(w_{\mathrm{P}})$ is of class C^∞.
ii) If $W_{\mathrm{P}} \cap W_{\mathrm{Q}} \neq \emptyset$,

$$h_{\mathrm{P}}(w_{\mathrm{P}}(a)) \left| \frac{dw_{\mathrm{Q}}}{dw_{\mathrm{P}}}(w_{\mathrm{P}}(a)) \right|^{-2} = h_{\mathrm{Q}}(w_{\mathrm{Q}}(a)), \qquad a \in W_{\mathrm{P}} \cap W_{\mathrm{Q}}.$$

Because of the above equation we write $h = h_{\mathrm{P}} dw_{\mathrm{P}} \cdot d\bar{w}_{\mathrm{P}} = h_{\mathrm{P}}|dw_{\mathrm{P}}|^2$. We consider $dw_{\mathrm{P}} \cdot d\bar{w}_{\mathrm{P}} = |dw_{\mathrm{P}}|^2$ to be transformed as

$$dw_{\mathrm{P}} \cdot d\bar{w}_{\mathrm{P}} = \left| \frac{dw_{\mathrm{P}}}{dw_{\mathrm{Q}}} \right|^2 dw_{\mathrm{Q}} \cdot d\bar{w}_{\mathrm{Q}}.$$

For a holomorphic mapping $f: X_1 \to X_2$, the *pull-back* f^*h of h by f is defined as follows. Take a disk neighborhood V (resp., W) of X_1 (resp., X_2) with holomorphic local coordinate z (resp., w) so that $f(V) \subset W$, and $h = h(w)|dw|^2$ in W. Set

$$f^*h = h \circ f \left| \frac{dw \circ f}{dz} \right|^2 |dz|^2. \tag{5.3.23}$$

It is clear that f^*h is independent of the choice of holomorphic local coordinates. If $dw \circ f/dz \neq 0$, then f^*h defines an hermitian metric there. If f is biholomorphic, then f^*h is an hermitian metric.

Let $\phi: [T_0, T_1] \to X_2$ be a piecewise continuously differentiable curve. For a $t \in [T_0, T_1]$ at which ϕ is differentiable, take a disk neighborhood W containing $\phi(t)$ and a holomorphic local coordinate w in W as in (5.3.23). Then the function

$$\sqrt{h(\phi(t))} \left| \frac{dw \circ \phi(t)}{dt} \right| \quad (\geqq 0)$$

is independent of the choice of w by (5.3.22), ii). We define the *length* $L_h(C)$ of the curve $C = C(\phi)$ with respect to h by

$$L_h(C) = \int_{T_0}^{T_1} \sqrt{h(\phi(t))} \left| \frac{dw \circ \phi(t)}{dt} \right| dt. \tag{5.3.24}$$

For a subset E of X_2 we define the *area* $A_h(E)$ with respect to h by

$$A_h(E) = \int_E h(w)dudv, \qquad w = u + iv. \tag{5.3.25}$$

Here it is easily checked in the same way as in the case of length that the right side of the above equation is locally well-defined independently from holomorphic local coordinates and hence globally, provided that the integration exists.

EXERCISE 8. i) On $\widehat{\mathbf{C}} \setminus \{\infty\} = \mathbf{C}$, we set $h = 4(1+|z|^2)^{-2}|dz|^2$. On $\widehat{\mathbf{C}} \setminus \{0\} = \mathbf{C}$, we set $h = 4(1+|\tilde{z}|^2)^{-2}|d\tilde{z}|^2$. Show that these define an hermitian metric on the Riemann sphere $\widehat{\mathbf{C}}$, which is called the *Fubini-Study metric.*

ii) Obtain the length of the real axis $\mathbf{R}$ with respect to the above Fubini-Study metric.

iii) Obtain the area of $\widehat{\mathbf{C}}$ with respect to the Fubini-Study metric.

EXERCISE 9. Assume that the composition $g \circ f$ of two holomorphic mappings f and g is defined. Let h be an hermitian metric defined on a domain containing the image of g. Show that $(g \circ f)^* h = f^*(g^* h)$.

Problems

1. Construct the Riemann surface of the multi-valued function $\sqrt{z(z-1)}$. What are the branch points and the branch orders?
2. Construct the Riemann surface of the inverse function of $f(z) = z^2 - 2z$ ($z \in \mathbf{C}$). What are the branch points and the branch orders?
3. Let X be a Riemann surface and let u be a real valued function of class C^2. We define u to be harmonic if for an arbitrary point $\mathrm{P} \in X$ and for a disk neighborhood V of P with holomorphic local coordinate $z_V = z = x + iy$, the function $u \circ z_V^{-1}(x,y)$ is harmonic in (x,y). Show that this definition is independent of the choice of a holomorphic local coordinate.
4. Let $D \subset \widehat{\mathbf{C}}$ be a simply connected Riemann surface and let f be a non-vanishing holomorphic function in D. Show that there is a holomorphic function g in D such that $f(z) = e^{g(z)}$.
5. Let f be a holomorphic function on a Riemann surface X. Show that if $|f|$ takes a maximum at a point of X, then f is constant.
6. Let u be a harmonic function on a Riemann surface X. Show that if u takes the maximum or the minimum at a point of X, then u is constant.
7. On $\Delta(R)(0 < R < \infty)$ we define an hermitian metric

$$h = \frac{4R^2}{(R^2 - |z|^2)^2}|dz|^2,$$

which is called the *Poincaré metric*. Obtain the length of the curve $C_{(r,\theta)}(\phi : [0,r] \ni t \to te^{i\theta} \in \Delta(R))$ $(0 \leqq r < R)$, and the area of the disk $\Delta(r)$ with respect to h.

8. Let $h = 2a(z)|dz|^2$ (local expression) be an hermitian metric on a Riemann surface X. Show that the function

$$K_h(z) = -\frac{4}{a(z)}\Delta \log a(z) = -\frac{1}{a(z)}\partial_z\overline{\partial}_z \log a(z)$$

is defined independently of the choice of the local holomorphic coordinate z. We call K_h the *Gaussian curvature* of h.

9. i) Let h_1 be the Fubini-Study metric on $\widehat{\mathbf{C}}$ (see §3, Exercise 8). Show that $K_{h_1} \equiv 1$.

ii) Show that for the hermitian metric $h_2 = |dz|^2$ (the Euclidean metric) on $\mathbf{C}$, $K_{h_2} \equiv 0$.

iii) Show that for the Poincaré metric h_3 (cf. problem 6) on $\Delta(R)$ with $0 < R < \infty$, $K_{h_3} \equiv -1$.

CHAPTER 6

Holomorphic Mappings

The aim of this chapter is to prove the Riemann mapping theorem and Picard's theorem, which are the two greatest theorems in the theory of complex functions of one variable. The former is a very specific and essential property of the one variable case, and cannot be extended to the several variable case as it is. It is an open and deep problem to generalize it in the higher dimensional case. On the other hand, Picard's theorem furnished the starting point for modern complex function theory, and was later developed to a quantitative theory, called the Nevanlinna theory (1925). Furthermore, introducing a geometric method, L. Ahlfors proved that the number 2 of the so-called exceptional values is the Euler number of $\widehat{\mathbf{C}}$, which is a topological invariant of the 2-dimensional sphere (1935). For this work Ahlfors was awarded the first Fields medal in 1936. Picard's theorem and the Nevanlinna theory have been extended to the several variable case in a number of ways, and interesting results have been obtained.

6.1. Linear Transformations

As defined in Chapter 2, §8, a linear transformation f is represented by $f(z) = (az+b)/(cz+d), ad-bc \neq 0$, which is a meromorphic function on $\widehat{\mathbf{C}}$, and also a holomorphic mapping $f : \widehat{\mathbf{C}} \to \widehat{\mathbf{C}}$.

(6.1.1) THEOREM. i) *A holomorphic transformation f of $\widehat{\mathbf{C}}$ is a linear transformation.*

ii) *A holomorphic transformation g of $\mathbf{C}$ is a linear transformation represented by $g(z) = az + b, a \neq 0$.*

PROOF. i) Set $c = f(\infty)$. If $c \neq \infty$, we consider $1/(f(z)-c)$; if this is a linear transformation, so is f itself. Thus we may assume that $c = \infty$. Restricting f to $\mathbf{C}$, we have a biholomorphic mapping $f|\mathbf{C} : \mathbf{C} \to \mathbf{C}$. Hence, it suffices to prove ii).

ii) Set $g(0) = b$. Since g^{-1} is continuous, $g^{-1}(\overline{\Delta(b;1)})$ is a bounded and closed subset of $\mathbf{C}$. Take $R > 0$ so that $g^{-1}(\overline{\Delta(b;1)}) \subset \Delta(0;R)$. Let

$$g(z) = \sum_{n=0}^{\infty} a_n z^n$$

be the Taylor expansion of g. The Laurent expansion of $h(\tilde{z}) = g \circ \tilde{z} = g(1/z)$, $\tilde{z} \in \Delta(1/R) \setminus \{0\}$, is given by

$$h(\tilde{z}) = \sum_{n=0}^{\infty} a_n \tilde{z}^{-n} = \sum_{-\infty}^{0} a_{-n} \tilde{z}^n.$$

Note that h restricted to $\Delta(1/R) \setminus \{0\}$ does not take any value in $\Delta(b;1)$. By Casorati-Weierstrass' Theorem (5.1.2) $\tilde{z} = 0$ is not an isolated essential singularity of h; that is, $\tilde{z} = 0$ is at most a pole of h. Therefore, there is a number n_0 such that $a_{n_0} \neq 0$ and $a_n = 0$ for all $n > n_0$, and so $g(z)$ is a polynomial

$$g(z) = b + \sum_{n=1}^{n_0} a_n z^n, \qquad a_{n_0} \neq 0.$$

Since g is biholomorphic, $g'(z)$ has no zero. The fundamental theorem of algebra, Theorem (3.5.23), implies that $n_0 = 1$, and hence $g(z) = a_1 z + b, g'(z) = a_1 \neq 0$. □

The next result is called Schwarz' lemma:

(6.1.2) LEMMA. *Let $f(z)$ be a holomorphic function on $\Delta(1)$ such that $f(0) = 0$ and $|f(z)| < 1$. Then*

$$|f(z)| \leqq |z|, \qquad |f'(0)| \leqq 1.$$

The first equality holds for some $z \neq 0$, or the second holds if and only if

$$f(z) = e^{i\theta} z, \qquad \theta \in \mathbf{R}.$$

PROOF. Set $f(z) = \sum_{n=1}^{\infty} a_n z^n$. Then the function

$$g(z) = \frac{f(z)}{z} = \sum_{n=0}^{\infty} a_{n+1} z^n$$

is defined on $\Delta(1)$ and holomorphic there. The assumption implies

$$\lim_{|z| \to 1} |g(z)| \leqq 1.$$

It follows from the maximum principle (Theorem (3.5.21)) that unless g is a constant such that $|g| = 1$, $|g(z)| < 1$ for $z \in \Delta(1)$. Since $g(0) = f'(0)$,

$$|f(z)| < |z|, \qquad 0 < |z| < 1,$$
$$|f'(0)| < 1.$$

If $g = e^{i\theta}$ ($\theta \in \mathbf{R}$), then $f(z) = e^{i\theta} z$; in this case, $|f(z)| \equiv |z|$ and $|f'(z)| \equiv 1$. □

Let $f(z)$ be a holomorphic function on $\Delta(R)$ such that $f(0) = 0$ and $|f(z)| < M$. Applying Lemma (6.1.2) to

$$g(z) = \frac{1}{M} f(Rz), \qquad |z| < 1,$$

we get

$$|f(z)| \leqq \frac{M}{R}|z|, \qquad |f'(0)| \leqq \frac{M}{R}. \tag{6.1.3}$$

(6.1.4) THEOREM. i) *A holomorphic transformation of a disk $\Delta(a; R)$ is a linear transformation.*

ii) *A holomorphic transformation of the upper half plane $\mathbf{H}$ is a linear transformation.*

PROOF. i) We may assume that $a = 0$ and $R = 1$. Let $f : \Delta(1) \to \Delta(1)$ be a holomorphic transformation, and set $a = f(0) \in \Delta(1)$. As in (2.8.10) we set

$$\phi_a(z) = \frac{z - a}{-\bar{a}z + 1}.$$

It follows from Theorem (2.8.11) that ϕ_a is a holomorphic transformation of $\Delta(1)$, and ϕ_a^{-1} is a linear transformation, too. It suffices to show that $g = \phi_a \circ f$ is a linear transformation. Since $g(0) = 0$ and $|g(z)| < 1$, Lemma (6.1.2) implies

$$|g'(0)| \leqq 1. \tag{6.1.5}$$

Here, the equality holds if and only if $g(z) = e^{i\theta}z, \theta \in \mathbf{R}$; it is a linear transformation. Applying Lemma (6.1.2) to g^{-1}, we have

$$|(g^{-1})(0)| = \left|\frac{1}{g'(0)}\right| \leqq 1.$$

It follows from this and (6.1.5) that $|g'(0)| = 1$; thus, $g(z) = e^{i\theta}z$.

ii) Let $f : \mathbf{H} \to \mathbf{H}$ be a holomorphic transformation. As in (2.8.12), we set

$$\psi(z) = \frac{z - i}{z + i},$$

which defines a biholomorphic mapping $\psi : \mathbf{H} \to \Delta(1)$. Then $\psi \circ f \circ \psi^{-1}$ is a holomorphic transformation of $\Delta(1)$, and hence a linear transformation by i). Therefore f is a linear transformation. □

By the above Theorems (6.1.1) and (6.1.4) the notations, $\mathrm{Aut}(\widehat{\mathbf{C}})$, $\mathrm{Aut}(\Delta(1))$, and $\mathrm{Aut}(\mathbf{H})$ used in Chapter 2, §8 may be taken for holomorphic transformation groups in the sense defined in Chapter 5, §3.

EXERCISE 1. Let D_i $(i = 1, 2)$ be a domain defined as the inside of a circle of $\mathbf{C}$ or one side of a line of $\mathbf{C}$. Show that a biholomorphic mapping $f : D_1 \to D_2$ is a linear transformation.

6.2. Poincaré Metric

We define an hermitian metric g_R on $\Delta(R)$ by

$$a_R(z) = \frac{2R^2}{(R^2 - |z|^2)^2}, \qquad (6.2.1)$$
$$g_R = 2a_R(z)|dz|^2.$$

This is called the *Poincaré metric* on $\Delta(R)$. Taking a biholomorphic mapping

$$f_R : \Delta(1) \ni z \to Rz \in \Delta(R),$$

we have by (5.3.23)

$$f_R^* g_R = g_1, \qquad (6.2.2)$$

and hence we may normalize, $R = 1$. The next theorem plays a basic role.

(6.2.3) THEOREM. *The Poincaré metric* g_1 *on* $\Delta(1)$ *is* $\mathrm{Aut}(\Delta(1))$*-invariant; i.e., for every* $f \in \mathrm{Aut}(\Delta(1))$, $f^* g_1 = g_1$.

PROOF. Since

$$f^* g_1 = \frac{4|f'(z)|^2}{(1 - |f(z)|^2)^2}|dz|^2 \qquad (6.2.4)$$

for $f \in \mathrm{Aut}(\Delta(1))$, it suffices to prove that

$$\frac{|f'(z)|^2}{(1 - |f(z)|^2)^2} = \frac{1}{(1 - |z|^2)^2}. \qquad (6.2.5)$$

This clearly holds for $f(z) = e^{i\theta} z$ ($\theta \in \mathbf{R}$). By Theorem (2.8.11) it is sufficient to prove (6.2.5) for

$$f(z) = \phi_a(z) = \frac{z - a}{-\bar{a}z + 1}, \qquad a \in \Delta(1). \qquad (6.2.6)$$

Since

$$f'(z) = \frac{-\bar{a}z + 1 + (z - a)\bar{a}}{(-\bar{a}z + 1)^2} = \frac{1 - |a|^2}{(\bar{a}z - 1)^2},$$

we have

$$\begin{aligned}
\frac{|f'(z)|}{1 - |f(z)|^2} &= \frac{1}{1 - \left|\frac{z-a}{-\bar{a}z+1}\right|^2} \frac{1 - |a|^2}{|-\bar{a}z + 1|^2} \\
&= \frac{1 - |a|^2}{|\bar{a}z - 1|^2 - |z - a|^2} \\
&= \frac{1 - |a|^2}{|\bar{a}z|^2 - \bar{a}z - a\bar{z} + 1 - |z|^2 + \bar{z}a + z\bar{a} - |a|^2} \\
&= \frac{1 - |a|^2}{1 - |z|^2 + |a|^2|z|^2 - |a|^2} = \frac{1}{1 - |z|^2}. \qquad \square
\end{aligned}$$

Let $C(\phi : [T_0, T_1] \to \Delta(1))$ be a piecewise continuously differentiable curve in $\Delta(1)$. As in (5.3.24) we define the length $L_{\Delta(1)}(C) = L_{g_1}(C)$ of C with respect to g_1 by

(6.2.7)
$$L_{\Delta(1)} = \int_{T_0}^{T_1} \sqrt{2g_1(\phi(t))} \left| \frac{d\phi}{dt}(t) \right| dt$$
$$= \int_{T_0}^{T_1} \frac{2}{1 - |\phi(t)|^2} \sqrt{\left| \frac{d\phi_1}{dt}(t) \right|^2 + \left| \frac{d\phi_2}{dt}(t) \right|^2} \, dt,$$

where $\phi(t) = \phi_1(t) + i\phi_2(t)$. Let $f \in \mathrm{Aut}(\Delta(1))$. Then it follows from Theorem (6.2.3) (cf. also (3.2.18)) that

(6.2.8)
$$L_{\Delta(1)}(f(C)) = \int_{T_0}^{T_1} \frac{2}{1 - |f(\phi(t))|^2} \left| \frac{df \circ \phi}{dt}(t) \right| dt$$
$$= \int_{T_0}^{T_1} \frac{2}{1 - |f(\phi(t))|^2} \cdot |f'(\phi(t))| \left| \frac{d\phi}{dt}(t) \right| dt$$
$$= \int_{T_0}^{T_1} \frac{2}{1 - |\phi(t)|^2} \left| \frac{d\phi}{dt}(t) \right| dt$$
$$= L_{\Delta(1)}(C).$$

Therefore we see that the length of C with respect to g_1 is invariant under holomorphic transformations of $\Delta(1)$. We call $L_{\Delta(1)}(C)$ the *hyperbolic length* of C.

For two arbitrary points $z_1, z_2 \in \Delta(1)$ we define the *hyperbolic distance* by

(6.2.9)
$$d_{\Delta(1)}(z_1, z_2) = \inf\{L_{\Delta(1)}(C); C \text{ is a piecewise continuous differentiable curve in } \Delta(1) \text{ connecting } z_1 \text{ and } z_2\}.$$

In fact, this satisfies the following axioms of metric:

(6.2.10) i) $d_{\Delta(1)}(z_1, z_2) = d_{\Delta(1)}(z_2, z_1)$,
ii) $d_{\Delta(1)}(z_1, z_2) + d_{\Delta(1)}(z_2, z_3) \geqq d_{\Delta(1)}(z_1, z_3)$,
iii) $d_{\Delta(1)}(z_1, z_2) = 0 \iff z_1 = z_2$.

The proofs of i) and ii) are clear, and iii) follows from the fact that $a_1(z)$ is a positive continuous function; but it will also be shown in the following argument. It follows from (6.2.8) that

(6.2.11)
$$d_{\Delta(1)}(f(z_1), f(z_2)) = d_{\Delta(1)}(z_1, z_2), \qquad f \in \mathrm{Aut}(\Delta(1)).$$

For this property, we say that $d_{\Delta(1)}$ is $\mathrm{Aut}(\Delta(1))$-invariant. Set

$$f(z) = e^{i\theta} \frac{z - z_1}{-\bar{z}_1 z + 1},$$

so that $z_2' = f(z_2) \in \mathbf{R}$ and $z_2' > 0$. Let $C(\phi : [0,1] \to \Delta(1))$ be a piecewise continuously differentiable curve with initial point 0 and terminal point z_2'. Set

$$\begin{aligned} &\phi(t) = \phi_1(t) + i\phi_2(t), \\ &\dot{\phi}(t) = \frac{d\phi}{dt}(t), \qquad \dot{\phi}_j(t) = \frac{d\phi_j}{dt}(t), \quad j = 1, 2. \end{aligned}$$

Since $|\dot{\phi}(t)| \geqq |\dot{\phi}_1(t)|$ and $1 - |\phi(t)|^2 \leqq 1 - |\phi_1(t)|^2$,

$$\frac{|\dot{\phi}(t)|^2}{(1 - |\phi(t)|^2)^2} \geqq \frac{\dot{\phi}_1(t)^2}{(1 - \phi_1(t)^2)^2}. \tag{6.2.12}$$

We have a curve $C_1 = C_1(\phi_1)$ given by ϕ_1. Then the initial (resp., terminal) point of C_1 is 0 (resp., z_2'), and by (6.2.12)

$$L_{\Delta(1)}(C) \geqq L_{\Delta}(C_1).$$

The equality holds if and only if $\phi_2(t) \equiv 0$. Furthermore, we have

$$\begin{aligned} L_{\Delta(1)}(C_1) &= \int_0^1 \frac{2|\dot{\phi}_1(t)|}{1 - \phi_1(t)^2} dt \\ &\geqq \int_0^1 \frac{2\dot{\phi}_1(t)}{1 - \phi_1(t)^2} dt = \int_0^{z_2'} \frac{2}{1 - x^2} dx \\ &= \int_0^{z_2'} \left(\frac{1}{1+x} + \frac{1}{1-x} \right) dx = \big[\log(1+x) - \log(1-x)\big]_0^{z_2'} \\ &= \log \frac{1 + z_2'}{1 - z_2'}. \end{aligned}$$

It is to be noted that the equality in the above inequality holds if and only if $\dot{\phi}_1(t) \geqq 0$; for such an example, we may take $\phi_1(t) = tz_2'$. Thus we see that $L_{\Delta(1)}(C) \geqq \log\{(1 + z_2')/(1 - z_2')\}$, and the equality holds if and only if C is a line segment $C_{(0,z_2')}$ connecting 0 and z_2'. It follows that

$$d_{\Delta(1)}(z_1, z_2) = d_{\Delta(1)}(0, z_2') = \log \frac{1 + \left| \frac{z_2 - z_1}{-\bar{z}_1 z_2 + 1} \right|}{1 - \left| \frac{z_2 - z_1}{-\bar{z}_1 z_2 + 1} \right|}. \tag{6.2.13}$$

From this, (6.2.10), iii) follows.

In general, a curve C connecting two points $z_1, z_2 \in \Delta(1)$ is called a *hyperbolic geodesic* if

$$d_{\Delta(1)}(z_1, z_2) = L_{\Delta(1)}(C).$$

The above $C_{(0,z_2')}$ is the unique geodesic connecting 0 and z_2', and $C_{(z_1,z_2)} = f^{-1}(C_{(0,z_2')})$ is the unique geodesic connecting z_1 and z_2. The mapping f is analytically continued to a neighborhood of $\overline{\Delta(1)}$. By the conformality of f (Theorem (3.1.15)) $C_{(z_1,z_2)}$ is a part of the circle (including the case of a line), passing through z_1 and z_2 and orthogonally crossing the boundary circle $C(0;1)$ of $\Delta(1)$. We denote by $L(z_1, z_2)$ the part of that circle contained in $\Delta(1)$. For any two points $w_1, w_2 \in L(z_1, z_2)$ the segment of $L(z_1, z_2)$ connecting w_1 and w_2 is the geodesic $C_{(w_1,w_2)}$ connecting w_1 and w_2. We also call $L(z_1, z_2)$ a hyperbolic geodesic. Take the third point $z_3 \in \Delta(1)$ outside $L(z_1, z_2)$. Then there are infinitely many hyperbolic geodesics passing through z_3 and having no intersection with $L(z_1, z_2)$ (see Figure 67). To confirm this, it is easier to set $z_3 = 0$ by a linear transformation of $\mathrm{Aut}(\Delta(1))$. If we associate a hyperbolic geodesic to a line in Euclidean geometry, the above property contradicts the axiom of parallels. The geometry in which hyperbolic geodesics in $\Delta(1)$ are considered as lines is called *non-Euclidean* or *hyperbolic.*

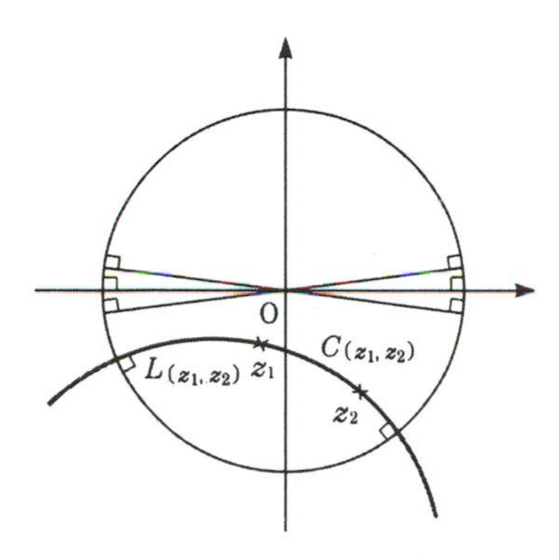

FIGURE 67

The metric $d_{\Delta(1)}$ defines the so-called metric-topology on $\Delta(1)$. That is, a sequence $\{z_n\}_{n=0}^{\infty}$ of points $z_n \in \Delta(1)$ is said to converge to $a \in \Delta(1)$ if

$$\lim_{n\to\infty} d_{\Delta(1)}(z_n, a) = 0.$$

We say that $d_{\Delta(1)}$ is *complete* if any Cauchy sequence $\{z_n\}_{n=0}^{\infty}$ with respect to $d_{\Delta(1)}$ (i.e., for an arbitrary $\epsilon > 0$ there is a number n_0 such that $d_{\Delta(1)}(z_n, z_m) < \epsilon$ for all $n, m \geqq n_0$) converges to a point of $\Delta(1)$. In fact we have the following theorem.

(6.2.14) THEOREM. i) *A sequence $\{z_n\}_{n=0}^{\infty}$ of points z_n of $\Delta(1)$ converges to a point $a \in \Delta(1)$ with respect to $d_{\Delta(1)}$ if and only if it converges to a as a sequence of points of* $\mathbf{C}$.

ii) *$d_{\Delta(1)}$ is a complete metric.*

PROOF. i) By (6.2.13) $d_{\Delta(1)}(z, w)$ is continuous in the two variables z and w. Therefore

$$\lim_{n\to\infty} |z_n - a| = 0 \Longrightarrow \lim_{n\to\infty} d_{\Delta(1)}(z_n, a) = 0.$$

Conversely, we assume $\lim_{n\to\infty} d_{\Delta(1)}(z_n, a) = 0$. By the transformation $\phi_a \in \mathrm{Aut}(\Delta(1))$ in (6.2.13) we may assume $a = 0$ without loss of generality. It follows from (6.2.13) that

$$d_{\Delta(1)}(z_n, 0) = \log \frac{1 + |z_n|}{1 - |z_n|} \to 0 \qquad (n \to \infty). \tag{6.2.15}$$

Hence, $\lim_{n\to\infty} |z_n| = 0$.

ii) Let $\{z_n\}_{n=0}^{\infty}$ be a Cauchy sequence in $\Delta(1)$ with respect to $d_{\Delta(1)}$; that is, for an arbitrary $\epsilon > 0$ there is a number n_0 such that

$$d_{\Delta(1)}(z_n, z_m) < \epsilon \tag{6.2.16}$$

for all $n, m \geqq n_0$. In particular, $d_{\Delta(1)}(z_n, z_{n_0}) < \epsilon$, $n \geqq n_0$, and so

$$d_{\Delta(1)}(0, z_n) \leqq d_{\Delta(1)}(0, z_{n_0}) + d_{\Delta(1)}(z_{n_0}, z_n) < d_{\Delta(1)}(0, z_{n_0}) + \epsilon.$$

Thus $\{d_{\Delta(1)}(0, z_n)\}_{n=0}^{\infty}$ is bounded. We set

$$M = \sup\{d_{\Delta(1)}(0, z_n)\}_{n=0}^{\infty}.$$

It follows from (6.2.15) that

$$|z_n| \leqq \frac{e^M - 1}{e^M + 1} < 1.$$

Therefore there is a subsequence $\{z_{n_\nu}\}_{\nu=0}^{\infty}$ of $\{z_n\}_{n=0}^{\infty}$ which converges to a point $a \in \overline{\Delta((e^M - 1)/(e^M + 1))} \subset \Delta(1)$ as a sequence of points of $\mathbf{C}$. It follows from i) that for $\epsilon > 0$ there is a number ν_0 such that

$$d_{\Delta(1)}(z_{n_\nu}, a) < \epsilon, \qquad \nu \geqq \nu_0. \tag{6.2.17}$$

We may assume that $n_{\nu_0} \geqq n_0$. Thus, by (6.2.16) and (6.2.17)

$$d_{\Delta(1)}(z_n, a) < 2\epsilon. \qquad \square$$

By i) we deduce that it is not necessary to distinguish the convergence with respect to $d_{\Delta(1)}$ and the convergence in $\mathbf{C}$. Because of ii) the Poincaré metric g_1 (and g_R as well) is also said to be *complete*.

The upper half plane $\mathbf{H}$ is biholomorphic to $\Delta(1)$ by (2.8.12):

$$\psi(z) = \frac{z - i}{z + i},$$
$$\psi : \mathbf{H} \to \Delta(1).$$

Set $h = \psi^* g_1$. A direct computation yields

$$h = \frac{1}{(\operatorname{Im} z)^2} |dz|^2, \tag{6.2.18}$$

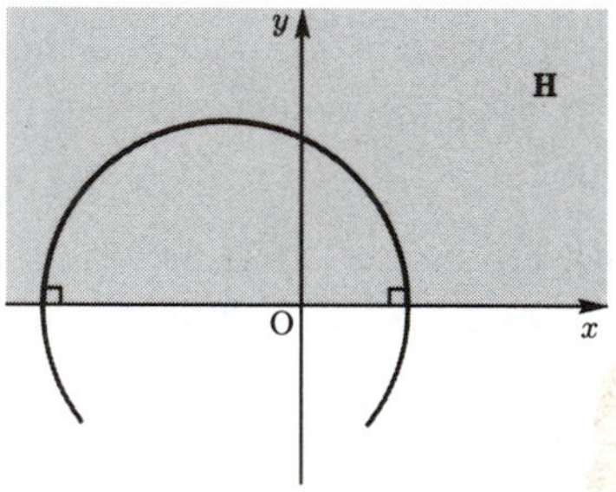

FIGURE 68

which is called the *Poincaré metric* on $\mathbf{H}$. We can define the length $L_{\mathbf{H}}(C)$ of a piecewise continuously differentiable curve C in $\mathbf{H}$, the hyperbolic metric $d_{\mathbf{H}}$, and hyperbolic geodesics of $\mathbf{H}$ in the same way as in $\Delta(1)$. By the correspondence of circle to circle of linear transformations (Chapter 2, §8), a hyperbolic geodesic of $\mathbf{H}$ is a part of a circle orthogonally crossing the real axis, contained in $\mathbf{H}$.

EXERCISE 1. Show that a hyperbolic geodesic of $\mathbf{H}$ is as described above.

EXERCISE 2. Let $\{z_n\}_{n=0}^{\infty}$ be a sequence of points of $\Delta(1)$. Assume that there is a point $a \in \Delta(1)$ such that $\{d_{\Delta(1)}(a, z_n)\}_{n=0}^{\infty}$ is bounded. Show that $\{z_n\}_{n=0}^{\infty}$ contains a subsequence which converges to a point of $\Delta(1)$.

EXERCISE 3. Compute $d_{\mathbf{H}}(z_1, z_2)$ for points $z_1, z_2 \in \mathbf{H}$.

6.3. Contraction Principle

If the universal covering of a domain $D \subset \widehat{\mathbf{C}}$ is biholomorphic to $\Delta(1)$, D is said to be *hyperbolic*. In this case, the universal covering mapping

$$\pi : \Delta(1) \to D$$

is holomorphic. The deck transformation group $\pi_1(D)$ consists of holomorphic transformations of $\mathrm{Aut}(\Delta(1))$. Thus $\pi_1(D)$ is a subgroup of $\mathrm{Aut}(\Delta(1))$, and its elements are linear transformations. For an arbitrary point $\mathrm{P} \in D$ we take a disk neighborhood $U_{\mathrm{P}} \subset D$. It follows from (5.3.6) that for every $z_1 \in \pi^{-1}(\mathrm{P})$ there is a neighborhood V_1 of z_1 such that the restriction

$$\pi|V_1 : V_1 \to U_{\mathrm{P}}$$

is biholomorphic. Now we define an hermitian metric $h_{U_{\mathrm{P}}}$ by

$$h_{U_{\mathrm{P}}} = ((\pi|V_1)^{-1})^*(g_1|V_1). \tag{6.3.1}$$

Here $g_1|V_1$ stands for the restriction of the Poincaré metric g_1 to V_1.

Take another $z_2 \in \pi^{-1}(\mathrm{P})$ and V_2 in the same way. Then, by (5.3.12) there is an element $\gamma \in \pi_1(D) \subset \mathrm{Aut}(\Delta(1))$ such that

$$\gamma(z_1) = z_2, \qquad \gamma(V_1) = V_2.$$

Since g_1 is $\mathrm{Aut}(\Delta(1))$-invariant (Theorem (6.2.3)), $\gamma^*(g_1|V_2) = g_1|V_1$. Thus it follows from Chapter 5, §3, Exercise 9 that

$$((\pi|V_1)^{-1})^*(g_1|V_1) = ((\pi|V_2)^{-1})^*(g_1|V_2).$$

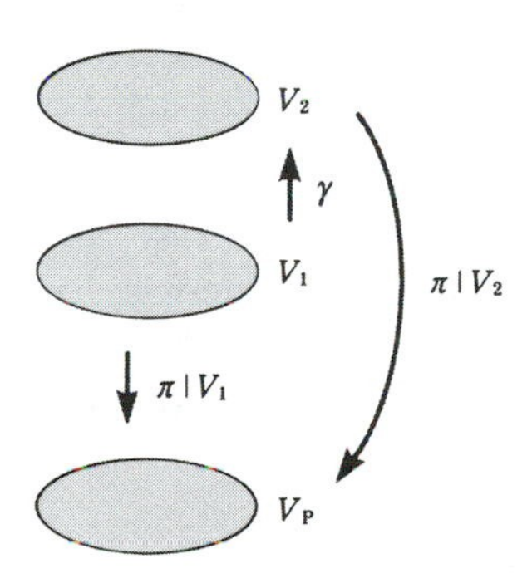

FIGURE 69

Therefore $h_{U_{\mathrm{P}}}$ is defined independently of the choice of V_1, and defines an hermitian metric h_D on D. We call h_D the *hyperbolic metric* on D. For a piecewise continuously differentiable curve C in D, we may define the length $L_D(C) = L_{h_D}(C)$ with respect to h_D, and then the metric d_D in the same way as $d_{\Delta(1)}$. We call $L_D(C)$ the *hyperbolic length* of C, and d_D the *hyperbolic metric*. Let $\tilde{C}$ be a lifting of C. Then by the definition

$$L_D(C) = L_{\Delta(1)}(\tilde{C}). \tag{6.3.2}$$

(6.3.3) THEOREM. *Let $D \subset \widehat{\mathbf{C}}$ be a hyperbolic domain, and let $\pi : \Delta(1) \to D$ be the universal covering. Then we have the following.*

i) *A sequence $\{z_n\}_{n=0}^{\infty}$ of points of D converges to $a \in D$ as points of $\widehat{\mathbf{C}}$ if and only if it converges to a with respect to d_D.*

ii) *The metric d_D is complete.*

PROOF. i) Let $\lim z_n = a$ as points of $\widehat{\mathbf{C}}$. Take a disk neighborhood U_a of a. We may assume that $z_n \in U_a, n = 0, 1, 2, \ldots$. Take a point $\hat{a} \in \pi^{-1}(a)$ and a neighborhood V of $\hat{a}$ so that $\pi(V) = U_a$. Take $\hat{z}_n \in V \cap \pi^{-1}(z_n), n = 0, 1, 2, \ldots$. It follows from (6.3.2) and the definition of d_D that

$$d_D(z_n, a) \leqq d_{\Delta(1)}(\hat{z}_n, \hat{a}). \tag{6.3.4}$$

By the choice of the points, $\lim_{n\to\infty} \hat{z}_n = \hat{a}$ as points of $\mathbf{C}$. It follows from Theorem (6.2.14) that $\lim_{n\to\infty} d_{\Delta(1)}(\hat{z}_n, \hat{a}) = 0$. By (6.3.4) we have $\lim d_D(z_n, a) = 0$.

Conversely, we assume that $\lim d_D(z_n, a) = 0$. Take the above U_a so that it is relatively compact in D. It follows from the above argument that $d_D(z, a)$ is continuous in $z \in D$. Hence

$$\epsilon_0 = \min\{d_D(z, a); z \in \partial U_a\} > 0.$$

Take an arbitrary point $z \in D$ with $d_D(z, a) < \epsilon_0$. There is a piecewise continuously differentiable curve C from a to z such that

$$d_D(z, a) \leqq L_D(C) < \epsilon_0. \tag{6.3.5}$$

If $z \notin U_a$, then $C \cap \partial U_a \neq \emptyset$; by the definition of ϵ_0, $\epsilon_0 \leqq L_D(C)$. By (6.3.5) we get a contradiction, $\epsilon_0 < \epsilon_0$. Therefore

$$\{z \in D; d_D(z, a) < \epsilon_0\} \subset U_a. \tag{6.3.6}$$

By the hypothesis we may assume that $z_n \in U_a$ and $d_D(z_n, a) < \epsilon_0$, $n = 0, 1, 2, \ldots$. Take a piecewise continuously differentiable curve C_n connecting a and z_n as in (6.3.5). Then (6.3.6) implies that $C \subset U_a$. Take a lifting of C_n with the initial point $\hat{a}$, and let $\hat{z}_n \in \pi^{-1}(z_n)$ be its terminal point. We get

$$L_D(C_n) \geqq d_{\Delta(1)}(\hat{z}_n, \hat{a}).$$

Letting $L_D(C_n) \to d_D(z_n, a)$, we see that $d_D(z_n, a) \geqq d_{\Delta(1)}(\hat{z}_n, \hat{a})$. Combining this with (6.3.4), we obtain

$$d_D(z_n, a) = d_{\Delta(1)}(\hat{z}_n, \hat{a}). \tag{6.3.7}$$

It follows again from Theorem (6.2.14) that $\lim \hat{z}_n = \hat{a}$ as points of $\mathbf{C}$, and that $\lim z_n = a$ as points of $\widehat{\mathbf{C}}$.

ii) For $r > 0$ we set

$$D_r = \{z \in D; d_D(z, \pi(0)) < r\}.$$

If D_r is relatively compact in D, then any Cauchy sequence with respect to d_D has a convergent subsequence with limit in D as points of $\widehat{\mathbf{C}}$, and so the convergence of the original sequence with respect to d_D follows as in the proof of Theorem (6.2.14), ii). Set

$$\Delta(1)_r = \{\tilde{z} \in \Delta(1); d_{\Delta(1)}(\hat{z}, 0) < r\}.$$

For an arbitrary $z \in D_r$ we take a piecewise continuously differentiable curve C from $\pi(0)$ to z such that

$$d_D(z, \pi(0)) \leqq L_D(C) < r.$$

Let $\hat{C}$ be the lifting of C with the initial point 0. By (6.3.2), $L_{\Delta(1)}(\hat{C}) < r$. Let $\hat{z}$ be the terminal point of $\hat{C}$. Then $\hat{z} \in \Delta(1)_r$, $\pi(\hat{z}) = z$, and

$$D_r \subset \pi(\Delta(1)_r) \subset \pi(\overline{\Delta(1)}_r).$$

By Exercise 2 of §2, $\overline{\Delta(1)}_r$ is compact, and so is $\pi(\overline{\Delta(1)}_r)$. Hence $\overline{D}_r$ is compact in D. □

We deduce the following from the proof of ii):

(6.3.8) COROLLARY. *Let $\pi : \Delta(1) \to D$ be as above, and fix a point $a \in D$. Then a sequence $\{z_n\}_{n=0}^{\infty}$ of points of D has no accumulation point in D (that is, no convergent subsequence) if and only if $\lim_{n\to\infty} d_D(z_n, a) = \infty$.*

Let $D \subset \widehat{\mathbf{C}}$ be a domain, and let $h_j, j = 1, 2$, be two hermitian pseudo-metrics on D. Write

$$h_j = b_j(z)|dz|^2 = \tilde{b}_j(\tilde{z})|d\tilde{z}|^2, \qquad j = 1, 2 \quad \left(\tilde{z} = \frac{1}{z}\right).$$

For $a \in D$ we define $h_1(a) \leqq h_2(a)$ if and only if

$$\text{(6.3.9)} \qquad \begin{aligned} b_1^*(a) &\leqq b_2(a) \qquad \text{if } a \neq \infty, \\ \tilde{b}_1^*(0) &\leqq \tilde{b}_2(0) \qquad \text{if } a = \infty. \end{aligned}$$

If $h_1(a) \leqq h_2(a)$ for all $a \in D$, we write $h_1 \leqq h_2$.

(6.3.10) THEOREM. *Let $D_j \subset \widehat{\mathbf{C}}, j = 1, 2$, be two hyperbolic domains with hyperbolic metrics h_{D_j}. Let $f : D_1 \to D_2$ be a holomorphic mapping. Then we have*

i) $f^* h_{D_2} \leqq h_{D_1}$*;*
ii) $d_{D_1}(a, b) \geqq d_{D_2}(f(a), f(b))$ *for all* $a, b \in D_1$.

This property of the hyperbolic metrics is called the *contraction principle.* For the proof we first deal with the case where $D_1 = D_2 = \Delta(1)$. The next is called *Schwarz-Pick's lemma*:

(6.3.11) LEMMA. *For an arbitrary holomorphic mapping $f : \Delta(1) \to \Delta(1)$ we have*

i) $f^* g_1 \leqq g_1$*, and if the equality holds at a point, then* $f \in \mathrm{Aut}(\Delta(1))$*;*
ii) $d_{\Delta(1)}(a, b) \geqq d_{\Delta(1)}(f(a), f(b))$ *for all* $a, b \in \Delta(1)$.

PROOF. i) We want to show that

$$f^*g_1(a) \leqq g_1(a) \tag{6.3.12}$$

at an arbitrary point $a \in \Delta(1)$. By making use of

$$\phi_{-a}(z) = \frac{z+a}{\bar{a}z+1} = \phi_a^{-1}(z),$$

we get $\phi_{-a}^* g_1(0) = g_1(a)$ from Theorem (6.2.3). Thus (6.3.12) is equivalent to

$$f^*\phi_{-a}^* g_1(0) \leqq \phi_{-a}^* g_1(0);$$

that is, $g_1(0) \geqq \phi_{-a}^{-1*} \circ f^* \circ \phi_{-a}^* g_1(0) = (\phi_{-a} \circ f \circ \phi_a)^* g_1(0)$. Therefore it suffices to prove (6.3.12) at $a = 0$ for an arbitrary holomorphic mapping $f : \Delta(1) \to \Delta(1)$:

$$f^*g_1(0) \leqq g_1(0). \tag{6.3.13}$$

Since

$$(\phi_{f(0)})^* g_1(0) = f^*(\phi_{f(0)}^* g_1)(0) = f^* g_1(0),$$

we may assume that $f(0) = 0$. By the definition of g_1, (6.3.13) with $f(0) = 0$ is equivalent to $|f'(0)| \leqq 1$. This was proved by Lemma (6.1.2).

Suppose that there is a point $a \in \Delta(1)$ such that $f^*g_1(a) = g_1(a)$. By the above arguments, we may assume that $a = 0 = f(0)$. Then it follows from Lemma (6.2.1) that $f \in \mathrm{Aut}(\Delta(1))$.

ii) Take a piecewise continuously differentiable curve C from a to b in $\Delta(1)$. It follows from i) that

$$L_{\Delta(1)}(C) \geqq L_{\Delta(1)}(f(C)) \geqq d_{\Delta(1)}(f(a), f(b)).$$

Thus $d_{\Delta(1)}(a, b) \geqq d_{\Delta(1)}(f(a), f(b))$. □

PROOF OF THEOREM (6.3.10). i) Let $\pi_j : \Delta(1) \to D_j, j = 1, 2$, be the universal coverings. Let $F : \Delta(1) \to \Delta(1)$ be a lifting of $f \circ \pi_1 : \Delta(1) \to D_2$:

$$\begin{array}{ccc} \Delta(1) & \xrightarrow[F]{} & \Delta(1) \\ \downarrow \pi_1 & & \downarrow \pi_2 \\ D_1 & \xrightarrow{f} & D_2 \end{array}$$

It follows from Lemma (6.3.11) that $F^*g_1 \leqq g_1$. Hence, the definition (6.3.1) implies that $f^*h_{D_2} \leqq h_{D_1}$.

ii) The proof is similar to that of Lemma (6.3.11), ii). □

The notion of hyperbolic distance was generalized for general complex manifolds by S. Kobayashi [18], and it has played an important role in complex analysis.

EXERCISE 1. Let D be a hyperbolic domain, and let h_D be the hyperbolic metric. Show that $f^*h_D = h_D$ for $f \in \mathrm{Aut}(D)$.

EXERCISE 2. Let D be as above, and let $\pi : \Delta(1) \to D$ be the universal covering. Show that for two points $w_1, w_2 \in D$

$$d_D(w_1, w_2) = \min \left\{ d_{\Delta(1)}(z_1, z_2); z_j \in \Delta(1), \pi(z_j) = w_j, j = 1, 2 \right\}.$$

6.4. The Riemann Mapping Theorem

Let $D \subset \widehat{\mathbf{C}}$ be a domain. Let $\{f_n\}_{n=0}^{\infty}$ be a sequence of holomorphic functions on D which converges uniformly on compact subsets to f. Then it follows from Theorem (3.5.19) that f is holomorphic, and that

$$(6.4.1) \qquad \begin{array}{cccll} \dfrac{df_n}{dz} & \to & \dfrac{df}{dz} & (n \to \infty) & \text{on } D \cap \mathbf{C}, \\ \dfrac{df_n \circ \tilde{z}}{d\tilde{z}} & \to & \dfrac{df \circ \tilde{z}}{d\tilde{z}} & (n \to \infty) & \text{on } D \cap \widehat{\mathbf{C}} \setminus \{0\}, \end{array}$$

where the convergence is uniform on compact subsets. Now let $\mathcal{F}$ be a family of holomorphic functions on D. We say that $\mathcal{F}$ is a *normal family* if any sequence of holomorphic functions in $\mathcal{F}$ contains a subsequence which converges uniformly on compact subsets. For normal families the following Montel's theorem is fundamental.

(6.4.2) THEOREM. *If a family $\mathcal{F}$ of holomorphic functions on D is uniformly bounded, then $\mathcal{F}$ is normal.*

PROOF. There is an $M > 0$ such that for every $f \in \mathcal{F}$

$$|f(a)| \leqq M, \qquad a \in D.$$

Take $a \in D$ such that $a \neq \infty$. Take a disk neighborhood $\Delta(a; r)$ with $0 < r < d(a; \partial D)$. It follows from (3.5.4) that

$$f'(z) = \frac{1}{2\pi i} \int_{C(a;r)} \frac{f(\zeta)}{(\zeta - z)^2} d\zeta, \qquad z \in \Delta(a; r).$$

Therefore, for $z \in \Delta(a; r/2)$

$$|f'(z)| \leqq \frac{1}{2\pi} \left(\frac{2}{r} \right)^2 M \cdot 2\pi r = \frac{4M}{r}.$$

For arbitrary $z, z' \in \Delta(a; r/2)$

$$|f(z) - f(z')| = \left| \int_z^{z'} f'(\zeta) d\zeta \right| \leqq \frac{4M}{r} |z - z'|.$$

Hence $\mathcal{F}$ is equicontinuous on $\Delta(a; r/2)$. If $a = \infty$, making use of the complex coordinate $\tilde{z}$ about ∞, we infer in the same way as above that $\mathcal{F}$ is equicontinuous on a neighborhood of ∞. Thus, $\mathcal{F}$ is equicontinuous on compact subsets of D. We see by Theorem (2.2.6) that $\mathcal{F}$ is normal. $\square$

It follows from the above argument that Theorem (6.4.2) holds if $\mathcal{F}$ is bounded on every compact subset of D.

EXERCISE 1. Show that a family $\mathcal{F}$ of holomorphic functions on D is normal if and only if for every $a \in D$ there is a neighborhood V of a such that the restriction $\mathcal{F}|V = \{f|V; f \in \mathcal{F}\}$ of $\mathcal{F}$ to V is normal.

(6.4.3) LEMMA. *For an arbitrary $a \in \Delta(1)$ there is a unique $f \in \mathrm{Aut}(\Delta(1))$ such that $f(0) = a$ and $f'(0) > 0$.*

PROOF. As in (2.8.10), we set

$$f(z) = \phi_{-a}(z) = \frac{z+a}{\bar{a}z+1}.$$

Then $f \in \mathrm{Aut}(\Delta(1))$, and $f(0) = a$. A simple computation yields

$$f'(0) = 1 - |a|^2 > 0.$$

Thus we have shown the existence of such $f \in \mathrm{Aut}(\Delta(1))$. Let $g \in \mathrm{Aut}(\Delta(1))$ satisfy the conditions. Then, $g^{-1} \circ f \in \mathrm{Aut}(\Delta(1))$ satisfies

$$g^{-1} \circ f(0) = 0, \qquad (g^{-1} \circ f)'(0) > 0.$$

It follows from Theorems (6.1.4) and (2.8.11) that $g^{-1} \circ f(z) = e^{i\theta}z$ with $\theta \in \mathbf{R}$. Since $(g^{-1} \circ f)'(0) = e^{i\theta} > 0$, $e^{i\theta} = 1$, and hence $g^{-1} \circ f(z) = z$; i.e., $g = f$. □

Our next theorem is called the *Riemann mapping theorem.*

(6.4.4) THEOREM. *Let $D \subset \widehat{\mathbf{C}}$ be a simply connected domain such that the boundary ∂D contains at least two distinct points. Then, for an arbitrarily given $a \in D$ there exists a unique biholomorphic mapping $f : D \to \Delta(1)$ such that*

$$\begin{aligned} f(a) &= 0, \\ f'(a) &> 0, \qquad a \neq \infty, \\ \frac{df \circ \tilde{z}}{d\tilde{z}}(0) &> 0, \qquad a = \infty. \end{aligned}$$

PROOF. We are going to reduce D to a bounded domain of $\widehat{\mathbf{C}}$. Take distinct points $b_1, b_2 \in \partial D$. We may assume without loss of generality that $b_1 \in \mathbf{C}$. We take a linear transformation

$$\phi_1 = \frac{z - b_1}{-\frac{z}{b_2} + 1}.$$

Here, if $b_2 = \infty$, we set $z/b_2 = 0$. Then D is biholomorphic to $\phi_1(D) \subset \mathbf{C} \setminus \{0\}$. We may assume that $D \subset \mathbf{C} \setminus \{0\}$. Since D is simply connected, one may take by Theorem (5.2.5), ii) a branch $\phi_2(z)$ of $\sqrt{z}$ on D. As shown in Example (5.2.7), $\phi_2 : D \to \phi_2(D)$ is biholomorphic, and

$$\phi_2(D) \cap (-\phi_2(D)) = \emptyset.$$

Thus, $\phi_2(D) \subset \mathbf{C}$ has an exterior point $b_3 \in \mathbf{C}$. Set

$$\phi_3(z) = \frac{1}{z - b_3}.$$

Then, $\phi_3 \circ \phi_2 : D \to \phi_3 \circ \phi_2(D) \subset \mathbf{C}$ is biholomorphic, and $\phi_3 \circ \phi_2(D)$ is bounded. Therefore one may assume that D is a bounded domain of $\mathbf{C}$.

Let $\mathcal{F}$ be the family of all univalent holomorphic functions f on D satisfying

$$f(a) = 0, \qquad f'(a) = 1.$$

Since $(z - a) \in \mathcal{F}$, $\mathcal{F} \neq \emptyset$. Set

$$\|f\|_\infty = \sup\{|f(z)|; z \in D\}, \qquad f \in \mathcal{F},$$
$$\rho = \inf\{\|f\|_\infty; f \in \mathcal{F}\}.$$

Since $(z - a) \in \mathcal{F}$,

$$\rho \leqq \max\{|z - a|; z \in \partial D\} < \infty.$$

Take a sequence $\{f_n\}_{n=0}^\infty$ in $\mathcal{F}$ so that

$$\lim_{n \to \infty} \|f_n\| = \rho.$$

We may assume that $\|f_n\|_\infty < \rho + 1$ for all f_n. Thus, $\{f_n\}_{n=1}^\infty$ is uniformly bounded. It follows from Theorem (6.4.2) that $\{f_n\}_{n=1}^\infty$ contains a subsequence converging uniformly on compact subsets of D. We rewrite $\{f_n\}_{n=1}^\infty$ for the subsequence, and set

$$f_0 = \lim_{n \to \infty} f_n.$$

Note that $f_0(a) = 0$, and by (6.4.1) $f_0'(a) = \lim f_n'(a) = 1$. Hence, f_0 is not a constant. Theorem (4.3.10) implies that f_0 is univalent, and so $f_0 \in \mathcal{F}$. By definition, $\rho \leqq \|f_0\|_\infty$. On the other hand, for an arbitrary $\epsilon > 0$ there is a number n_0 such that for all $n \geqq n_0$, $\|f_n\|_\infty < \rho + \epsilon$. Letting $n \to \infty$, we have $\|f_0\| \leqq \rho$, and hence $\|f_0\| = \rho$. It suffices to show that

$$f_0(D) = \Delta(\rho). \tag{6.4.5}$$

By the maximum principle (Theorem (3.5.21)) it is clear that $f_0(D) \subset \Delta(\rho)$. Assume that $f_0(D) \neq \Delta(\rho)$. Taking $c \in \Delta(\rho) \cap \partial(f_0(D))$, we set

$$\phi_4(w) = \frac{\rho(w - c)}{-\bar{c}w + \rho^2}.$$

The mapping $\phi_4 : \Delta(\rho) \to \Delta(1)$ is biholomorphic, and $\phi_4(c) = 0$. Taking a branch $\sqrt{\phi_4 \circ f_0(z)}$ on D, we set

$$\phi_5 = \rho\sqrt{\phi_4 \circ f_0(z)} \in \Delta(\rho), \qquad z \in D.$$

It follows that ϕ_5 is univalent, and $\phi_5(a) = \sqrt{-\rho c}$. Moreover, set

$$\phi_6 = \frac{\rho^2(\phi_5(z) - \phi_5(a))}{-\overline{\phi_5(a)}\phi_5(z) + \rho^2} \in \Delta(\rho), \qquad z \in D.$$

The mapping ϕ_6 is univalent, too, and

$$\phi_6(a) = 0, \qquad \|\phi_6\|_\infty \leqq \rho.$$

A simple computation yields

$$\phi_6'(a) = \frac{\rho + |c|}{2\sqrt{-\rho c}}.$$

Since $0 < |c| < \rho$, $|\phi_6'(a)| > 1$. Setting

$$\phi_7(z) = \frac{1}{\phi_6'(a)}\phi_6(z), \qquad z \in D,$$

we have $\phi_7 \in \mathcal{F}$, and

$$\|\phi_7\|_\infty \leqq \frac{\rho}{|\phi_6'(a)|} < \rho.$$

This contradicts the choice of ρ. Thus, (6.4.5) is proved.

To show the uniqueness, we take a biholomorphic mapping $g : D \to \Delta(1)$ such that $g(a) = 0$ and $g'(0) > 0$. Then $f \circ g^{-1} \in \mathrm{Aut}(\Delta(1))$, $f \circ g^{-1}(0) = 0$, and $(f \circ g^{-1})'(0) > 0$. Lemma (6.4.3) implies that $f \circ g^{-1}(z) = z$, and so $f = g$. □

(6.4.6) THEOREM. *An arbitrary simply connected domain D of $\widehat{\mathbf{C}}$ is biholomorphic to $\widehat{\mathbf{C}}$ itself, $\mathbf{C}$, or $\Delta(1)$; furthermore, these three domains are not biholomorphic to each other.*

PROOF. If $\partial D = \emptyset$, $D = \widehat{\mathbf{C}}$. If ∂D consists of only one point $a \in \widehat{\mathbf{C}}$, we may assume $a = \infty$ by a linear transformation; hence, $D = \mathbf{C}$. If ∂D contains more than one point, it follows from Theorem (6.4.4) that D is biholomorphic to $\Delta(1)$.

The latter assertion easily follows from Theorems (3.5.22) and (3.7.8). □

It is known that the above Theorem (6.4.6) holds in fact for a general simply connected Riemann surface. It is called the *uniformization theorem*, and was proved by Koebe (1907), and independently by Poincaré (1908). The Riemann mapping theorem was first claimed in his doctoral dissertation (1851), and a compact simply connected Riemann surface with boundary was dealt with. Riemann's proof depended on the existence of a solution of a variational problem which had not been established, and was criticized by Weierstrass. Riemann did not answer this criticism, but Hilbert did later (1904). Theorem (6.4.4) was first completely proved in its current form by Osgood (1900).

An arbitrary domain or a Riemann surface, even if it is not simply connected, carries the universal covering which is simply connected. It is biholomorphic to $\widehat{\mathbf{C}}$, $\mathbf{C}$ or $\Delta(1)$ by the uniformization theorem. Thus the domain or the Riemann surface is $\widehat{\mathbf{C}}$ itself, or is represented as a quotient space of $\mathbf{C}$ or $\Delta(1)$ by the deck transformation group (cf. (5.3.11)).

EXERCISE 2. Let $f : D \to \Delta(1)$ be a biholomorphic mapping with $f(z_0) = 0$ ($z_0 \in D$). Let g be a holomorphic function on D such that $|g(z)| < 1$ on D, and $g(z_0) = 0$. Show that $|g'(z)| \leqq |f'(z_0)|$.

EXERCISE 3. Show in Theorem (6.4.4) that if $a \in \mathbf{R}$ and D is symmetric with respect to the real axis, then $f(\bar{z}) = \overline{f(z)}$.

EXERCISE 4. Obtain a biholomorphic mapping from $D = \{z = x + iy; y^2 > 4c^2(x + c^2)\}$ $(c > 0)$ to the upper half plane $\mathbf{H}$.

6.5. Boundary Correspondence

The boundary of the upper half plane $\mathbf{H}$ in $\widehat{\mathbf{C}}$ consists of the real axis $\mathbf{R}$ and the infinity ∞. The two ends, $-\infty$ and $+\infty$, of $\mathbf{R}$ coincide with ∞.

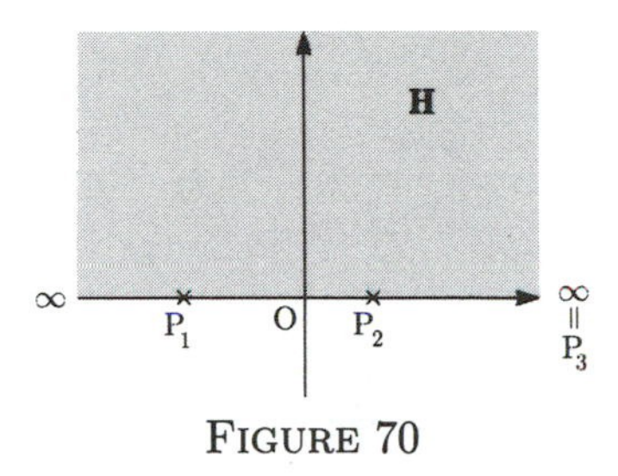

FIGURE 70

We endow $\partial\mathbf{H}$ with the orientation from $-\infty$ to $+\infty$ via 0. We take an ordered triple $(\mathrm{P}_1, \mathrm{P}_2, \mathrm{P}_3)$ of distinct points of $\partial\mathbf{H}$. Suppose that it is one of the following:

(6.5.1) i) $\mathrm{P}_3 = \infty$, $\mathrm{P}_1 < \mathrm{P}_2$;
ii) $\mathrm{P}_2 = \infty$, $\mathrm{P}_3 < \mathrm{P}_1$;
iii) $\mathrm{P}_1 = \infty$, $\mathrm{P}_2 < \mathrm{P}_3$;
iv) $\infty \neq \mathrm{P}_1 < \mathrm{P}_2 < \mathrm{P}_3 \neq \infty$.

In this case we say that the triple $(\mathrm{P}_1, \mathrm{P}_2, \mathrm{P}_3)$ has the *positive orientation.* By Theorem (6.1.4), ii) and Theorem (2.8.13) an arbitrary $f \in \mathrm{Aut}(\mathbf{H})$ is represented by

$$(6.5.2) \qquad f(z) = \frac{az+b}{cz+d}, \qquad \begin{pmatrix} a & b \\ c & d \end{pmatrix} \in SL(2, \mathbf{R}).$$

The restriction

$$f|\partial\mathbf{H} : \partial\mathbf{H} \to \partial\mathbf{H}$$

is injective and homeomorphic in the sense of the topology induced from $\widehat{\mathbf{C}}$. Since

$$f'(z) = \frac{ad-bc}{(cz+d)^2} = \frac{1}{(cz+1)^2},$$
$$f'(z) > 0, \qquad z \in \mathbf{R},$$

it follows that $f|\mathbf{R}$ is monotone increasing. Therefore, a triple $(\mathrm{P}_1, \mathrm{P}_2, \mathrm{P}_3)$ of positive orientation is mapped to a triple $(f(\mathrm{P}_1), f(\mathrm{P}_2), f(\mathrm{P}_3))$ of positive orientation. Hence we say that f *preserves the orientation* of $\partial\mathbf{H}$.

Now, cases ii)~iv) of (6.5.1) are reduced to case i) by a linear transformation

$$f(z) = \frac{-1}{z - \mathrm{P}_3} \in \mathrm{Aut}(\mathbf{H}).$$

Moreover, case i) is reduced to

$$(\mathrm{P}_1, \mathrm{P}_2, \mathrm{P}_3) = (0, 1, \infty) \tag{6.5.3}$$

by a linear transformation

$$f(z) = \frac{\frac{1}{\sqrt{\mathrm{P}_2 - \mathrm{P}_1}} z - \frac{\mathrm{P}_1}{\sqrt{\mathrm{P}_2 - \mathrm{P}_1}}}{\sqrt{\mathrm{P}_2 - \mathrm{P}_1}} = \frac{1}{\mathrm{P}_2 - \mathrm{P}_1}(z - \mathrm{P}_1) \in \mathrm{Aut}(\mathbf{H}).$$

That is, every triple $(\mathrm{P}_1, \mathrm{P}_2, \mathrm{P}_3)$ of positive orientation is mapped to $(0, 1, \infty)$ by the boundary correspondence of some $f \in \mathrm{Aut}(\mathbf{H})$. Conversely, assume that $f \in \mathrm{Aut}(\mathbf{H})$ fixes the triple $(0, 1, \infty)$:

$$f(0) = 0, \quad f(1) = 1, \quad f(\infty) = \infty.$$

As in (6.5.2), write

$$f(z) = \frac{az + b}{cz + d}, \qquad \begin{pmatrix} a & b \\ c & d \end{pmatrix} \in SL(2, \mathbf{R}).$$

Since $f(\infty) = \infty$, $c = 0$; $f(0) = 0$ implies $b = 0$. Since $f(1) = 1$, $a/d = 1$. On the other hand, $ad = 1$. Thus, $a = d = \pm 1$, and so $f(z) = z$. That is, if $f \in \mathrm{Aut}(\mathbf{H})$ fixes the triple $(0, 1, \infty)$ of positive orientation, then f must be the identity. Summarizing the above, we see that for two arbitrary triples $(\mathrm{P}_1, \mathrm{P}_2, \mathrm{P}_3)$ and $(\mathrm{P}'_1, \mathrm{P}'_2, \mathrm{P}'_3)$ of positive orientation, there exists a unique $f \in \mathrm{Aut}(\mathbf{H})$ such that

$$f(\mathrm{P}_1) = \mathrm{P}'_1, \quad f(\mathrm{P}_2) = \mathrm{P}'_2, \quad f(\mathrm{P}_3) = \mathrm{P}'_3.$$

The disk $\Delta(1)$ is biholomorphic to $\mathbf{H}$ by

$$\psi(z) = \frac{z - i}{z + i},$$
$$\psi : \mathbf{H} \to \Delta(1), \qquad \psi(i) = 0.$$

By ψ the boundary $\partial \mathbf{H} = \mathbf{R} \cup \{\infty\}$ corresponds to $\partial \Delta(1) = C(0; 1)$ as follows:

$$\psi(\infty) = 1, \qquad \psi(\mathbf{R}) = C(0; 1) \setminus \{1\}.$$

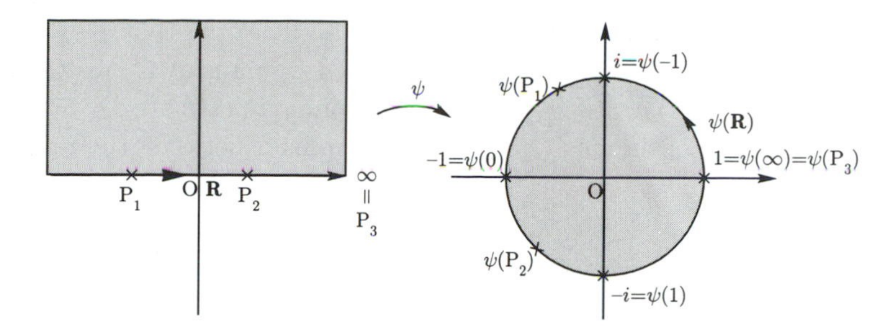

FIGURE 71

A triple (P_1, P_2, P_3) of positive orientation with $P_j \in \partial\mathbf{H}, 1 \leqq j \leqq 3$, is mapped by ψ to a triple $(\psi(P_1), \psi(P_2), \psi(P_3))$ of points of $C(0;1)$ in anti-clockwise order. Let (Q_1, Q_2, Q_3) be a triple of points of $C(0;1)$. We say that the triple (Q_1, Q_2, Q_3) of $C(0;1)$ is of positive orientation if they are in anti-clockwise order. A triple (Q_1, Q_2, Q_3) of positive orientation of $C(0;1)$ is mapped by ψ^{-1} to a triple $(\psi^{-1}(Q_1), \psi^{-1}(Q_2), \psi^{-1}(Q_3))$ of positive orientation of $\partial\mathbf{H}$. Thus we have proved the following theorem.

(6.5.4) THEOREM. *Let D be $\mathbf{H}$ or $\Delta(1)$. Then every $f \in \mathrm{Aut}(D)$ preserves the orientation of ∂D, and for arbitrary triples (P_1, P_2, P_3) and (P'_1, P'_2, P'_3) of positive orientation, there exists a unique $f \in \mathrm{Aut}(D)$ such that*

$$f(P_1) = P'_1, \quad f(P_2) = P'_2, \quad f(P_3) = P'_3.$$

Lemma (6.4.3) gives the uniqueness of holomorphic transformations at an interior point, and Theorem (6.5.4) does it by the boundary correspondence.

EXERCISE 1. Obtain $f \in \mathrm{Aut}(\mathbf{H})$, mapping $(0, 1, \infty)$ to $(-1, 0, 1)$.

EXERCISE 2. Obtain a biholomorphic mapping $f : \mathbf{H} \to \Delta(1)$, mapping $(0, 1, \infty)$ to $(1, i, -1)$.

(6.5.5) LEMMA. *Let $E_j \subset \widehat{\mathbf{C}}$, $j = 1, 2$, be subsets, and let $F : E_1 \to E_2$ be an injective continuous mapping. If E_1 is closed, then F is a homeomorphism.*

PROOF. It suffices to show the continuity of $F^{-1} : E_2 \to E_1$. Let $Q \in E_2$ be an arbitrary point, and let $\{Q_n\}_{n=0}^{\infty}$ be an arbitrary sequence converging to Q. Set $P = f^{-1}(Q)$, and $P_n = f^{-1}(Q_n)$, $n = 1, 2, \ldots$. Note that E_1 is compact. Let P' be an accumulation point of $\{P_n\}_{n=0}^{\infty}$. It is sufficient to show that $P' = P$. Let $\{P_{n_\nu}\}_{\nu=0}^{\infty}$ be a subsequence of $\{P_n\}_{n=0}^{\infty}$ which converges to P. Since F is continuous,

$$F(P') = \lim_{\nu\to\infty} F(P_{n_\nu}) = \lim_{\nu\to\infty} Q_{n_\nu} = Q = F(P).$$

The injectivity of F implies $P' = P$. □

Let $D \subset \widehat{\mathbf{C}}$ be a simply connected domain, and assume that $\overline{D} \neq \widehat{\mathbf{C}}$; that is, D has an exterior point. If D is not a bounded domain of $\mathbf{C}$, we take a point $c \in \mathbf{C} \backslash \overline{D}$, and set $f(z) = 1/(z-c)$. Then D is biholomorphic to $D_1 = f(D) \Subset \mathbf{C}$, and f is extended continuously to a homeomorphism between the boundaries:

$$f : \overline{D} \to \overline{D}_1. \tag{6.5.6}$$

(6.5.7) THEOREM. *Let $D \subset \widehat{\mathbf{C}}$ be a simply connected domain such that $\overline{D} \neq \widehat{\mathbf{C}}$. Assume that the boundary ∂D is given by a closed Jordan curve $C(\phi : [0,1] \to \widehat{\mathbf{C}})$ (that is, $\partial D = \phi([0,1])$). Then every biholomorphic mapping $f : \Delta(1) \to D$ is extended to a homeomorphism from $\overline{\Delta(1)}$ to $\overline{D}$.*

PROOF. By (6.5.6) we may assume that D is a bounded domain of $\mathbf{C}$. Take an arbitrary point $\zeta \in C(0;1)$. For $\rho > 0$ we set

$$\begin{aligned} \Omega_\rho &= \Delta(1) \cap \Delta(\zeta;\rho), & \Gamma_\rho &= C(\zeta;\rho), \\ \omega_\rho &= f(\Omega_\rho), & \gamma_\rho &= f(\Gamma), \\ A(\rho) &= \text{the area of } \omega_\rho, & L(\rho) &= \text{the length of } \gamma_\rho. \end{aligned}$$

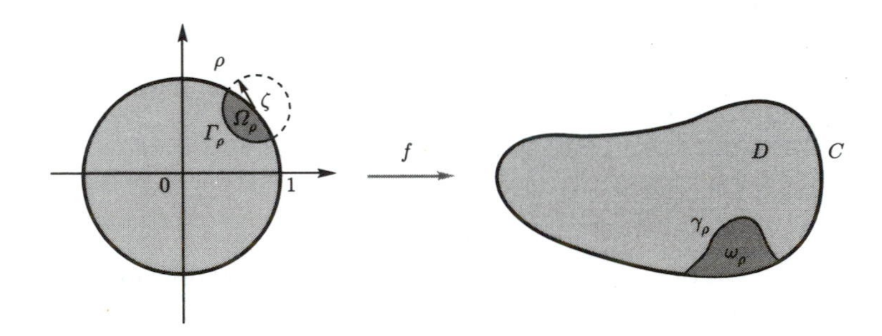

FIGURE 72

It is sufficient to prove that ω_ρ accumulates to one point of C as $\rho \to 0$. Furthermore, it is sufficient to prove it for a sequence $\{\rho_n\}_{n=1}^\infty$ ($\rho_n > 0$), converging to 0. Since $\lim_{\rho \to 0} A(\rho) = 0$, for an arbitrary $\epsilon > 0$ there exists $\delta(\epsilon) > 0$ such that

$$A(\rho) < \epsilon, \qquad 0 < \rho < \delta(\epsilon).$$

By Schwarz' inequality we have

$$\begin{aligned} \int_0^\delta L(\rho)d\rho &= \iint_{\Omega_\delta} |f'(\zeta + te^{i\theta})| t dt d\theta \\ &\leqq \left(\iint_{\Omega_\delta} |f'(\zeta + te^{i\theta})|^2 t dt d\theta \right)^{1/2} \left(\iint_{\Omega_\delta} t dt d\theta \right)^{1/2} \\ &< \sqrt{A(\delta)}\sqrt{\pi\delta^2} \\ &< \delta\sqrt{\pi\epsilon}, \qquad 0 < \delta < \delta(\epsilon). \end{aligned}$$

Therefore, for an arbitrary $0 < \delta < \delta(\epsilon)$, there is some $\rho \in (0, \delta)$ such that

$$L(\rho) < \sqrt{\pi\epsilon}$$

Hence, for a positive sequence $\{\epsilon_n\}_{n=1}^{\infty}$ converging to 0, there exists a monotone decreasing positive sequence $\{\rho_n\}_{n=1}^{\infty}$ converging to 0 such that

$$A(\rho_n) < \epsilon_n, \qquad L(\rho_n) < \sqrt{\pi\epsilon_n}. \tag{6.5.8}$$

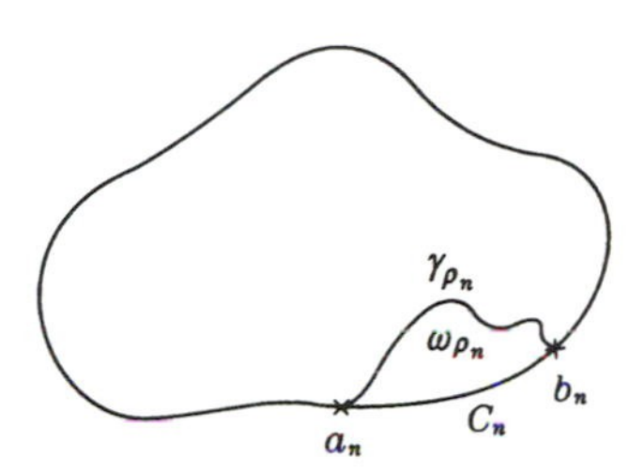

FIGURE 73

As a point $z \in \Gamma_{\rho_n}$ approaches on Γ_{ρ_n} to its end points $\alpha_n, \beta_n \in \overline{\Gamma}_{\rho_n} \cap C(0;1)$, $f(z) \in \gamma_{\rho_n}$ converges to points $a_n, b_n \in \overline{\gamma}_{\rho_n} \cap C$, respectively; otherwise, γ_{ρ_n} should oscillate infinitely times between some positive distance, and hence the length of γ_{ρ_n} would be infinite. This contradicts (6.5.8). It follows again from (6.5.8) that

$$|a_n - b_n| < \sqrt{\pi\epsilon}. \tag{6.5.9}$$

Set $C_n = \overline{\omega}_{\rho_n} \cap C$ $(\ni a_n, b_n)$. Then

$$\partial\omega_{\rho_n} = \gamma_{\rho_n} \cup C_n, \qquad C_n \supset C_{n+1}. \tag{6.5.10}$$

The sets C_n are closed subsets of C, and connected. Assume that $a_{n_1} = b_{n_1}$ for some n_1.

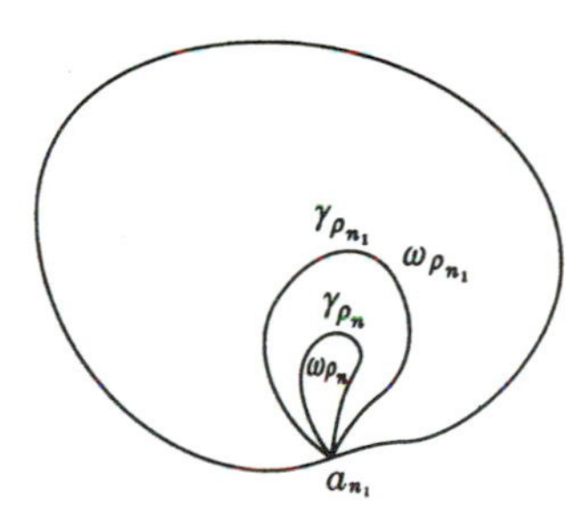

FIGURE 74

For $n \geqq n_1$ we have

$$C_n = \{a_{n_1}\}, \qquad \partial\omega_{\rho_n} = \gamma_{\rho_n} \cup \{a_{n_1}\}. \tag{6.5.11}$$

Therefore, the diameter of ω_{ρ_n} is less than $L(\rho_n)$, and by (6.5.8)

$$\omega_{\rho_n} \longrightarrow a_{n_1}, \qquad n \to \infty.$$

This means that for an arbitrary $\epsilon > 0$ there is a number n_0 such that for every $n \geqq n_0$, $\omega_{\rho_n} \subset \Delta(a_{n_1}; \epsilon)$.

Assume that $a_n \neq b_n$ for any n. We write S^1 for $C(0;1)$. There is an injective and surjective continuous mapping

$$\phi : S^1 \to \partial D = C.$$

It follows from Lemma (6.5.5) that the inverse mapping $\phi^{-1} : C \to S^1$ is continuous. Hence, $\phi^{-1}(C_n) = I_n$ is a connected closed subset of S^1. The points $a_n, b_n \in C$ divide C into two curves C'_n, C''_n. One of them is C_n. Let s_n, t_n be the end points of I_n. Then it follows from (6.5.9) and (6.5.10) that

$$|s_n - t_n| \to 0, \qquad I_n \supset I_{n+1} \supset \cdots.$$

Therefore, $\bigcap_{n=1}^{\infty} I_n = \{\tau\}$ with $\tau \in S^1$, and so

$$C_n \to c = \phi(\tau), \qquad n \to \infty.$$

Since $\lim_{n\to\infty} a_n = \lim_{n\to\infty} b_n = c$, and $\lim_{n\to\infty} L(\gamma_{\rho_n}) = 0$, we have by the above and (6.5.10) that

$$\omega_{\rho_n} \to c, \qquad n \to \infty.$$

Thus, f is extended to a continuous mapping $f : \overline{\Delta(1)} \to \overline{D}$, and $f(\overline{\Delta(1)}) = \overline{D}$.

In the above proof, the case of (6.5.11) does not occur. For, if $a_{n_1} = b_{n_1}$, f takes the value a_{n_1} constantly on the arc $S^1 \cap \Delta(\zeta; \rho_{n_1})$. It follows from Chapter 3, §5, Exercise 5 that $f(z) \equiv a_{n_1}$, $z \in \Delta(1)$. This is a contradiction. Therefore we see that f is injective on the boundary, and so by Lemma (6.5.5) $f : \overline{\Delta(1)} \to \overline{D}$ is a homeomorphism. □

Let D be as in the above theorem, and let $f_0 : \Delta(1) \to D$ be a biholomorphic mapping. Extend it to a homeomorphism $f_0 : \overline{\Delta(1)} \to \overline{D}$. We say that a triple $(\mathrm{Q}_1, \mathrm{Q}_2, \mathrm{Q}_3)$ of points of ∂D is of *positive orientation* if $(f_0^{-1}(\mathrm{Q}_1), f_0^{-1}(\mathrm{Q}_2), f_0^{-1}(\mathrm{Q}_3))$ is of positive orientation. This is a property independent of the choice of f_0 (Theorem (6.5.4)). We immediately have the following by Theorems (6.4.4), (6.5.4), and (6.5.7)

(6.5.12) THEOREM. *Let $D \subset \widehat{\mathbf{C}}$ be a simply connected domain such that $\overline{D} \neq \widehat{\mathbf{C}}$ and the boundary ∂D is given by a Jordan closed curve. Let $(\mathrm{P}_1, \mathrm{P}_2, \mathrm{P}_3)$ be a triple of positive orientation of $\partial\Delta(1)$, and let $(\mathrm{Q}_1, \mathrm{Q}_2, \mathrm{Q}_3)$ be a triple of positive orientation of ∂D. Then there exists a unique biholomorphic mapping $f : \Delta(1) \to D$ such that $f(\mathrm{P}_i) = \mathrm{Q}_i, 1 \leqq i \leqq 3$.*

REMARK 1. The above theorem holds with $\Delta(1)$ replaced by $\mathbf{H}$.

REMARK 2. If Jordan's Theorem (3.3.8) is admitted, it is not necessary to assume the simple connectedness of D or $\overline{D} \neq \widehat{\mathbf{C}}$ in Theorems (6.5.4) and (6.5.12).

For an arbitrary simply connected domain $D \subsetneq \mathbf{C}$, Carathéodory topologically defined the notion of boundary elements, and proved that every boundary element corresponds to one point of $\partial\Delta(1)$ through a biholomorphic mapping $f : \Delta(1) \to D$; this gives a complete description of the boundary correspondence of f.

EXERCISE 3. Let D be as in Theorem (6.5.7). Let $h(z)$ be an arbitrary real-valued continuous function on ∂D. Show that there exists a unique continuous function $u(z)$ on $\overline{D}$ such that $u(z)$ is harmonic in D and has the boundary value $h(z)$.

6.6. Universal Covering of $\mathbf{C}\setminus\{0,1\}$

As shown in Chapter 5, §3, for an arbitrary domain $D \subset \widehat{\mathbf{C}}$ there exists the universal covering $\pi : X \to D$ of D. The space X has a structure of Riemann surface such that π gives rise to a holomorphic mapping. Here we are going to construct X for $D = \mathbf{C}\setminus\{0,1\}$.

Let C_1 be the half-line $\{\infty\} \cup \{z \in \mathbf{C}; \operatorname{Re} z = 0, \operatorname{Im} z \geqq 0\}$ from ∞ to 0, let C_2 be the half-circle $C(1/2;1/2) \cap \{\operatorname{Im} z \geqq 0\}$ from 0 to 1, and let C_3 be the half-line $\{z \in \mathbf{C}; \operatorname{Re} z = 1, \operatorname{Im} z \geqq 0\} \cup \{\infty\}$ from 1 to ∞. Let L_1 be the half-line $\{\infty\} \cup \{z \in \mathbf{C}; \operatorname{Re} z \leqq 0, \operatorname{Im} z = 0\}$ from ∞ to 0, let L_2 be the real line segment from 0 to 1, and let L_3 be the half-line $\{z \in \mathbf{C}; \operatorname{Re} z \geqq 1, \operatorname{Im} z = 0\} \cup \{\infty\}$ from 1 to ∞.

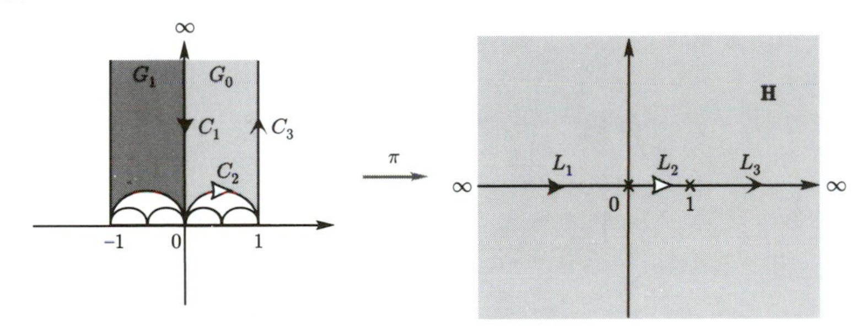

FIGURE 75

Let G_0 be the domain bounded by C_1, C_2 and C_3. By Theorem (6.5.12) there exists a unique biholomorphic mapping $\pi : G_0 \to \mathbf{H}$ with the homeomorphic extension $\pi : \overline{G_0} \to \overline{\mathbf{H}}$ such that $\pi(\infty) = \infty$, $\pi(0) = 0$, and $\pi(1) = 1$. Therefore we have

$$\pi(C_1) = L_1, \quad \pi(C_2) = L_2, \quad \pi(C_3) = L_3. \tag{6.6.1}$$

We shall obtain the universal covering $\pi : \mathbf{H} \to \mathbf{C}\setminus\{0,1\}$ by analytic continuation of π. We denote by G_1 the domain which is the reflection of G_0 with respect to C_1. By the reflection principle, Theorem (5.1.7), π analytically extends, crossing the set $\overset{\circ}{C}_1 = \{z \in \mathbf{C}; \operatorname{Re} z = 0, \operatorname{Im} z > 0\}$ of the interior points of C_1, to a holomorphic mapping from G_1 to the lower half-plane $\mathbf{L} = \{z \in \mathbf{C}; \operatorname{Im} z < 0\}$. Thus we get a holomorphic mapping $\pi : G_0 \cup \overset{\circ}{C}_1 \cup G_1 \to \mathbf{H} \cup \overset{\circ}{L}_1 \cup \mathbf{L}$. We carry out this analytic continuation for the other boundaries, C_2 and C_3, and again do the same for the resulting images of $C_j, 1 \leqq j \leqq 3$. Let $G_n, n = 1, 2, \ldots$, denote all the domains obtained by the above process of reflections of G_0. Then, we get

$$\mathbf{H} = \bigcup_{n=0}^{\infty} (\overline{G}_n \cap \mathbf{H}),$$

and a holomorphic mapping

$$\lambda : \mathbf{H} \to \mathbf{C} \setminus \{0, 1\}. \tag{6.6.2}$$

It is clear that the above λ satisfies condition (5.3.3) of a covering mapping. Since $\mathbf{H}$ is simply connected, (6.6.2) is the universal covering. Set $E = (\overline{G}_1 \cap \mathbf{H}) \cup G_0$. Then, the set of the interior points of E forms a domain, $\lambda(E) = \mathbf{C} \setminus \{0, 1\}$, and for every $z \in E$

$$\lambda^{-1}(\lambda(z)) \cap E = \{z\}.$$

We call such E a *fundamental domain* of the universal covering (6.6.2). Thus, we see that $\mathbf{C} \setminus \{0, 1\}$ is a hyperbolic domain. It follows from Theorem (6.3.3) that

(6.6.3) $\mathbf{C} \setminus \{0, 1\}$ carries a hyperbolic metric $h_{\mathbf{C} \setminus \{0,1\}}$, which is complete.

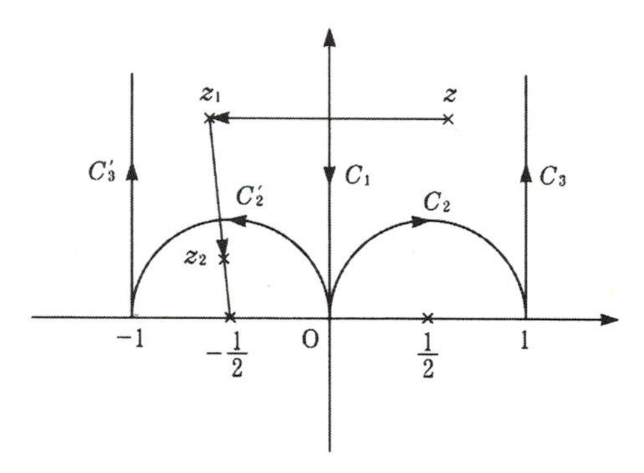

FIGURE 76

The deck transformation group $\pi_1(\mathbf{C} \setminus \{0, 1\})$ of the universal covering (6.6.2) is a subgroup of $\operatorname{Aut}(\mathbf{H}) = PSL(2, \mathbf{R})$. We determine what it is. Consider the transformation for $w \in \mathbf{C} \setminus \{0, 1\}$ obtained by the reflections with respect to $L_j, 1 \leqq j \leqq 3$. Then the point w is mapped to itself only by an even number of repetitions of such reflections. Thus we see that an element of $\pi_1(\mathbf{C} \setminus \{0, 1\}) \subset \operatorname{Aut}(\mathbf{H})$ is a transformation obtained by an even number of repetitions of reflections with respect to $C_j, 1 \leqq j \leqq 3$. For example, we take the reflection of G_0 with respect to C_1, and then the reflection with respect to the image C_2' of C_2, and denote it by T_1 (cf. Figure 76):

the reflection with respect to C_1	$z = x + iy \to z_1 = -x + iy = -\overline{z}$,
the image C_2' of C_2	$\left(x + \frac{1}{2}\right)^2 + y^2 = \left(\frac{1}{2}\right)^2$,
the reflection with respect to C_2'	$z_1 \to z_2$,
	$\left(\overline{z}_1 + \frac{1}{2}\right)\left(z_2 + \frac{1}{2}\right) = \left(\frac{1}{2}\right)^2$.

Therefore we have

$$\begin{aligned} z_2 + \frac{1}{2} &= \frac{1}{4} \cdot \frac{1}{\overline{z}_1 + \frac{1}{2}} = \frac{1}{4} \cdot \frac{1}{-z + \frac{1}{2}}, \\ z_2 &= \frac{1}{-4z + 2} - \frac{1}{2} = \frac{1}{2}\left(\frac{1}{-2z+1} - 1\right) \\ &= \frac{1}{2}\left(\frac{2z}{-2z+1}\right) = \frac{z}{-2z+1} \\ &= T_1(z). \end{aligned}$$

Take a point $\mathrm{P}_0 \in \mathbf{C} \setminus \{0,1\}$; for instance, $\mathrm{P}_0 = i$. Then, the transformation T_1 corresponds to a Jordan closed curve $\alpha_1 \subset \mathbf{C} \setminus \{0,1\}$ from P_0 with the anti-clockwise orientation, crossing $\mathring{L}_1$ and $\mathring{L}_2$, and then coming back to P_0.

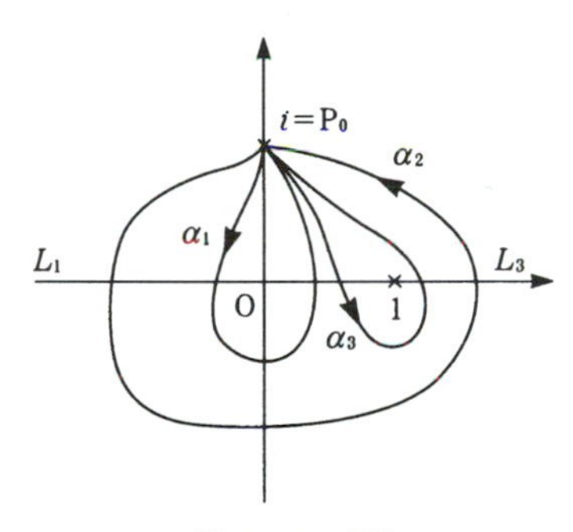

FIGURE 77

In the same way, let C_3' be the reflection image of C_3 with respect to C_1, and let T_2 denote the composition of the reflections with respect to C_1 and C_3'. Then

$$T_2(z) = z - 2.$$

The transformation T_2 corresponds to a Jordan curve $\alpha_2 \subset \mathbf{C} \setminus \{0,1\}$ from P_0 which crosses $\mathring{L}_1$, goes into $\mathbf{L}$, crosses $\mathring{L}_3$, and comes back to P_0. If α_3 denotes a Jordan curve from P_0 which crosses $\mathring{L}_2$ and $\mathring{L}_3$, going around 1 anti-clockwisely, we have

$$\{\alpha_2\} = \{\alpha_1\} + \{\alpha_3\}.$$

Note that T_1 is given by $A_1 = \left(\begin{smallmatrix} 1 & 0 \\ -2 & 1 \end{smallmatrix}\right) \in SL(2, \mathbf{R})$, and T_2 by $A_2 = \left(\begin{smallmatrix} 1 & -2 \\ 0 & 1 \end{smallmatrix}\right) \in SL(2, \mathbf{R})$. Hence, we get

$$A_1, A_2 \in \Gamma(2).$$

(6.6.4) LEMMA. *The group $\Gamma(2)$ is generated by $\pm A_1$ and $\pm A_2$; that is, every $A \in \Gamma(2)$ is expressed as $A = \pm A_1^{m_1} A_2^{n_1} A_1^{m_2} A_2^{n_2} \cdots A_1^{m_l} A_2^{n_l}$, $m_j, n_j \in \mathbf{Z}$, $1 \leqq j \leqq l < \infty$.*

PROOF. We have that $A_1^{-1} = \left(\begin{smallmatrix} 1 & 0 \\ 2 & 1 \end{smallmatrix}\right)$, $A_2^{-1} = \left(\begin{smallmatrix} 1 & 2 \\ 0 & 1 \end{smallmatrix}\right)$, and

$$A_1^n = \begin{pmatrix} 1 & 0 \\ -2n & 1 \end{pmatrix}, \quad A_2^n = \begin{pmatrix} 1 & -2n \\ 0 & 1 \end{pmatrix}, \qquad n \in \mathbf{Z}.$$

Take an arbitrary $A = \left(\begin{smallmatrix} a & b \\ c & d \end{smallmatrix}\right) \in \Gamma(2)$. We have

$$AA_2^n = \begin{pmatrix} a & b - 2na \\ c & d - 2nc \end{pmatrix} = \begin{pmatrix} a & b' \\ c & d' \end{pmatrix}.$$

By the condition, $b' = b - 2na \neq \pm a$, $n \in \mathbf{Z}$. One may take $n = n_1 \in \mathbf{Z}$ so that

$$|b'| < |a|. \tag{6.6.5}$$

Next, we get

$$AA_2^{n_1} A_1^m = \begin{pmatrix} a & b' \\ c & d' \end{pmatrix} \begin{pmatrix} 1 & 0 \\ -2m & 1 \end{pmatrix} = \begin{pmatrix} a - 2mb' & b' \\ c - 2md' & d' \end{pmatrix} = \begin{pmatrix} a' & b' \\ c' & d' \end{pmatrix}.$$

If $b' \neq 0$, one may take $m = m_1 \in \mathbf{Z}$ so that

$$|a'| < |b'|. \tag{6.6.6}$$

Set $A^{(0)} = A, A^{(1)} = AA_2^{n_1} A_1^{m_1} = \left(\begin{smallmatrix} a^{(1)} & b^{(1)} \\ c^{(1)} & d^{(1)} \end{smallmatrix}\right)$. Inductively, one obtains

$$A^{(k)} = \begin{pmatrix} a^{(k)} & b^{(k)} \\ c^{(k)} & d^{(k)} \end{pmatrix} = A^{(k-1)} A_2^{n_k} A_1^{m_k},$$
$$|a^{(k)}| < |b^{(k)}| < |a^{(k-1)}| < |b^{(k-1)}|, \qquad k = 2, 3, \ldots$$

(cf. (6.6.5) and (6.6.6)). The sequence $\{|b^{(k)}|\}_{k=0}^{\infty}$ is strictly decreasing, and one may continue this process so long as $b^{(k)} \neq 0$. Hence, there is a number $l \in \mathbf{Z}$ with $b_l = 0$, so that

$$A^{(l)} = \begin{pmatrix} a^{(l)} & 0 \\ c^{(l)} & d^{(l)} \end{pmatrix} = \pm A_1^{-c^{(l)}/2}.$$

Therefore, one has an expression $A = \pm A_1^{r_1} A_2^{s_1} A_1^{r_2} A_2^{s_2} \cdots A_1^{r_l} A_2^{s_l}$, $r_j, s_j \in \mathbf{Z}$, $1 \leqq j \leqq l$. $\square$

Summarizing the above, we have the following:

(6.6.7) THEOREM. *The universal covering of* $\mathbf{C}\setminus\{0,1\}$ *is the upper half plane* $\mathbf{H}$ *(or the disk* $\Delta(1)$*), and the deck transformation group* $\pi_1(\mathbf{C}\setminus\{0,1\})$ *is isomorphic to* $\Gamma(2)/\{\pm\left(\begin{smallmatrix}1&0\\0&1\end{smallmatrix}\right)\}$*;*

$$\mathbf{C}\setminus\{0,1\}=\mathbf{H}/\Gamma(2).$$

In particular, $\mathbf{C}\setminus\{0,1\}$ *carries a complete hyperbolic metric.*

The function $w=\lambda(z)$ given by (6.6.2) is called the *lambda function*, and is one of those called modular functions. It has a close relationship with elliptic functions which will be dealt with in Chapter 7.

EXERCISE 1. Let D and $\tilde{D}$ be domains of $\widehat{\mathbf{C}}$, and let $\pi:\tilde{D}\to D$ be the universal covering. Show that if D is not biholomorphic to $\widehat{\mathbf{C}}$, $\mathbf{C}$, or $\mathbf{C}^*$, then $\tilde{D}$ is biholomorphic to $\Delta(1)$.

6.7. The Little Picard Theorem

If a rational function $f(z)$ is not a constant, for any $\alpha\in\mathbf{C}$ the equation $f(z)=\alpha$ has a solution $z\in\mathbf{C}$, and if f has a pole, then f has the value ∞ there. However, this is not the case in general, if f is a holomorphic or meromorphic function on $\mathbf{C}$. For instance, the entire function $f(z)=e^z$ does not take the value 0, nor, of course, the infinity ∞. Thus, the number of points which f misses is two. In general, if a meromorphic function f, regarded as a holomorphic mapping into $\widehat{\mathbf{C}}$, misses a point of $\widehat{\mathbf{C}}$, that point is called an *exceptional value.* The next theorem is called the *little Picard theorem*, and shows that the above "two" is the maximum.

(6.7.1) THEOREM. *The number of exceptional values of a non-constant meromorphic function* f *on* $\mathbf{C}$ *is at most two.*

PROOF. Assume that the number of exceptional values of f is more than two. Then we are going to show that f reduces to a constant. Let three distinct points a, b, and c of $\widehat{\mathbf{C}}$ be exceptional values of f. By a linear transformation, we may assume that $\{a,b,c\}=\{0,1,\infty\}$. Thus we have a holomorphic mapping

$$f:\mathbf{C}\to\mathbf{C}\setminus\{0,1\}.$$

It follows from Theorem (6.6.7) that the universal covering of $\mathbf{C}\setminus\{0,1\}$ is the unit disk $\Delta(1)$. Let $\pi:\Delta(1)\to\mathbf{C}\setminus\{0,1\}$ be the universal covering mapping. Since $\mathbf{C}$ is simply connected, it follows from Theorem (5.3.18) and (5.3.21) that f has a holomorphic lifting

$$\hat{f}:\mathbf{C}\to\Delta(1),\qquad \pi\circ\hat{f}(z)=f(z).$$

Liouville's Theorem (3.5.22) implies that $\hat{f}$ is constant, so that f is constant. $\square$

By the same idea as above, Montel's Theorem (6.4.2) can be greatly improved, as Montel himself showed. We first generalize the notion of normal families.

Let $D \subset \widehat{\mathbf{C}}$ be a domain, and let $f_n : D \to \widehat{\mathbf{C}}$, $n = 1, 2, \ldots$, be a sequence of holomorphic mappings. We say that $\{f_n\}$ converges uniformly on every compact subset to $f : D \to \widehat{\mathbf{C}}$ if for any point $\mathrm{P} \in D$ there are a neighborhood $U \subset D$ of P and a number n_0 such that

$$(6.7.2) \qquad f_n(U) \subset \widehat{\mathbf{C}} \setminus \{\infty\}, \qquad n \geqq n_0,$$

or

$$(6.7.3) \qquad f_n(U) \subset \widehat{\mathbf{C}} \setminus \{0\}, \qquad n \geqq n_0,$$

and that in the case of (6.7.2) (resp., (6.7.3)) the sequence $\{f_n|U\}$ (resp., $\{\tilde{w} \circ f_n|U\} = \{1/f_n|U\}$) of holomorphic functions converges uniformly on every compact subset of U to f (resp., $\tilde{w} \circ f$). In this case, f is, of course, a holomorphic mapping. A family $\mathcal{F}$ of holomorphic mappings from D into $\widehat{\mathbf{C}}$ is said to be *normal* if each sequence in $\mathcal{F}$ has a subsequence which converges uniformly on every compact subset.

(6.7.4) REMARK. Lemma (2.2.7) holds for a domain $D \subset \widehat{\mathbf{C}}$. In fact, if $D = \widehat{\mathbf{C}}$, D is itself compact; if $D \neq \widehat{\mathbf{C}}$, by a linear transformation, D is homeomorphic to a domain of $\mathbf{C}$ for which Lemma (2.2.7) is valid.

(6.7.5) LEMMA. *Let $\mathcal{F}$ be a family of holomorphic mappings from D into $\widehat{\mathbf{C}}$. For $\mathcal{F}$ to be normal, it is necessary and sufficient that for an arbitrary point $\mathrm{P} \in D$ there exists a neighborhood $U \subset D$ of P such that the restriction $\mathcal{F}|U = \{f|U; f \in \mathcal{F}\}$ of $\mathcal{F}$ to U is normal.*

PROOF. The necessity is clear. By Remark (6.7.4) we take $U_n \subset D$, $n = 1, 2, \ldots$, as in Lemma (2.2.7). Take a sequence $\{f_\nu\}$ in $\mathcal{F}$. For a point $\mathrm{P} \in \overline{U}_n$ there is a neighborhood $V_{\mathrm{P}} \subset D$ of P such that $\{f_\nu|V_{\mathrm{P}}\}$ contains a subsequence which converges uniformly on V_{P}. Since $\overline{U}_n$ is compact, there are a finite number of $\mathrm{P}_j \in \overline{U}_n$ and $V_j = V_{\mathrm{P}_j}$, $1 \leqq j \leqq l$, such that

$$\overline{U}_n \subset \bigcup_{j=1}^{l} V_j.$$

Let $\{f_{1\nu}\}$ be a subsequence of $\{f_\nu\}$ which converges uniformly on V_1. Let $\{f_{2\nu}\}$ be a subsequence of $\{f_{1\nu}\}$ which converges uniformly on V_2. Inductively, we get a subsequence $\{f_{l\nu}\}_{\nu=1}^{\infty}$ of $\{f_\nu\}$ which converges uniformly on $\overline{U}_n$. For $n = 1$, we denote by $\{f_{(1)\nu}\}$ the subsequence of $\{f_\nu\}$ obtained as above. For $n = 2$ we similarly take a subsequence $\{f_{(2)\nu}\}$ of $\{f_{(1)\nu}\}$. Inductively, we get $\{f_{(n)\nu}\}_{\nu=1}^{\infty}$, $n = 1, 2, \ldots$. Then, the subsequence $\{f_{(\nu)\nu}\}$ of $\{f_\nu\}$ converges uniformly on every compact subset of D. □

The next theorem is due to *Montel.*

(6.7.6) THEOREM. *Let $\mathcal{F}$ be a family of holomorphic mappings from D into $\widehat{\mathbf{C}}$. If there are three distinct points $a, b, c \in \widehat{\mathbf{C}}$ which are exceptional values for all $f \in \mathcal{F}$, then $\mathcal{F}$ is normal.*

PROOF. We take an arbitrary point $\mathrm{P} \in D$. By Lemma (6.7.5) it suffices to show that there is a neighborhood $U \subset D$ of P such that $\mathcal{F}|U$ is normal. Let $\{f_n\}_{n=1}^{\infty}$ be a sequence in $\mathcal{F}$. Since $\widehat{\mathbf{C}}$ is compact, we may assume that $\{f_n(\mathrm{P})\}$ converges to a point $\mathrm{Q} \in \widehat{\mathbf{C}}$. By a linear transformation we may also assume that $\{a, b, c\} = \{0, 1, \infty\}$. It follows from Theorems (6.6.7) and (6.3.3) that $\widehat{\mathbf{C}} \setminus \{0, 1, \infty\} = \mathbf{C} \setminus \{0, 1\}$ carries a complete hyperbolic metric $d_{\mathbf{C}\setminus\{0,1\}}$. We assume that $\mathrm{P} \in \mathbf{C}$; if $\mathrm{P} = \infty$ we make the same arguments using the complex coordinate $\tilde{z}$. For the sake of simplicity, we may assume after a translation that $\mathrm{P} = 0$. Take $r > 0$ so that $\Delta(r) \subset D$. For a point $z \in \overline{\Delta(r/2)}$ we take a line segment $C_z(\phi : t \in [0, 1] \to tz)$. Then by (6.2.13) we have

$$\begin{aligned} d_{\Delta(r)}(0, z) &= L_{\Delta(r)}(C_z) \\ &= \log \frac{r + |z|}{r - |z|} \leqq \log 3. \end{aligned}$$

It follows from Theorem (6.3.10) that

$$d_{\mathbf{C}\setminus\{0,1\}}(f_n(0), f_n(z)) \leqq d_{\Delta(r)}(0, z) \leqq \log 3. \tag{6.7.7}$$

(6.7.8) The case of $\mathrm{Q} = \lim_{n\to\infty} f_n(0) \in \{0, 1, \infty\}$:

For instance, we suppose that $\mathrm{Q} = 0$. Since $d_{\mathbf{C}\setminus\{0,1\}}$ is complete, for an arbitrary $\epsilon > 0$ we have

$$\begin{aligned} d_{\mathbf{C}\setminus\{0,1\}}(f_n(0), C(0; \epsilon)) &= \min\{d_{\mathbf{C}\setminus\{0,1\}}(f_n(0), w); w \in C(0; \epsilon)\} \\ &\to +\infty \qquad (n \to \infty). \end{aligned}$$

It follows from (6.7.7) that there is a number n_0 such that

$$f_n(\overline{\Delta(r/2)}) \subset \Delta(0; \epsilon), \qquad n \geqq n_0.$$

Thus $\{f_n|\Delta(r/2)\}$ converges uniformly to 0. The other cases of $\mathrm{Q} = 1, \infty$ are discussed similarly.

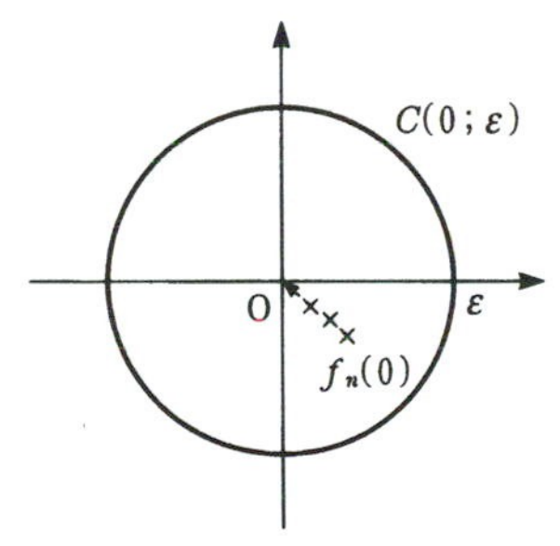

FIGURE 78

(6.7.9) The case of $Q \in \mathbf{C} \setminus \{0, 1\}$:

There is a number n_0 such that

$$d_{\mathbf{C}\setminus\{0,1\}}(f_n(0), Q) < 1, \qquad n \geqq n_0.$$

Thus, by (6.7.7)

$$d_{\mathbf{C}\setminus\{0,1\}}(Q, f_n(z)) < 1 + \log 3, \qquad z \in \Delta(r/2), \quad n \geqq n_0.$$

Since $d_{\mathbf{C}\setminus\{0,1\}}$ is complete, the set $\{w \in \mathbf{C} \setminus \{0, 1\}; d_{\mathbf{C}\setminus\{0,1\}}(Q, w) \leqq 1 + \log 3\}$ is compact; in particular, it is a bounded set of $\mathbf{C}$. Therefore, $\{f_n|\Delta(r/2)\}$ is uniformly bounded, and Theorem (6.4.2) implies that $\{f_n|\Delta(r/2)\}$ contains a subsequence which converges uniformly on compact subsets. □

EXERCISE 1. Let $\mathcal{F}$ be a family of holomorphic functions f on a domain $D \subset \widehat{\mathbf{C}}$ such that for some fixed number $\alpha \in \mathbf{R}$, $\operatorname{Im} f(z) > \alpha$. Show that $\mathcal{F}$ is normal.

EXERCISE 2. Prove the little Picard Theorem (6.7.1) by making use of Montel's Theorem (6.7.6).

EXERCISE 3. Show by an example that the number three of exceptional values in Montel's Theorem (6.7.6) cannot be smaller.

6.8. The Big Picard Theorem

In this section we write

$$\Delta^*(r) = \Delta(r) \setminus \{0\} \quad (r > 0), \qquad \Delta^* = \Delta^*(1).$$

The universal covering of Δ^* is given by

(6.8.1) $$\pi : \mathbf{H} \ni z \to e^{2\pi i z} \in \Delta^*.$$

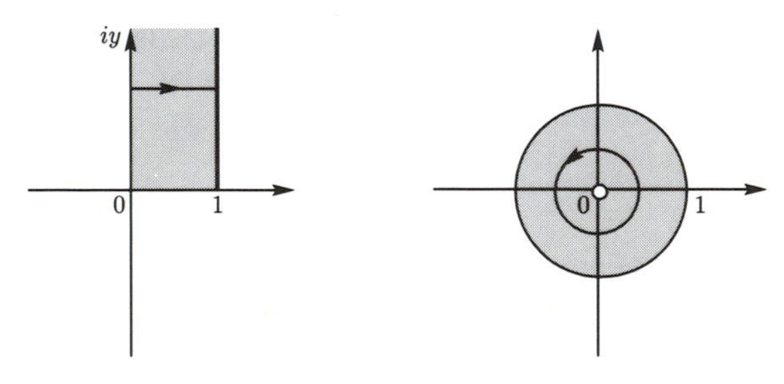

FIGURE 79

A fundamental domain is

$$\{z = x + iy; 0 \leqq x < 1, y > 0\}.$$

As described in §3, Δ^* carries the complete hyperbolic metric h_{Δ^*}, which is, in view of (6.2.18) and (6.8.1), expressed by

$$h_{\Delta^*}(z) = \frac{4}{|z|^2(\log|z|^2)^2}|dz|^2, \qquad z \in \Delta^*. \tag{6.8.2}$$

The hyperbolic length $L_{\Delta^*}(C(0;r))$ of the circle $C(0;r)$ $(0 < r < 1)$ with respect to h_{Δ^*} (cf. (5.3.24)) is given by

$$L_{\Delta^*}(C(0;r)) = \int_0^{2\pi} \frac{2}{r|\log r^2|} r d\theta = \frac{2\pi}{|\log r|}. \tag{6.8.3}$$

The area $A_{\Delta^*}(\Delta^*(r))$ of $\Delta^*(r)$ with respect to h_{Δ^*} (cf. (5.3.25)) is given by

$$\begin{aligned} A_{\Delta^*}(\Delta^*(r)) &= \int_{\Delta^*(r)} \frac{4}{t^2(\log t^2)^2} t dt d\theta \\ &= 2\pi \int_0^r \frac{1}{(\log t)^2} d\log t = 2\pi \left[-\frac{1}{\log t}\right]_0^r \\ &= \frac{2\pi}{|\log r|}. \end{aligned} \tag{6.8.4}$$

We generalize the notion of isolated essential singularities of holomorphic functions for meromorphic functions. Let $\mathrm{P} \in \widehat{\mathbf{C}}$ be a point, and let U be a neighborhood of P. Let f be a meromorphic function in $U \setminus \{\mathrm{P}\}$. If f cannot be extended meromorphically to U, P is called an *isolated essential singularity* of f. As f is regarded as a holomorphic mapping $f : U \setminus \{\mathrm{P}\} \to \widehat{\mathbf{C}}$, this is equivalent to saying that f cannot be extended to a holomorphic mapping from U into $\widehat{\mathbf{C}}$. The next theorem is called the *big Picard theorem.*

(6.8.5) THEOREM. *If a meromorphic function f on $U \setminus \{\mathrm{P}\}$ has an isolated essential singularity at* P, *then f assumes all points of $\widehat{\mathbf{C}}$ infinitely many times in an arbitrary neighborhood of* P *with* P *deleted, except for at most two points of* $\widehat{\mathbf{C}}$.

PROOF. We may assume that $\mathrm{P} = 0$, and f is defined on $\Delta^*(R)$ $(R > 0)$. Suppose that there are three distinct points $a, b, c \in \widehat{\mathbf{C}}$ such that $f^{-1}(\{a,b,c\})$ is a finite set. Then there is a $0 < R_0 < R$ such that

$$f^{-1}(\{a,b,c\}) \cap \Delta^*(R_0) = \emptyset.$$

We show that in this case f will be meromorphically extended to $\Delta^*(R_0)$. By the change of variable z/R_0, f may be assumed to be defined on Δ^*, and $f(\Delta^*) \cap \{a,b,c\} = \emptyset$ as well. We regard f as a holomorphic mapping

$$f : \Delta^* \to \mathbf{C} \setminus \{0, 1\}. \tag{6.8.6}$$

By Theorem (6.6.7) $\mathbf{C} \setminus \{0,1\}$ carries the complete hyperbolic metric $h_{\mathbf{C}\setminus\{0,1\}}$. Take a sequence $\{z_n\}_{n=1}^{\infty}$ of Δ^* so that

$$
\begin{aligned}
(6.8.7) \qquad |z_n| &> |z_{n+1}|, \qquad \lim_{n\to\infty} z_n = 0,\\
\alpha_n &= f(z_n), \qquad \lim_{n\to\infty} \alpha_n = \alpha \in \widehat{\mathbf{C}}.
\end{aligned}
$$

Take a disk neighborhood $U(\alpha)$ of α. If for a positive number $r < 1$

$$
(6.8.8) \qquad f(\Delta^*(r)) \subset U(\alpha),
$$

by Riemann's extension Theorem (5.1.1) f extends to a holomorphic mapping from $\Delta(r)$ into $\widehat{\mathbf{C}}$. Assume that (6.8.8) does not hold. Then, there is a sequence $\{w_n\}_{n=1}^{\infty}$ of Δ^* such that

$$
\lim_{n\to\infty} w_n = 0, \qquad \beta_n = f(w_n) \in \widehat{\mathbf{C}} \setminus U(\alpha).
$$

Since $\widehat{\mathbf{C}} \setminus U(\alpha)$ is compact, by taking a subsequence one may assume that

$$
\beta = \lim_{n\to\infty} f(w_n) \in \widehat{\mathbf{C}} \setminus U(\alpha).
$$

Moreover, taking subsequences of $\{z_n\}$ and $\{w_n\}$ and renumbering them, one may assume that

$$
|z_n| > |w_n| > |z_{n+1}|, \qquad n = 1, 2, \ldots .
$$

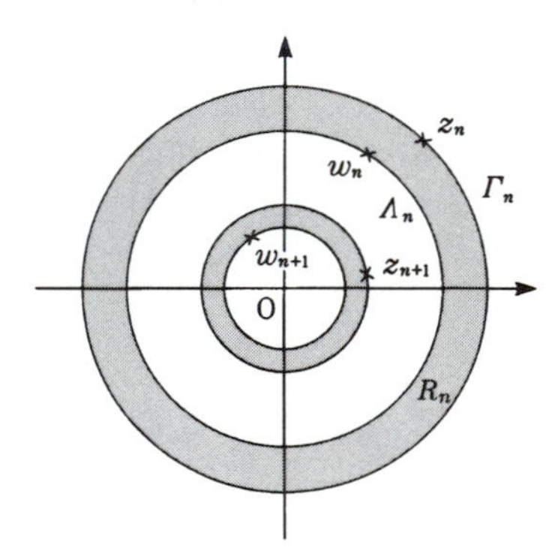

FIGURE 80

Set

$$
\begin{aligned}
R_n &= \{z \in \mathbf{C}; |w_n| < |z| < |z_n|\},\\
\Gamma_n &= C(0; |z_n|) \ni z_n,\\
\Lambda_n &= C(0; |w_n|) \ni w_n.
\end{aligned}
$$

Then, $\partial R_n = \Gamma_n \cup \Lambda_n$. For an arbitrary point $z \in \Gamma_n$ we have by the contraction principle, Theorem (6.3.10), and (6.8.3)

$$\begin{aligned} d_{\mathbf{C}\setminus\{0,1\}}(f(z), \alpha_n) &\leqq L_{\mathbf{C}\setminus\{0,1\}}(f(\Gamma_n)) \\ &\leqq L_{\Delta^*}(\Gamma_n) = \frac{2\pi}{|\log|z_n||} \to 0 \qquad (n \to \infty). \end{aligned}$$

Since $\lim_{n\to\infty} \alpha_n = \alpha$, $f(\Gamma_n) \to \alpha$ $(n \to \infty)$ if $\alpha \in \mathbf{C} \setminus \{0,1\}$. If $\alpha \in \{0,1,\infty\}$, similarly to (6.7.8) one sees that $f(\Gamma_n) \to \alpha$ $(n \to \infty)$. Therefore one always has

$$f(\Gamma_n) \to \alpha \qquad (n \to \infty). \tag{6.8.9}$$

In the same way, one sees that

$$f(\Lambda_n) \to \alpha \qquad (n \to \infty). \tag{6.8.10}$$

Since one of α and β is in $\mathbf{C}$, we assume that $\alpha \in \mathbf{C}$.

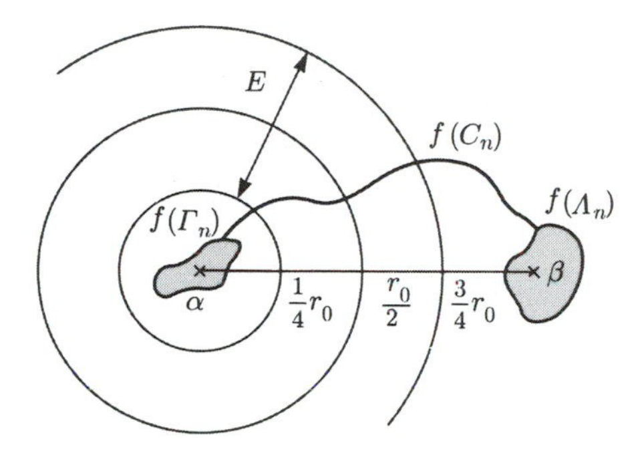

FIGURE 81

Set

$$r_0 = \begin{cases} 1, & \beta = \infty, \\ |\alpha - \beta|, & \beta \neq \infty, \end{cases}$$

and

$$E = \overline{\Delta\left(\alpha; \frac{3}{4} r_0\right)} \setminus \Delta\left(\alpha; \frac{r_0}{4}\right).$$

It follows from (6.8.9) and (6.8.10) that there is a number n_0 such that

$$E \cap (f(\Gamma_n) \cup f(\Lambda_n)) = \emptyset, \qquad n \geqq n_0. \tag{6.8.11}$$

Take a curve C_n in $\overline{R}_n$ from z_n to w_n. Then $f(C_n)$ intersects $C(\alpha; r_0/2)$. By the Prinzip von der Gebietstreue, Corollary (4.3.14), $f(R_n)$ is a domain, and its boundary is a subset of $f(\Gamma_n) \cup f(\Lambda_n)$. Thus, it follows from (6.8.11) that

$$f(R_n) \supset E. \tag{6.8.12}$$

We consider the area of E with respect to $h_{\mathbf{C}\setminus\{0,1\}}$. It is clear that

$$A_{\mathbf{C}\setminus\{0,1\}}(E) > 0. \tag{6.8.13}$$

On the other hand, by the contraction principle, Theorem (6.3.10), and (6.8.4)

$$\begin{aligned} A_{\mathbf{C}\setminus\{0,1\}}(E) &\leqq \int_{R_n} f^* h_{\mathbf{C}\setminus\{0,1\}} \leqq \int_{R_n} h_{\Delta^*} \\ &< A_{\Delta^*}(\Delta^*(|z_n|)) = \frac{2\pi}{|\log|z_n||} \to 0 \qquad (n \to \infty). \end{aligned}$$

This contradicts (6.8.13). □

If a meromorphic function f on $\mathbf{C}$ has an isolated essential singularity at ∞, f is called a *transcendental* meromorphic function; that is, if f is not transcendental, f is rational. The following is a direct consequence of Theorem (6.8.5).

(6.8.14) COROLLARY. *A transcendental meromorphic function on* $\mathbf{C}$ *takes infinitely many times all values of* $\widehat{\mathbf{C}}$ *except for at most two values.*

For example, $f(z) = e^z$ misses 0 and ∞, and in fact takes all other values infinitely many times. Hence the above "two" cannot be made smaller.

The proof of Theorem (6.8.5) essentially depended on the fact that the universal covering of $\widehat{\mathbf{C}} \setminus \{0, 1, \infty\}$ is the upper half plane. This fact was proved in §6 by making use of the Riemann mapping Theorem (6.6.7). Hence it may be said that the Riemann mapping Theorem (6.6.7) played a key role in the proof of the big Picard Theorem (6.8.14). The Riemann mapping theorem is a deep result, but is a speciality of the one variable case. It is, however, sufficient for the proof to construct a complete hermitian metric h on $\widehat{\mathbf{C}} \setminus \{0, 1, \infty\}$ for which the contraction principle holds. There is a quantity $K(h)$, the so-called Gaussian curvature of h, and the contraction principle for h holds if there is a constant $C_0 > 0$ such that

$$K(h) \leqq -C_0.$$

From this viewpoint, the big Picard theorem can be generalized to the higher dimensional case. For this, cf. [18] and [9].

Problems

1. Calculate $d_{\Delta(1)}\left(\frac{1}{2}, -\frac{1}{2}\right)$, and $d_{\Delta(1)}\left(\frac{1}{2}, \frac{i}{2}\right)$.
2. Show in Theorem (6.3.10) that if $f^* h_{D_2}(a) = h_{D_1}(a)$ holds at a point $a \in D_1$, then the equality holds at all points of D_1.
3. Let D be a hyperbolic domain. Show that for arbitrary points $z_i \in D$, $i = 1, 2$, there exists a continuously differentiable curve C from z_1 to z_2 such that

$$d_D(z_1, z_2) = L_D(C).$$

4. Let f be an entire function, and let N be a natural number. Prove that if the cardinality of $f^{-1}(w)$ is not more than N for all $w \in \mathbf{C}$, then f is a polynomial of degree at most n.

5. Obtain a biholomorphic mapping from $D = \{z \in \Delta(1); \operatorname{Im} z > 0\}$ to the upper half plane $\mathbf{H}$.

6. Obtain a biholomorphic mapping from $D = \{z = re^{i\theta}; 0 < r < 1, 0 < \theta < \alpha\}$ $(0 < \alpha < 2\pi)$ to $\Delta(1)$.

7. Show that by $w = z + 1/z$, $\Delta(1)$ is univalently mapped onto $D = \widehat{\mathbf{C}} \setminus [-2, 2]$. Where is $\widehat{\mathbf{C}} \setminus \Delta(1)$ mapped to?

8. Show that by $w = z(1 + \epsilon z)^2$ with $\epsilon = -e^{-i\alpha}, \alpha \in \mathbf{R}$ (this is called *Koebe's function*), $\Delta(1)$ is univalently mapped onto $\{w \in \mathbf{C}; w \neq r\epsilon, r \geqq 1/4\}$.

9. For a meromorphic function f on a domain $D \subset \mathbf{C}$, set

$$\{f; z\} = \frac{f'''(z)}{f'(z)} - \frac{3}{2}\left(\frac{f''(z)}{f'(z)}\right)^2.$$

This is called the *Schwarzian derivative* of f. Show the following.

(a) $\{f; z\} = 0$ if and only if $f(z) = (az + b)/(cz + d)$.

(b) For $\left(\begin{smallmatrix} a & b \\ c & d \end{smallmatrix}\right) \in SL(2, \mathbf{C})$, $\left\{\frac{af+b}{cf+d}; z\right\} = \{f; z\}$.

(c) Set $w = f(z)$, and assume that $f'(z) \neq 0$. Then,

$$\{w; z\} = -\left(\frac{dw}{dz}\right)\{z; w\}.$$

(d) Set $w = f(z)$, and let $g(w)$ be a meromorphic function in w. Then,

$$\{g \circ f; z\} = \{g; w\}\left(\frac{dw}{dz}\right)^2 + \{w; z\}.$$

10. (Vitali) Let E be a subset of a domain D which has an accumulation point in D. Let $\{f_n\}$ be a sequence of holomorphic functions which is uniformly bounded. Prove that if $\{f_n\}$ converges on E, then $\{f_n\}$ converges uniformly on compact subsets in D.

11. Let $\{f_n\}$ be a sequence of holomorphic functions on a domain D that converges at a point $z_0 \in D$. Prove that if $\{\operatorname{Re} f_n(z)\}$ converges uniformly on compact subsets in D, then so does $\{f_n\}$ itself.

12. Let D be a hyperbolic domain, and let $f : D \to D$ be a holomorphic mapping. Show that if there is a point $z_0 \in D$ such that $f(z_0) = z_0$ and $f^* h_D(z_0) = h_D(z_0)$, then $f \in \operatorname{Aut}(D)$.

13. Let D_i, $i = 1, 2$, be rectangle domains of $\mathbf{C}$, and let $f : D_1 \to D_2$ be a biholomorphic mapping. Show that f extends continuously up to the boundary of D_1. Moreover, show that if each vertex of D_1 is mapped by f to a vertex of D_2, then D_1 is similar to D_2.

14. Let D be the interior of an n-gon in $\mathbf{C}$, and let $f : \mathbf{H} \to D$ be a biholomorphic mapping. Suppose that for $-\infty < a_1 < \cdots < a_n \leqq +\infty$, $b_j = f(a_j)$ are vertices of D. Let $\alpha_j\pi$ $(0 < \alpha_j < 2)$ be the inner angles of the vertices b_j. Prove that f can be expressed as follows:

$$f(z) = C_1 \int^z \prod_{j=1}^{n} (z - a_j)^{\alpha_j - 1} dz + C_2 \qquad (a_n \neq +\infty),$$

$$f(z) = C_1 \int^z \prod_{j=1}^{n-1} (z - a_j)^{\alpha_j - 1} dz + C_2 \qquad (a_n = +\infty),$$

where C_1 $(\neq 0)$ and C_2 are constants. (The above formula is called *Schwarz-Christoffel's formula.*)
Hint 1. Suppose that $a_n \neq +\infty$. Then, by Schwarz' reflection principle the function $f(z)$ is analytically continued to a multi-valued holomorphic function on $\widehat{\mathbf{C}} \setminus \{a_j\}_{j=1}^n$.
Hint 2. By making use of a branch of $t = (f(z) - b_j)^{1/\alpha_j}$, one sees that t is analytically continued to the lower half plane in a neighborhood of a_j, and has a zero of order 1. Thus, one gets an expansion

$$f(z) - b_j = c_0(z - a_j)^{\alpha_j}(1 + c_1(z - a_j) + \cdots)^{\alpha_j},$$

and sees that the principal part of $f''(z)/f'(z)$ is $(\alpha_j - 1)/(z - a_j)$.
Hint 3. Note that the multi-valuedness of $f(z)$ is caused by the even number of repetitions of reflections with respect to line segments. Hence, letting $f^*(z)$ denote another local branch of $f(z)$, one gets

$$f^*(z) = Af(z) + B \qquad (A\ (\neq 0) \text{ and } B \text{ are constants}),$$

and $f''(z)/f'(z)$ is invariant, so one-valued, and at ∞ holomorphic with value 0. It follows that

$$\frac{f''(z)}{f'(z)} = \sum_{j=1}^{n} \frac{\alpha_j - 1}{z - a_j}.$$

Now, integrate this.
Hint 4. In the case when $a_n = +\infty$, take a linear transformation $\phi \in \operatorname{Aut}(\mathbf{H})$ such that $\phi(a_n) \neq +\infty$. Transform the formula obtained above by ϕ^{-1}.

15. Replace $\mathbf{H}$ by $\Delta(1)$ in problem 14. Then show that $f(z)$ can be expressed as

$$f(z) = C_1 \int^z \prod_{j=1}^{n} (z - a_j)^{\alpha_j - 1} dz + C_2.$$

Here, $|a_j| = 1$ and $b_j = f(a_j)$.

16. Set

$$f(z) = \int_0^z \frac{1}{\sqrt{z(1 - z^2)}} dz.$$

Prove that $f(z)$ maps $\mathbf{H}$ biholomorphically to the interior of a square in the first quadrant with one edge a real interval from 0 to $\Gamma(1/4)^2/1\sqrt{2\pi}$.

17. Let $P(X)$ be a polynomial of degree $\geqq 2$ without double root, and let $f(z)$ be a holomorphic function on $\Delta^*(1)$ (resp., $\mathbf{C}$). Show that if $P(f(z))$ has only finitely many zeros (resp., no zero), $f(z)$ has at most a pole at 0 (resp., f is a constant).

18. Show that the number of exceptional values of a non-constant rational or meromorphic function on $\mathbf{C}^*$ is at most 2.

19. Prove the big Picard Theorem (6.8.5) by making use of Montel's Theorem (6.7.6).

CHAPTER 7

Meromorphic Functions

In this chapter we investigate more deeply the properties of meromorphic functions. Beginning with an approximation theorem, we prove two kinds of existence theorems, the introductory part of value distribution of meromorphic functions on **C**, the infinite product expression of meromorphic functions, and the basic part of elliptic functions. These lead to function theory of several complex variables, the value distribution theory of one and several variables, the theory of Riemann surfaces, and the theory of elliptic curves and automorphic forms.

7.1. Approximation Theorem

Let $U \subset \widehat{\mathbf{C}}$ be an open set. In the present chapter we mainly deal with the case of $U \neq \widehat{\mathbf{C}}$. By a linear transformation we may assume $U \subset \mathbf{C}$ without loss of generality.

(7.1.1) LEMMA. *Assume that $U \subset \mathbf{C}$ is a bounded open subset. Let $K \subset U$ be a compact subset, and let f be a holomorphic function on U. Then, for an arbitrary $\epsilon > 0$ there is a rational function F such that all poles are contained in ∂U and*

$$|f(z) - F(z)| < \epsilon, \qquad z \in K.$$

That is, f can be approximated uniformly on K by such F.

PROOF. Let n be a natural number, and let k and j be integers. Set

(7.1.2)

$$E_n(j,k) = \left\{ z = x + iy; \frac{j}{2^n} \leqq x \leqq \frac{j+1}{2^n}, \frac{k}{2^n} \leqq y \leqq \frac{k+1}{2^n} \right\},$$

$$E_n = \bigcup_{E_n(j,k) \subset U} E_n(j,k).$$

Let U_n denote the interior of E_n. Then

$$U_n \subset U_{n+1}, \qquad U = \bigcup_{n=1}^{\infty} U_n.$$

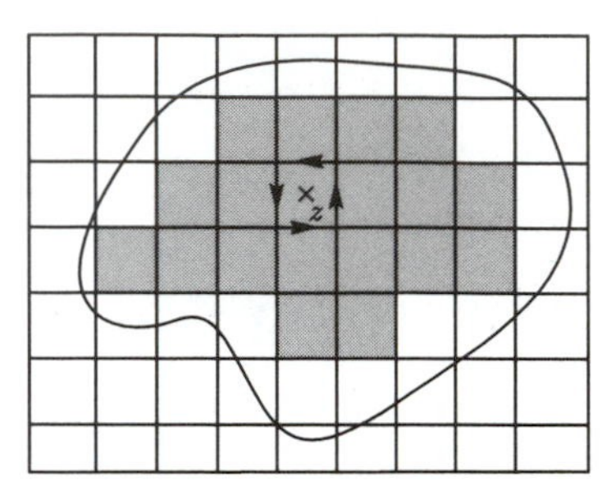

Figure 82

Since $K \subset U$ is compact, there is a number n_0 such that

$$K \subset U_n, \qquad n \geqq n_0. \tag{7.1.3}$$

Take $z \in K$ and $n \geqq n_0$ arbitrarily. If z is not a boundary point of $E_n(j,k)$'s, then there is a unique (j_0, k_0) with $z \in E_n(j_0, k_0)$. By Cauchy's integral formula, Theorem (3.5.16), and Cauchy's integral Theorem (3.4.14) we see that for $E_n(j,k) \subset U$

$$\frac{1}{2\pi i}\int_{\partial E_n(j,k)} \frac{f(z)}{\zeta - z} d\zeta = \begin{cases} f(z), & (j,k) = (j_0, k_0), \\ 0, & \text{otherwise.} \end{cases}$$

Here, the orientation of $\partial E_n(j,k)$ is anti-clockwise. If $E_n(j,k)$ and $E_n(j',k')$ share an edge, then the sum of the curvilinear integrals over the edge with respect to $\partial E_n(j,k)$ and $\partial E_n(j',k')$ is zero. Therefore we obtain

$$f(z) = \frac{1}{2\pi i}\int_{\partial U_n} \frac{f(z)}{\zeta - z} d\zeta. \tag{7.1.4}$$

Both sides of the above equation are continuous in $z \in K$, and so it holds for all $z \in K$. Since K is compact,

$$\begin{aligned} d(K; \partial U_n) &= \inf\{d(z; \partial U_n); z \in K\} \\ &\qquad = \min\{d(z; \partial U_n); z \in K\} > 0, \\ d(K; \partial U_n) &\geqq d(K; \partial U_{n-1}). \end{aligned}$$

Set $\delta_0 = d(K; \partial U_{n_0})$. Choose and fix $n \geqq n_0$ so that

$$\frac{1}{2^n} \leqq \frac{\delta_0}{2}. \tag{7.1.5}$$

Let $\gamma_1, \ldots, \gamma_{l_n}$ denote the line segments of length 2^{-n} that form the boundary ∂U_n. The rectangle with an edge γ_ν which lies on the opposite side of $E_n(j,k) \subset$

U with the edge γ_ν intersects ∂U. Let ξ_ν be a point of the intersection. It follows from (7.1.5) that for $z \in K$ and $\zeta \in \gamma_\nu$

$$|\zeta - \xi_\nu| \leqq \frac{\sqrt{2}}{2^n} \leqq \sqrt{2}\frac{\delta_0}{2} \leqq \frac{1}{\sqrt{2}}|z - \xi_\nu|. \tag{7.1.6}$$

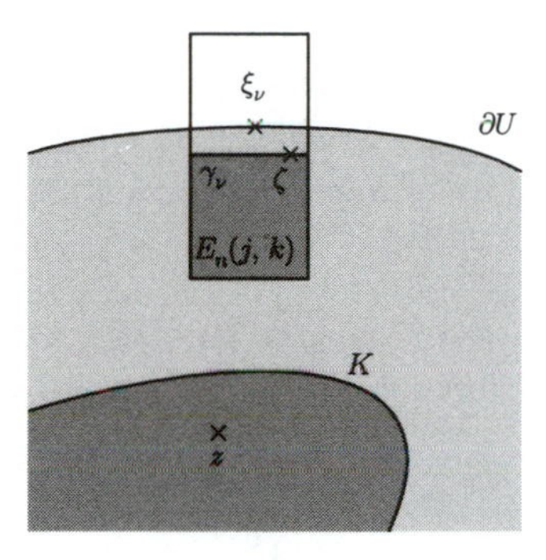

Figure 83

For these z and ζ we have

$$\begin{aligned}\frac{1}{\zeta - z} &= -\frac{1}{z - \xi_\nu} \cdot \frac{1}{1 - \frac{\zeta - \xi_\nu}{z - \xi_\nu}} \\ &= -\frac{1}{z - \xi_\nu}\sum_{j=0}^{\infty}\left(\frac{\zeta - \xi_\nu}{z - \xi_\nu}\right)^j.\end{aligned} \tag{7.1.7}$$

By (7.1.6) this power series converges uniformly and absolutely in $z \in K$ and $\zeta \in \gamma_\nu$. Therefore, for an arbitrary $\epsilon > 0$ there is a number N_ν such that

$$\left|\frac{1}{\zeta - z} + \sum_{j=0}^{N_\nu}\frac{(\zeta - \xi_\nu)^j}{(z - \xi_\nu)^{j+1}}\right| < \epsilon.$$

We set

$$\begin{aligned}F_\nu(z) &= \frac{1}{2\pi i}\int_{\gamma_\nu}\sum_{j=0}^{N_\nu}\frac{(\zeta - \xi_\nu)^j}{(z - \xi_\nu)^{j+1}}f(\zeta)d\zeta \\ &= \sum_{j=0}^{N_\nu}\frac{1}{2\pi i}\int_{\gamma_\nu}(\zeta - \xi_\nu)^j f(\zeta)d\zeta \cdot \frac{1}{(z - \xi_\nu)^{j+1}}, \\ F(z) &= -\sum_{\nu=1}^{l_n}F_\nu(z), \\ M &= \max\{|f(\zeta)|; \zeta \in \partial U_n\}.\end{aligned}$$

Then, we have

$$|f(z) - F(z)| \leqq \sum_{\nu=1}^{l_n} \left| \frac{1}{2\pi i} \int_{\gamma_\nu} \frac{f(\zeta)}{\zeta - z} d\zeta + F_\nu(z) \right|$$
$$\leqq \frac{l_n M}{2^{n+1}\pi}\epsilon. \qquad \square$$

(7.1.8) LEMMA. *Let $K \subset \mathbf{C}$ be a compact subset, and let $a, b \in \mathbf{C} \setminus K$ be two points in the same connected component of $\mathbf{C} \setminus K$. Let F be a rational function which possibly has a pole only at a. Then F is uniformly approximated on K by rational functions which possibly have a pole only at b.*

PROOF. By the assumption we have that $F(z) = \sum_{k=-N}^{N} c_k (z-a)^k$ with $0 \leqq N < \infty$ and $c_k \in \mathbf{C}$. It is sufficient to prove the assertion for every $1/(z-a)^k$ with $k > 0$. Take a curve from a to b which does not intersect K.

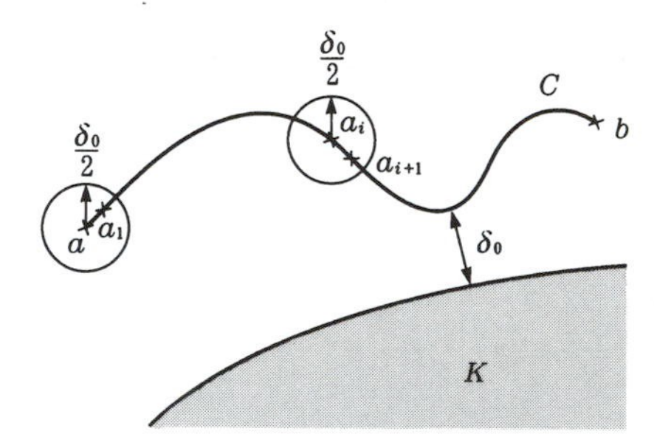

Figure 84

Set $\delta_0 = \min\{|z - w|; z \in K, w \in C\} > 0$, and take points of C:

$$a = a_0, a_1, \ldots, a_n = b, \qquad |a_{i+1} - a_i| < \frac{\delta_0}{2}.$$

It suffices to prove that $1/(z - a_i)^k$ $(k > 0)$ is uniformly approximated on K by rational functions with a pole only at a_{i+1}. It follows that for $z \in K$

$$\left| \frac{a_i - a_{i+1}}{z - a_{i+1}} \right| < \frac{1}{2}.$$

As in (7.1.7), we get

$$\frac{1}{z - a_i} = \sum_{j=0}^{\infty} \frac{(a_i - a_{i+1})^j}{(z - a_{i+1})^{j+1}},$$

which converges uniformly in a neighborhood of K. By Theorem (3.5.19) the $(k-1)$-th differentiation of both sides of the above equation yields

$$\frac{1}{(z - a_i)^k} = \sum_{j=0}^{\infty} \frac{(j+1)\cdots(j+k)(a_i - a_{i+1})^j}{(k-1)!\,(z - a_{i+1})^{j+k+1}},$$

which converges uniformly on K. Thus, we obtain our assertion. $\square$

The next is called *Runge's theorem.*

(7.1.9) THEOREM. *Let $D \subset \mathbf{C}$ be a domain and let $K \subset D$ be a compact subset. Let f be a holomorphic function in a neighborhood of K. If any connected component of $D \setminus K$ is not relatively compact in D, then f is uniformly approximated on K by holomorphic functions on D.*

PROOF. Take a bounded open set U so that $K \subset U \Subset D$ and f is holomorphic in U. It follows from Lemma (7.1.1) that f is uniformly approximated on K by rational functions with poles at finitely many points $a_\nu \in \partial U$. Take one $a_\nu \in \partial U \subset D \setminus K$, and denote by V the connected component of $D \setminus K$ containing a_ν. If V is bounded, then $\overline{V} \cap \partial D \neq \emptyset$. Take a point $b_\nu \in \overline{V} \cap \partial D$. Then a_ν and b_ν belong to the same connected component of $\mathbf{C} \setminus K$. It follows from Lemma (7.1.8) that a rational function with a pole only at a_ν is uniformly approximated on K by rational functions with a pole at b_ν. If V is not bounded, we take $b_\nu \in V$ so that $K \subset \Delta(|b_\nu|)$. A rational function with a pole at b_ν is expanded to a Taylor series about 0, and hence approximated uniformly on K by polynomials. Therefore, we deduce that f is uniformly approximated on K by rational functions with poles at most on ∂D. $\square$

Applying the above theorem to the case of $D = \mathbf{C}$, we have

(7.1.10) COROLLARY. *Let $K \subset \mathbf{C}$ be a compact subset such that $D \setminus K$ is connected. Then, a holomorphic function in a neighborhood of K is uniformly approximated on K by polynomials.*

(7.1.11) LEMMA. *Let $U \subset \mathbf{C}$ be an open subset, and let $K \subset U$ be a compact subset. Let $\{V_\alpha\}_{\alpha \in \Lambda}$ be the set of connected components of $U \setminus K$ which are relatively compact in U. Then, Λ is at most countable, and the set*

$$\tilde{K} = K \cup \bigcup_{\alpha \in \Lambda} V_\alpha$$

is compact.

PROOF. Since $U \setminus K$ is open, every V_α contains rational points. The set of rational points is countable. Hence, Λ is at most countable.

We first show that $\tilde{K}$ is bounded. Otherwise, there is a point $a_\alpha \in V_\alpha$ such that $K \subset \Delta(|a_\alpha|/2)$. Note that $\overline{V}_\alpha$ has no intersection with ∂U. Since $\mathbf{C} \setminus \overline{\Delta(|a_\alpha|/2)}$ $(\subset \mathbf{C} \setminus K)$ is connected and contains a_α, $V_\alpha \supset \mathbf{C} \setminus \overline{\Delta(|a_\alpha|/2)}$. Thus, V_α is unbounded. This is a contradiction, and hence $\tilde{K}$ is bounded. Take an arbitrary sequence of points $z_n \in \tilde{K}$, $n = 1, 2, \ldots$, which converges to a point $z_0 \in \mathbf{C}$. Suppose that $z_0 \notin \tilde{K}$. If $z_0 \in U$, then there is a connected component V of $U \setminus K$ which contains z_0 and is not relatively compact in U. For a sufficiently large n, $z_n \in V$. There is some V_α containing z_n. Then, $V_\alpha = V$. This contradicts the choice of V_α. Therefore, $z_0 \in \partial U$. Set

$$\delta_0 = \min\{d(z; \partial U); z \in K\} > 0.$$

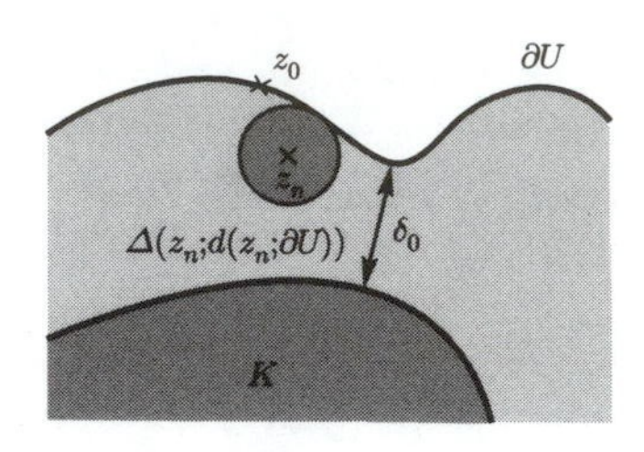

Figure 85

For a sufficiently large n, $d(z_n;\partial U) < \delta_0/2$, and

$$\Delta(z_n;d(z_n;\partial U)) \subset U \setminus K.$$

Hence, for V_α such that $V_\alpha \ni z_n$,

$$\Delta(z_n;d(z_n;\partial U)) \subset V_\alpha.$$

Since $\Delta(z_n;d(z_n;\partial U))$ is not relatively compact in U, neither is V_α. This is again absurd. Thus, $\tilde{K}$ is closed and bounded, and so compact. □

Let $D \subset \mathbf{C}$ be a domain and let $D = \bigcup_{n=1}^{\infty} U_n$ be an open covering. We say that $\{U_n\}_{n=1}^{\infty}$ is a *Runge increasing covering* if the following conditions are satisfied:

(7.1.12) i) $U_n \Subset U_{n+1}$, $n = 1, 2, \ldots$.
ii) Every holomorphic function in a neighborhood of $\overline{U}_n$ is uniformly approximated on $\overline{U}_n$ by holomorphic functions in D.

By Lemma (2.2.7) there exists an increasing sequence of open subsets V_n, $n = 1, 2, \ldots$, such that $V_n \Subset V_{n+1}$, and $D = \bigcup_{n=1}^{\infty} V_n$. Applying Lemma (7.1.11) to $\overline{V}_n$, we obtain $\tilde{\overline{V}}_n$. Let W_n denote the set of interior points of $\tilde{\overline{V}}_n$. It follows that

$$\overline{W}_n = \tilde{\overline{V}}_n, \quad V_n \subset W_n, \qquad n = 1, 2, \ldots. \tag{7.1.13}$$

Set $\overline{U}_1 = W_1$ and $n_1 = 1$. Since $\overline{U}_1 \Subset D$, there is a V_{n_2}, $n_2 \geqq n_1$, such that $\overline{U}_1 \subset V_{n_2}$. Set $U_2 = W_{n_2}$. Repeating this inductively, we obtain $\{U_n\}_{n=1}^{\infty}$, which is by Theorem (7.1.9) a Runge increasing covering. This shows the following.

(7.1.14) THEOREM. *Every domain $D \subset \mathbf{C}$ has a Runge increasing covering.*

7.2. Existence Theorems

Let $D \subset \widehat{\mathbf{C}}$ be a domain and let f be a meromorphic function on D. Let $a \in D$ be a pole of f. Then, f is expanded to a Laurent series about a:

$$\begin{aligned} f(z) =& \frac{c_{-k}}{(z-a)^k} + \frac{c_{-k+1}}{(z-a)^{k-1}} + \cdots + \frac{c_{-1}}{z-a} \\ &+ \sum_{n=0}^{\infty} c_n (z-a)^n, \qquad k > 0. \end{aligned}$$

Here, we call

$$Q_a(z) = \frac{c_{-k}}{(z-a)^k} + \frac{c_{-k+1}}{(z-a)^{k-1}} + \cdots + \frac{c_{-1}}{z-a} \tag{7.2.1}$$

the *principal part* of f at a. If $a = \infty$, f is expanded to

$$\begin{aligned} f(z) =& c_k z^k + \cdots + c_1 z + \sum_{n=0}^{\infty} c_{-n} z^{-n} \\ =& \frac{c_k}{\tilde{z}^k} + \cdots + \frac{c_1}{\tilde{z}} + \sum_{n=0}^{\infty} c_{-n} \tilde{z}^n \qquad \left(\tilde{z} = \frac{1}{z}\right). \end{aligned}$$

The *principal part* of f at ∞ is given by

$$Q_\infty(z) = c_k z^k + \cdots + c_1 z = \frac{c_k}{\tilde{z}^k} + \cdots + \frac{c_1}{\tilde{z}}. \tag{7.2.2}$$

It is a fundamental question whether there exists a meromorphic function which has a given principal part at every point of a given discrete subset of D. The following result, *Mittag-Leffler's theorem,* answers this question.

(7.2.3) THEOREM. *Let $\{a_n\}_{n=1}^N$ ($N \leqq +\infty$) be a series of distinct points of D without accumulation point in D. Suppose that at every a_n a rational function $Q_{a_n}(z)$ such as (7.2.1) or (7.2.2) is given. Then, there exists a meromorphic function on D which has the principal part Q_{a_n} at every point a_n.*

PROOF. If $D = \widehat{\mathbf{C}}$, then $N < +\infty$. Thus, we set

$$f(z) = \sum_{n=1}^{N} Q_{a_n}(z), \tag{7.2.4}$$

which clearly has the prescribed property. Assume that $D \neq \widehat{\mathbf{C}}$. By a linear transformation, we may assume that $D \subset \mathbf{C}$. It follows from Theorem (7.1.14) that D has a Runge increasing covering $\{U_\nu\}_{\nu=1}^{\infty}$. By the assumption each U_ν contains at most finitely many a_n's. Set

$$Q_\nu(z) = \sum_{a_n \in \overline{U}_\nu} Q_{a_n}(z), \qquad \nu = 1, 2, \ldots .$$

For ν such that $\overline{U}_\nu \cap \{a_n\} = \emptyset$, we set $Q_\nu = 0$. First we set $f_1 = Q_1$. Then, $Q_2 - f_1$ is holomorphic in a neighborhood of $\overline{U}_1$. Take a holomorphic function g_2 on D such that

$$|g_2 + Q_2 - f| < 1.$$

Set $f_2 = g_2 + Q_2$. Suppose that f_ν ($\nu \geqq 2$) was taken so that $f_\nu - Q_\nu$ is holomorphic in a neighborhood of $\overline{U}_\nu$ and

$$|f_\nu(z) - f_{\nu-1}(z)| < \frac{1}{2^{\nu-2}}, \qquad z \in \overline{U}_{\nu-1}. \tag{7.2.5}$$

Since $Q_{\nu+1} - f_\nu$ is holomorphic in a neighborhood of $\overline{U}_\nu$, there is a holomorphic function $g_{\nu+1}$ on D such that

$$|g_{\nu+1}(z) + Q_{\nu+1}(z) - f_\nu(z)| < \frac{1}{2^{\nu-1}}, \qquad z \in \overline{U}_\nu.$$

Set $f_{\nu+1} = g_{\nu+1} + Q_{\nu+1}$. Inductively, we take f_ν, $\nu = 1, 2, \ldots$. It follows from (7.2.5) that

$$\begin{aligned} f(z) &= f_1(z) + \sum_{\nu=1}^{\infty}(f_{\nu+1}(z) - f_\nu(z)) \\ &= f_\lambda(z) + \sum_{\nu=\lambda}^{\infty}(f_{\nu+1}(z) - f_\nu(z)), \qquad z \in D, \end{aligned}$$

defines a meromorphic function on D. By the construction, f clearly has the prescribed property. □

EXAMPLE. Let R be a meromorphic function on $\widehat{\mathbf{C}}$, that is, a rational function. Let $\{a_n\}_{n=1}^{N}$ ($N < +\infty$) be the set of poles of R, and let Q_{a_n} be the principal part of R at a_n. As in (7.2.4), $R(z) - \sum_{n=1}^{\infty} Q_{a_n}(z)$ is holomorphic on $\widehat{\mathbf{C}}$, and hence a constant by Theorem (3.7.8). Thus

$$R(z) = c_0 + \sum_{n=1}^{\infty} Q_{a_n}(z), \qquad c_0 \in \mathbf{C}. \tag{7.2.6}$$

The above right side is called the expansion by partial fractions.

EXAMPLE. Let $D = \mathbf{C}$, and set $f(z) = 1/\sin z$. Then the poles of f are $n\pi$, $n \in \mathbf{Z}$, the orders are 1, and the residues are $(-1)^n$. Set

$$g(z) = \sum_{n=1}^{\infty}(-1)^n \left(\frac{1}{z - n\pi} + \frac{1}{n\pi}\right) + (-1)^n \left(\frac{1}{z + n\pi} - \frac{1}{n\pi}\right).$$

Since

$$\left|\frac{1}{z \mp n\pi} \pm \frac{1}{n\pi}\right| = \frac{|z|}{n\pi|z \mp n\pi|} \leqq \frac{|z|}{\pi^2\left(1 - \frac{|z|}{n\pi}\right)} \cdot \frac{1}{n^2},$$

$g(z)$ converges absolutely and uniformly on every compact subset K of $\mathbf{C}$ except for a finite number of terms. Thus $g(z)$ is meromorphic function on $\mathbf{C}$. Set

$$h(z) = \frac{1}{z} + g(z) = \cdots + (-1)\left(\frac{1}{z+\pi} - \frac{1}{\pi}\right) + \frac{1}{z} + (-1)\left(\frac{1}{z-\pi} + \frac{1}{\pi}\right) + \cdots + (-1)^n\left(\frac{1}{z-n\pi} + \frac{1}{n\pi}\right) + \cdots .$$

Noting the absolute convergence, we have

$$h(z-\pi) = -h(z), \qquad h(z-2\pi) = h(z).$$

Thus, 2π is a period of $h(z)$, and hence

$$F(z) = \frac{1}{\sin z} - h(z), \qquad z \in \mathbf{C},$$

is an entire function with period 2π. For $z = x + iy$

$$\frac{1}{|\sin z|} = \frac{2}{|e^{ix-y} - e^{-ix+y}|} \leqq \frac{2}{e^{-y} - e^{y}} \to 0, \qquad |y| \to +\infty. \tag{7.2.7}$$

Here the convergence is uniform. On the other hand, $h(z)$ is written as

$$h(z) = \frac{1}{z} + \sum_{n=1}^{\infty} (-1)^n \frac{2z}{z^2 - n^2\pi^2}.$$

The above right hand side converges absolutely and uniformly on every compact subset except for a finite number of terms. We take its partial sum:

$$\begin{aligned} s_m(z) &= \sum_{n=1}^{2m} (-1)^n \frac{2z}{z^2 - n^2\pi^2} = \sum_{n=1}^{m} \left\{ \frac{-2z}{z^2 - (2n-1)^2\pi^2} + \frac{2z}{z^2 - 4n^2\pi^2} \right\} \\ &= \sum_{n=1}^{m} \frac{2\pi^2 z(4n-1)}{\{z^2 - (2n-1)^2\pi^2\}(z^2 - 4n^2\pi^2)}. \end{aligned}$$

For $z = x + iy$ with $0 \leqq x \leqq 2\pi$ and $|y| \geqq 4\pi$, we have

$$|s_m(z)| \leqq \sum_{n=1}^{m} \frac{c_2|y|n}{(y^2 + c_1 n^2)^2},$$

where $c_i > 0, i = 1, 2$, are constants. For $t \geqq 0$, the function

$$\phi(t) = \frac{c_2|y|t}{(y^2 + c_1 t^2)^2}$$

takes the maximum value $\phi(t_0) = 9c_2/16\sqrt{3c_1}y^2$ at $t_0 = |y|/\sqrt{3c_1}$. Therefore, we have

$$\sum_{n=1}^{m} \frac{c_2|y|n}{(y^2 + c_1 n^2)^2} \leqq c_2|y| \int_1^{\infty} \frac{t}{(y^2 + c_1 t^2)^2} dt + \frac{9c_2}{16\sqrt{3c_1}y^2}.$$

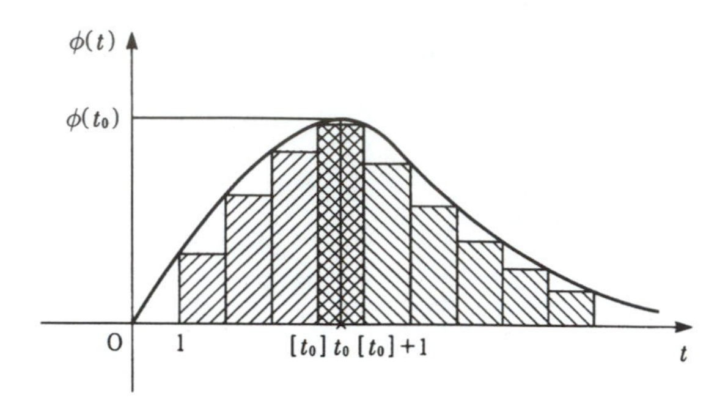

Figure 86

We compute the integral of the right side:

$$c_2|y|\int_1^\infty \frac{t}{(y^2+c_1t^2)^2}dt = \frac{c_2|y|}{2(y^2+c_1)}.$$

Thus, there is a constant $c_3 > 0$ such that

$$|s_m(z)| \leqq \frac{c_3}{|y|}.$$

Since $|g(z)| = |\lim_{m\to\infty} s_m(z)| \leqq c_3/|y|$, we get

$$|h(z)| \leqq \left|\frac{1}{z}\right| + |g(z)| \leqq \frac{1+c_3}{|y|}, \tag{7.2.8}$$
$$z = x+iy, \quad 0 \leqq x \leqq 2\pi, \quad |y| \geqq 4\pi.$$

It follows from (7.2.7) and (7.2.8) that $F(z)$ is bounded on $\{z = x+iy; 0 \leqq x \leqq 2\pi\}$. By the periodicity, $F(z)$ is a bounded holomorphic function on $\mathbf{C}$, and so by Liouville's Theorem (3.5.22) $F(z)$ is constant. It follows from (7.2.7) and (7.2.8) that $\lim_{|y|\to\infty} F(iy) = 0$, and hence $F \equiv 0$. Thus, one gets the following identity:

$$\begin{aligned}\frac{1}{\sin z} &= \frac{1}{z} + \sum_{n=-\infty}^{\infty}{}' (-1)^n\left(\frac{1}{z+n\pi} - \frac{1}{n\pi}\right)\\ &= \frac{1}{z} + \sum_{n=1}^{\infty}(-1)^n\frac{2z}{z^2-n^2\pi^2},\end{aligned}$$

where $\sum'$ stands for the sum with the term for $n=0$ omitted.

EXERCISE 1. Show that $\pi\cot\pi z = \dfrac{1}{z} + \displaystyle\sum_{n=-\infty}^{\infty}{}'\left(\frac{1}{z-n} + \frac{1}{n}\right)$.

EXERCISE 2. Show that $\dfrac{\pi^2}{\sin^2\pi z} = \displaystyle\sum_{n=-\infty}^{\infty}\frac{1}{(z-n)^2}$.

We are next going to show the existence of a holomorphic function possessing prescribed orders of zeros at prescribed discrete points in a domain D. This is called *Weierstrass' theorem*:

(7.2.9) THEOREM. *Let $D \subset \mathbf{C}$ be a domain, and let $\{a_n\}_{n=1}^{N}$ $(N \leqq \infty)$ be a discrete set of points of D. Let ν_n, $1 \leqq n \leqq N$, be integers. Then there exists a holomorphic function $f(z)$ which has zeros of order ν_n at a_n, $n = 1, \dots, N$, and no zeros elsewhere.*

PROOF. By Theorem (7.1.14) there is a Runge increasing covering $\{U_n\}_{n=1}^{\infty}$ of D. Because of the construction of U_n (cf. (7.1.13)), the following holds:

(7.2.10) Any connected component of $D \setminus \overline{U}_n$ is not relatively compact in D.

For every n we set

$$P_n(z) = \prod_{a_j \in \overline{U}_n} (z - a_j)^{\nu_j}.$$

For $a_j \in \overline{U}_{n+1} \setminus \overline{U}_n$ we take a point $b_j \in D \setminus \overline{U}_{n+1}$ which belongs to the connected component V of $D \setminus \overline{U}_n$ containing a_j. Then, the function $\log(z - a_j)/(z - b_j)$ has a one-valued branch in a neighborhood of $\overline{U}_n$. To show this, we take a piecewise linear curve C from a_j to b_j in V. The curve C consists of line segments L_j from z_k to z_{k+1}, where

$$a_j = z_0, z_1, \dots, z_l = b_j.$$

The function $(z - z_k)/(z - z_{k+1})$ is holomorphic and non-vanishing in $\widehat{\mathbf{C}} \setminus L_k$. Since $\widehat{\mathbf{C}} \setminus L_k$ is simply connected, $\log(z - z_k)/(z - z_{k+1})$ has a branch in $\widehat{\mathbf{C}} \setminus L_k \supset \overline{U}_n$. Therefore, the function

$$\log \frac{z - a_j}{z - b_j} = \sum_{k=0}^{l-1} \log \frac{z - z_k}{z - z_{k+1}} \tag{7.2.11}$$

has a branch in a neighborhood of $\overline{U}_n$. We take $n = 1$ and set $f_1(z) = P_1(z)$. For $a_j \in \overline{U}_2 \setminus \overline{U}_1$, we take $b_j \in D \setminus \overline{U}_2$ as above. Set

$$g_2(z) = \frac{P_2(z)}{\prod (z - b_j)^{\nu_j}} \cdot \frac{1}{f_1(z)} = \prod_{a_j \in \overline{U}_2 \setminus \overline{U}_1} \left(\frac{z - a_j}{z - b_j} \right)^{\nu_j}.$$

Then $\log g_2(z)$ has a branch in a neighborhood of $\overline{U}_1$. For an arbitrary $\epsilon_2 > 0$, there is a holomorphic function $h_2(z)$ in D such that

$$|\log g_2(z) + h_2(z)| < \epsilon_2, \qquad z \in \overline{U}_1. \tag{7.2.12}$$

Set

$$f_2(z) = \frac{P_2(z)}{\prod (z - b_j)^{\nu_j}} e^{h_2(z)}.$$

Then, in a neighborhood of $\overline{U}_1$ the function

$$\log \frac{f_2(z)}{f_1(z)} = \log g_2(z) + h_2(z)$$

has a branch, and by (7.2.12) it satisfies

$$\left| \log \frac{f_2(z)}{f_1(z)} \right| < \epsilon_2. \tag{7.2.13}$$

Here, we take $\epsilon_2 = 1/2$. Repeating this process, we construct holomorphic functions $f_n(z)$ in neighborhoods of $\overline{U}_n$ such that

(7.2.14)

i) $\dfrac{f_n(z)}{P_n(z)} \neq 0, \infty, \qquad z \in \overline{U}_n,\ n = 1, 2, \dots,$

ii) $\log \dfrac{f_{n+1}(z)}{f_n(z)}$ has a branch in a neighborhood of $\overline{U}_n$, and

$$\left| \log \frac{f_{n+1}(z)}{f_n(z)} \right| < \frac{1}{2^n}, \qquad z \in \overline{U}_n.$$

Set

$$f(z) = f_1(z) \prod_{n=1}^{\infty} \frac{f_{n+1}(z)}{f_n(z)} = f_m(z) \prod_{n=m}^{\infty} \frac{f_{n+1}(z)}{f_n(z)}. \tag{7.2.15}$$

On every $\overline{U}_m$ we have

$$\prod_{n=m}^{\infty} \frac{f_{n+1}(z)}{f_n(z)} = \exp\left(\sum_{n=m}^{\infty} \log \frac{f_{n+1}(z)}{f_n(z)} \right),$$

$$\sum_{n=m}^{\infty} \left| \log \frac{f_{n+1}(z)}{f_n(z)} \right| \leqq \sum_{n=m}^{\infty} \frac{1}{2^n} = \frac{1}{2^{m-1}}.$$

Therefore, the infinite product (7.2.15) converges absolutely and uniformly on compact subsets of D, and so $f(z)$ is a holomorphic function in D. It follows from (7.2.14) and (7.2.15) that $f(z)$ satisfies the required conditions. □

In the proofs of Mittag-Leffler's and Weierstrass' Theorems (7.2.3) and (7.2.9) the approximation theorem due to Runge (Theorem (7.1.9)) played an important role. In the case of several variables, an analogue of Mittag-Leffler's theorem holds on a domain where the approximation theorem holds (this is called the first Cousin problem, and was solved by K. Oka in 1936~1937). The analogue of Weierstrass' theorem (this is called the second Cousin problem), however, does not follow. When we applied the approximation theorem in (7.2.12), we took $\log g_2(z)$. In the case of several variables, this argument does not work, and K. Oka proved that some topological condition on D is necessary. This fact is referred as Oka's principle.

If a holomorphic function f on a domain D cannot be analytically continued to a neighborhood of any point of ∂D, ∂D is called a *natural boundary* of f, and D is called the *domain of existence* of f. This makes sense only when $D \neq \widehat{\mathbf{C}}$, and so we assume in the following that $D \subset \mathbf{C}$.

(7.2.16) THEOREM. *There is a holomorphic function f on D such that D is the domain of existence of f.*

PROOF. For positive integers n, we set

$$D_n = \left\{ z \in D; d(z; \partial D) > \frac{1}{n} \right\} \cap \Delta(z),$$
$$\Gamma_n = \partial D_n.$$

As in (7.2.12), we cover $\mathbf{C}$ by squares $E_n(j,k)$ with sides of length $1/2^n$.

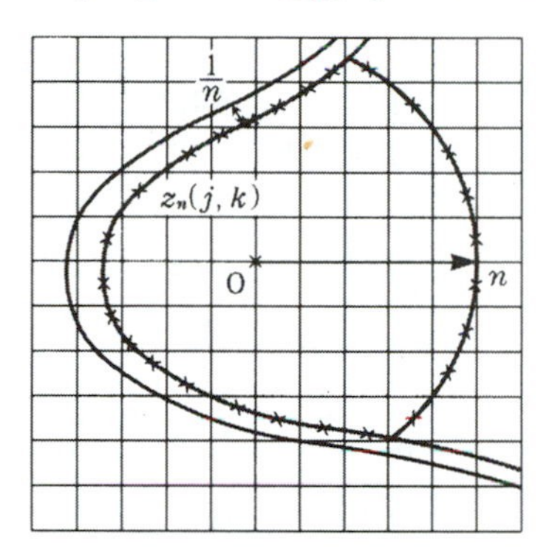

Figure 87

Take a point $z_n(j,k)$ from every $E_n(j,k) \cap \Gamma_n$ with $E_n(j,k) \cap \Gamma_n \neq \emptyset$. Let z_{nj}, $j = 1, \ldots, k_n$, be distinct points of $z_n(j,k)$'s, and order them as

$$z_{11}, \ldots, z_{1k_1}, z_{21}, \ldots, z_{2k_2}, z_{31}, \ldots .$$

Denote this sequence by $\{z_\nu\}$. Then $\{z_\nu\}$ is discrete in D and every point of ∂D is an accumulation point of $\{z_\nu\}$. It follows from Theorem (7.2.9) that there is a holomorphic function $f(z)$ in D which has precisely zeros of order one at the z_ν's. By the identity Theorem (2.4.14), ∂D is the natural boundary of f. $\square$

EXERCISE 3. Show that $f(z) = \sum_{n=0}^{\infty} z^{n!}$ is a holomorphic function whose domain of existence is $\Delta(1)$.

EXERCISE 4. Show that the real axis is the natural boundary of $\lambda(z)$ in (6.6.2)

The next theorem is most fundamental in the interpolation problem:

(7.2.17) THEOREM. *Let $\{a_n\}_{n=1}^{\infty}$ be a discrete set of points of D, and let*

$$P_n(z) = c_{n0} + c_{n1}(z - a_n) + \cdots + c_{nd_n}(z - a_n)^{d_n}$$

be arbitrary polynomials. Then, there is a holomorphic function $f(z)$ in D such that about every a_n, $f(z)$ is expanded to

$$f(z) = P_n(z) + \textit{higher order term.}$$

PROOF. By Weierstrass' Theorem (7.2.9) there is a holomorphic function $F(z)$ such that locally about every a_n

$$F(z) = (z - a_n)^{d_n+1} F_n(z),$$

where $F_n(z)$ is a holomorphic function in a neighborhood of a_n. By Mittag-Leffler's Theorem (7.2.3), there is a meromorphic function $G(z)$ in D such that locally about a_n

$$G(z) - \frac{P_n(z)}{F(z)} = H_n(z)$$

is a holomorphic function. Set

$$f(z) = F(z)G(z).$$

Then, about a_n we have

$$f(z) = P_n(z) + F(z)H_n(z)$$

and hence $f(z)$ satisfies the required property. □

7.3. Riemann-Stieltjes' Integral

For use in the next section we here describe necessary facts on Riemann-Stieltjes' integral. Let $\phi(t)$ be a real valued function on a bounded closed interval $[a, b] \subset \mathbf{R}$. For a partition $(d) : a = t_0 \leqq t_1 \leqq \cdots \leqq t_l = b$, we set

$$V(\phi; (d)) = \sum_{i=1}^{l} |\phi(t_i) - \phi(t_{i-1})|,$$
$$V(\phi) = \sup\{V(\phi; (d)); (d) \text{ is a partition of } [a, b]\} \leqq \infty.$$

We call $V(\phi)$ the *total variation* of ϕ; if $V(\phi) < \infty$, ϕ is said to be of bounded variation.

For $c \in [a, b]$ we have

(7.3.1)
$$\begin{aligned} \big|2|\phi(c)| - |\phi(a)| - |\phi(b)|\big| &\leqq |\phi(a) - \phi(c)| + |\phi(c) - \phi(b)| \\ &\leqq V(\phi|[a, c]) + V(\phi|[c, d]) = V(\phi). \end{aligned}$$

Specially, if $V(\phi) < \infty$, ϕ is bounded. For $x \in \mathbf{R}$, we set

$$x^+ = \max\{x, 0\}, \qquad x^- = \max\{-x, 0\}.$$

It follows that

$$|x| = x^+ + x^-, \qquad x = x^+ - x^-.$$

Set

$$V^{\pm}(\phi;(d)) = \sum_{i=1}^{l}(\phi(t_i)-\phi(t_{i-1}))^{\pm} \leqq V(\phi;(d)),$$
$$V^{\pm}(\phi) = \sup\{V^{\pm}(\phi;(d))\} \leqq V(\phi).$$

Then we see that

$$\begin{aligned} &V^{+}(\phi)+V^{-}(\phi)=V(\phi), \\ &V^{\pm}(\phi|[a,c])+V^{\pm}(\phi|[c,b])=V^{\pm}(\phi), \qquad a \leqq c \leqq b, \\ &\phi(b)-\phi(a)=V^{+}(\phi)-V^{-}(\phi). \end{aligned} \tag{7.3.2}$$

We call $V^{+}(\phi)$ (resp., $V^{-}(\phi)$) the *positive* (resp. *negative*) *variation* of ϕ.

EXERCISE 1. Prove (7.3.2).

(7.3.3) THEOREM. i) *A function ϕ on $[a,b]$ is of bounded variation if and only if ϕ is written as a difference of monotone increasing functions.*

ii) *If $V(\phi)<\infty$, there exist limits $\lim_{t\to\xi\pm 0}\phi(t)$ for $\xi\in[a,b]$. Here, at the end points $\xi=a,b$, the one-sided limits exist.*

iii) *If ϕ is continuously differentiable in a neighborhood of $[a,b]$, then*

$$V(\phi)=\int_a^b|\phi'(t)|dt.$$

PROOF. i) It is clear that a monotone increasing function is of bounded variation. Since $V(\phi\pm\psi)\leqq V(\phi)+V(\psi)$, the difference of two monotone increasing functions is of bounded variation. Conversely, assume that $V(\phi)<\infty$. We define $v^{\pm}(t)=V^{\pm}(\phi|[a,t])$ for $t\in[a,b]$. Clearly, $v^{\pm}(t)$ are monotone increasing, and by (7.3.2)

$$\phi(t)=\phi(a)+v^{+}(t)-v^{-}(t).$$

ii) This is clear.

iii) The mean value theorem implies this. □

Let $\psi:[a,b]\to\mathbf{R}$ be a function. Take a partition (d) of $[a,b]$ as above, and take points $t_{i-1}\leqq\xi\leqq t_i$, $1\leqq i\leqq l$. If the limit

$$\lim_{|(d)|\to 0}\sum_{i=1}^{l}\psi(\xi_i)(\phi(t_i)-\phi(t_{i-1}))$$

exists, we denote it by

$$\int_a^b\psi(t)d\phi(t),$$

which is called *Riemann-Stieltjes' integral* of ψ with respect to ϕ. This is just the curvilinear integral of ψ when ϕ is considered as a curve in $\mathbf{C}$.

EXERCISE 2. Let $\phi(t)=\phi_1(t)+i\phi_2(t)$, $a\leqq t\leqq b$, be a curve in $\mathbf{C}$. Show that the curve ϕ has finite length if and only if ϕ_i, $i=1,2$, are of bounded variation.

(7.3.4) THEOREM. i) *If $V(\phi) < \infty$ and ψ is continuous, then $\int_a^b \psi(t)d\phi(t)$ exists.*

ii) *Under the same assumption as above, $\int_a^b \phi(t)d\psi(t)$ exists, and*

$$\int_a^b \phi(t)d\psi(t) = \left[\phi(t)\psi(t)\right]_a^b - \int_a^b \psi(t)d\phi(t).$$

PROOF. i) This is a special case of (3.2.8) applied to $\phi : [a, b] \to \mathbf{R} \subset \mathbf{C}$.

ii) Take a partition (d) and points ξ_i as

$$a = t_0 = \xi_1 \leqq t_1 \leqq \xi_2 \leqq t_2 \leqq \cdots \leqq \xi_l = t_l = b.$$

Then, one obtains

$$\begin{aligned}&\sum_{i=1}^{l} \phi(\xi_i)(\psi(t_i) - \psi(t_{i-1}))\\ &= \left[\phi(t)\psi(t)\right]_a^b - \sum_{i=1}^{l-1} \psi(t_i)(\phi(\xi_{i+1}) - \phi(\xi_i)).\end{aligned}$$

As $|(d)| \to 0$, the required equality follows. □

(7.3.5) EXAMPLE. Set $\phi(t) = 0$ for $t < 0$, and $\phi(t) = 1$ for $t \geqq 0$. Let $\psi(t)$ be a continuous function on $[-\delta_0, \delta_0]$ $(\delta_0 > 0)$. Then

$$\int_{-\delta}^{\delta} \psi(t)d\phi(t) = \psi(0), \qquad 0 < \delta \leqq \delta_0.$$

To see this, we take a partition, $-\delta = t_0 < t_1 < \cdots < t_l = b$. Then, there is a t_{i_0} such that

$$t_{i_0-1} < 0 \leqq t_{i_0}.$$

Take $\xi \in [t_{i-1}, t_i]$ so that $\xi_{i_0} = 0$. Then

$$\sum_{i=1}^{l} \psi(\xi_i)(\phi(t_i) - \phi(t_{i-1})) = \psi(\xi_{i_0})(\phi(t_{i_0}) - \phi(t_{i_0-1})) = \psi(0).$$

It suffices to let $|(d)| \to 0$.

7.4. Meromorphic Functions on C

Let f be a meromorphic function on $\mathbf{C}$. Let $\{a_\nu\}_{\nu=1}^{\infty}$ (resp., $\{b_\mu\}_{\mu=1}^{\infty}$) be the zeros (resp., poles) of f counting multiplicities (e.g., if a_1 is a zero of order k, then $a_1 = a_2 = \cdots = a_k \neq a_j$, $j > k$). The following is called *Jensen's formula.*

(7.4.1) LEMMA. *Assume that* $f(z) = z^{k_0}(c_{k_0} + c_{k_0+1}z + \cdots)$ *with* $c_{k_0} \neq 0$ *and* $k_0 \in \mathbf{Z}$. *Then we have*

(7.4.2)

$$\begin{aligned}\frac{1}{2\pi}\int_0^{2\pi} &\log|f(re^{i\theta})|d\theta \\ &= \sum_{0<|a_\nu|<r} \log\frac{r}{|a_\nu|} - \sum_{0<|b_\mu|<r} \log\frac{r}{|b_\mu|} + k_0\log r + \log|c_{k_0}|.\end{aligned}$$

PROOF. If the case of $k_0 = 0$ is proved, we apply it to $f(z)z^{-k_0}$. Then equation (7.4.2) follows. Thus, we may assume that $f(z)$ has no zero nor pole at the origin 0. Assume first that there is no a_ν nor b_μ on $C(0;r)$. For $\alpha \in \Delta(r)$, set

$$E(z) = \frac{\frac{z}{r} - \frac{\alpha}{r}}{-\frac{\overline{\alpha}}{r}\frac{z}{r} + 1} = \frac{r(z-\alpha)}{-\overline{\alpha}z + r^2}.$$

For $z \in \Delta(r)$, $|E(z)| < 1$, and $|E(z)| = 1$ for $z \in C(0;r)$. Set

$$h(z) = f(z)\prod_{|b_\mu|<r}\frac{r(z-b_\mu)}{-\overline{b}_\mu z + r^2}\cdot\prod_{|a_\nu|<r}\frac{-\overline{a}_\nu z + r^2}{r(z-a_\nu)}.$$

Then, $h(z)$ is holomorphic and non-vanishing in a neighborhood of $\overline{\Delta(r)}$. Therefore, $\log|h(z)|$ is harmonic there, and $|f(z)| = |h(z)|$ on $C(0;r)$. By the mean value Theorem (3.6.10), iii)

$$\begin{aligned}\frac{1}{2\pi}\int_0^{2\pi}\log|f(re^{i\theta})|d\theta &= \frac{1}{2\pi}\int_0^{2\pi}\log|h(re^{i\theta})|d\theta = \log|h(0)| \\ &= \log|f(0)| + \sum_{|a_\nu|<r}\log\frac{r}{|a_\nu|} - \sum_{|b_\mu|<r}\log\frac{r}{|b_\mu|}.\end{aligned}$$

Now we consider the case where f has zeros or poles on $C(0;r)$. The right side of (7.4.2) is continuous in r. Therefore, it suffices to show the continuity of the integral of the left side of (7.4.2) in r. Suppose that $f(re^{i\theta_0}) = 0$, or ∞. By the parameter change $\theta - \theta_0$, one may assume that $\theta_0 = 0$. For θ (resp., t) close to 0 (resp., r) one has

$$f(te^{i\theta}) = f(r + te^{i\theta} - r) = (te^{i\theta} - r)^k g(te^{i\theta}),$$

where $k \in \mathbf{Z}$ and $g(te^{i\theta}) \neq 0$.

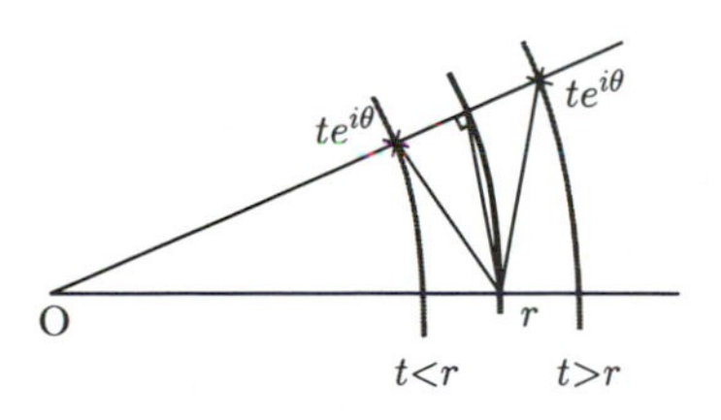

Figure 88

It is our aim to estimate $\int_{-\delta}^{\delta}\log|f(te^{i\theta})|d\theta$, so that it suffices to estimate

$$\left|\int_{-\delta}^{\delta}\log|te^{i\theta}-r|d\theta\right|.$$

As in the above figure, it follows that

$$\int_{-\delta}^{\delta}\log|re^{i\theta}-r|d\theta<\int_{-\delta}^{\delta}\log|te^{i\theta}-r|d\theta<0.$$

Thus, it is sufficient to estimate $\int_{-\delta}^{\delta}\log|e^{i\theta}-1|d\theta$:

$$\int_{-\delta}^{\delta}\log|e^{i\theta}-1|d\theta=\int_{-\delta}^{\delta}\log\left|\frac{i\theta}{1!}+\frac{(i\theta)^2}{2!}+\cdots\right|d\theta=\int_{-\delta}^{\delta}\log|\theta|d\theta+o(\delta),$$

where $|o(\delta)/\delta|\to 0$ as $\delta\to 0$. It follows that

$$\int_{-\delta}^{\delta}\log|\theta|d\theta=2\delta\log\delta-2\delta.$$

Therefore, for an arbitrary $\epsilon>0$ there is a $\delta_0>0$ such that

$$\left|\int_{-\delta}^{\delta}\log|f(te^{i\theta})|d\theta\right|<\epsilon,\qquad |t-r|<\delta_0.$$

Hence, there is a finite union I of open intervals of $[0,2\pi]$ such that I contains $\arg a_\nu$, $\arg b_\mu$ with $a_\nu, b_\mu\in C(0;r)$ and

$$\left|\int_I\log|f(te^{i\theta})|d\theta\right|<\epsilon,\qquad |t-r|<\delta_0.$$

Since

$$\frac{1}{2\pi}\int_{[0,2\pi]\setminus I}\log|f(te^{i\theta})|d\theta\to\frac{1}{2\pi}\int_{[0,2\pi]\setminus I}\log|f(re^{i\theta})|d\theta,\qquad t\to r,$$

the following holds:

$$\lim_{t\to r}\frac{1}{2\pi}\int_0^{2\pi}\log|f(te^{i\theta})|d\theta=\frac{1}{2\pi}\int_0^{2\pi}\log|f(re^{i\theta})|d\theta.\qquad\square\tag{7.4.3}$$

We set

$$\log^+x=\log\max\{x,1\}\ (\geqq 0),\qquad x\geqq 0.$$

It follows that

$$\begin{aligned}&\log x=\log^+x-\log^+\frac{1}{x},\quad |\log x|=\log^+x+\log^+\frac{1}{x},\\ &\log^+\prod_{i=1}^n x_i\leqq\sum_{i=1}^n\log^+x_i,\\ &\log^+\sum_{i=1}^n x_i\leqq\sum_{i=1}^n\log^+x_i+\log n.\end{aligned}\tag{7.4.4}$$

EXERCISE 1. Prove (7.4.4).

We set

$$m(r,f) = \frac{1}{2\pi}\int_0^{2\pi} \log^+ |f(re^{i\theta})| d\theta,$$

which is called the *proximity function* of f. As in the proof of (7.4.3), we see that $m(r,f)$ is continuous in $r > 0$. It follows from (7.4.4) that

$$\frac{1}{2\pi}\int_0^{2\pi} \log |f(re^{i\theta})| d\theta = m(r,f) - m\left(r, \frac{1}{f}\right). \tag{7.4.5}$$

Set

$$\begin{aligned} n(r,f) &= \text{the number of } b_\mu\text{'s in } \Delta(r), \\ n(0,f) &= \text{the number of } b_\mu\text{'s such that } b_\mu = 0, \\ N(r,f) &= \int_0^r \frac{n(t,f) - n(0,f)}{t} dt + n(0,f)\log r. \end{aligned} \tag{7.4.6}$$

We call $n(r,f)$ and $N(r,f)$ the *counting functions* (of poles) of f.

(7.4.7) LEMMA. *Let the notation be as above. Then we have*

$$\sum_{0<|b_\mu|<r} \log \frac{r}{|b_\mu|} = \int_0^r \frac{n(t,f)-n(0,f)}{t} dt.$$

PROOF. First, note that $n(r,f)$ is a monotone increasing function in r. By Theorem (7.3.4) and Example (7.3.5) one has

$$\begin{aligned} \sum_{0<|b_\mu|<r} \log \frac{r}{|b_\mu|} &= \int_0^r \log \frac{r}{t}\, d\{n(t,f) - n(0,f)\} \\ &= \int_0^r \frac{n(t,f)-n(0,f)}{t} dt. \qquad \square \end{aligned}$$

We define Nevanlinna's *order function* or *characteristic function* of f by

$$T(r,f) = N(r,f) + m(r,f).$$

Let g be a meromorphic function on **C**, and let $a \in \mathbf{C}$. Then, it follows from (7.4.4) that

$$\begin{aligned} T(r, f-a) &\leqq T(r,f) + \log^+|a| + \log 2, \\ T(r, f+g) &\leqq T(r,f) + T(r,g) + \log 2 \\ T(r, f\cdot g) &\leqq T(r,f) + T(r,g). \end{aligned} \tag{7.4.8}$$

By Lemma (7.4.1), (7.4.4), (7.4.5), and Lemma (7.4.7), we have *Nevanlinna's first main theorem*:

(7.4.9) THEOREM. *Let $f(z)$ be a meromorphic function on $\mathbf{C}$ such that about 0 $f(z) = c_k z^k + c_{k+1} z^{k+1} + \cdots$ with $c_k \neq 0$ and $k \in \mathbf{Z}$. Then*

$$T(r, f) = T\left(r, \frac{1}{f}\right) + \log |c_k|.$$

(7.4.10) COROLLARY. *For $a \in \mathbf{C}$*

$$N\left(r, \frac{1}{f-a}\right) \leqq T(r, f) + O(1),$$

where $O(1)$ stands for a bounded term as $r \to \infty$.

PROOF. It follows from (7.4.8) and Theorem (7.4.9) that

$$\begin{aligned} N\left(r, \frac{1}{f-a}\right) &\leqq N\left(r, \frac{1}{f-a}\right) + m\left(r, \frac{1}{f-a}\right) \\ &= T\left(r, \frac{1}{f-a}\right) = T(r, f-a) + O(1) \\ &= T(r, f) + O(1). \qquad \square \end{aligned}$$

The inequality of Corollary (7.4.10) is called *Nevanlinna's inequality.* As seen in the above proof, if the number of a-points of f is larger, i.e., $N(r, 1/(f-a))$ has a larger growth as $r \to \infty$, then that of the proximity function $m(r, 1/(f-a))$ is smaller. In the Nevanlinna theory the order function $T(r, f)$ is considered as the standard, and the quantitative properties are investigated by comparisons of other quantities such as $N(r, 1/(f-a))$ with $T(r, f)$.

EXAMPLE. For coprime polynomials $P(z)$ and $Q(z)$ we have

$$T\left(r, \frac{P}{Q}\right) = \deg \frac{P}{Q} \cdot \log r + O(1), \tag{7.4.11}$$

where $\deg P$ (resp., $\deg Q$) denotes the degree of P (resp., Q), and

$$\deg \frac{P}{Q} = \max\{\deg P, \deg Q\}.$$

For we may assume without loss of generality by the first main Theorem (7.4.9) that $\deg P \leqq \deg Q$. In this case, we get

$$\begin{aligned} m\left(r, \frac{P}{Q}\right) &= O(1), \\ N\left(r, \frac{P}{Q}\right) &= \deg Q \cdot \log r + O(1). \end{aligned}$$

Thus, (7.4.11) follows.

For an entire function f we define the *maximum function* of f by

$$M(r, f) = \max\{|f(z)|; z \in \overline{\Delta(r)}\} = \max\{|f(z)|; z \in C(0; r)\}.$$

(7.4.12) THEOREM. *Let f be an entire function. Then*

$$m(r,f) \leqq \log M(r,f) \leqq \frac{R+r}{R-r} m(R,f), \qquad 0 < r < R.$$

PROOF. By definition the first inequality is trivial. We show the second. By the change of variable Rz, we may assume that $R = 1$. Take $z_0 \in C(0;r)$ so that

$$\log |f(z_0)| = M(r,f).$$

Apply Lemma (7.4.1) to $f \circ \phi_{z_0}^{-1}$, where $\phi_{z_0}(z) = (z - z_0)/(-\overline{z}_0 z + 1)$. Then one gets

$$\log |f(z_0)| = \log |f \circ \phi_{z_0}^{-1}(0)| \leqq \frac{1}{2\pi} \int_0^{2\pi} \log |f \circ \phi_{z_0}^{-1}(e^{i\theta})| d\theta.$$

Applying the computation used in (3.6.2), one obtains

$$\begin{aligned} \log |f(z_0)| &\leqq \frac{1}{2\pi} \int_0^{2\pi} \log |f(e^{i\theta})| \cdot \frac{1 - r^2}{|e^{i\theta} - z_0|^2} d\theta \\ &\leqq \frac{1}{2\pi} \int_0^{2\pi} \log |f(e^{i\theta})| \cdot \frac{1 - r^2}{(1 - r^2)} d\theta \leqq \frac{1+r}{1-r} M(1,f). \qquad \square \end{aligned}$$

We use the notation

$$T(r,f) = O(\log r)$$

for the property that $T(r,f)/\log r$ is bounded as $r \to \infty$.

(7.4.13) LEMMA. *An entire function f is a polynomial if and only if $T(r,f) = O(\log r)$.*

PROOF. By (7.4.11) the "only if" part is clear. Suppose that there are positive numbers C_1 and r_0 such that $T(r,f) \leqq C_1 \log r$ for $r > r_0$. Since $T(r,f) = m(r,f)$, it follows from Theorem (7.4.12) that

$$\log M(r,f) \leqq 3m(2r,f) \leqq 3C_1(\log r + \log 2), \qquad r > r_0/2.$$

Therefore, $M(r,f) \leqq 2^{3C_1} r^{3C_1}$ for $r > r_0/2$. The coefficient estimate of Theorem (3.5.20), i) implies that f is a polynomial. $\square$

(7.4.14) LEMMA. *Let f be a meromorphic function on* **C**. *Then, the number of poles of f is finite if and only if $N(r,f) = O(\log r)$.*

PROOF. The "only if" part is trivial. Suppose as above that

$$N(r,f) \leqq C_1 \log r, \qquad r > r_0.$$

Since $n(r,f)$ is a monotone increasing function in r, for $r > r_0$ we have

$$\begin{aligned} n(r,f) &\leqq \frac{1}{\log r} \int_r^{r^2} \frac{n(t,f)}{t} dt \leqq \frac{1}{\log r} N(r^2, f) \\ &\leqq \frac{1}{\log r} C_1 \log r^2 \leqq 2C_1. \qquad \square \end{aligned}$$

(7.4.15) THEOREM. *A meromorphic function f on $\mathbf{C}$ is a rational function if and only if $T(r,f) = O(\log r)$.*

PROOF. The "only if" part follows from (7.4.11). Suppose that $T(r,f) = O(\log r)$. Then, $N(r,f) = O(\log r)$, so that the number of poles of f is finite by Lemma (7.4.14). Denote them by $\{b_\mu\}_{\mu=1}^{l}$, counting orders, and set

$$g(z) = f(z) \prod_{\mu=1}^{l} (z - b_\mu).$$

Then, $g(z)$ is an entire function, and by (7.4.8) it satisfies

$$\begin{aligned} T(r,g) &\leqq T(r,f) + T\left(r, \prod_{\mu=1}^{l}(z - b_\mu)\right) \\ &= O(\log r). \end{aligned}$$

Then, Lemma (7.4.13) implies that g is a polynomial, and so f is a rational function. □

We define the *order* of a meromorphic function f on $\mathbf{C}$ by

$$\rho_f = \varlimsup_{r\to\infty} \frac{\log T(r,f)}{\log r}. \tag{7.4.16}$$

When f is entire, Theorem (7.4.12) implies

$$\rho_f = \varlimsup_{r\to\infty} \frac{\log\log M(r,f)}{\log r}. \tag{7.4.17}$$

(7.4.18) EXAMPLE. Let $P(z)$ be a polynomial of degree n, and set $f(z) = e^{P(z)}$. It follows from (7.4.17) that $\rho_f = n$.

Historically, until the time of Hadamard the order of entire functions f was defined by (7.4.17), and the value distribution theory, i.e., the quantitative theory of the solutions of $f(z) = a$ had been developed. This treatment faced a difficulty in dealing with meromorphic functions. The definition of (7.4.16) is due to R. Nevanlinna. For more details, cf. [15], [16], and [17]. For the generalization to higher dimensions, cf. [19], [20], and [9].

Let $\{a_\nu\}_{\nu=1}^{\infty}$ be a discrete sequence in $\mathbf{C}$ such that $0 < |a_\nu| \leqq |a_{\nu+1}|$. We define the *exponent* λ *of convergence* of $\{a_\nu\}_{\nu=1}^{\infty}$ by

$$\lambda = \inf\left\{\mu > 0; \sum_{\nu=1}^{\infty} \frac{1}{|a_\nu|^\mu} < \infty\right\}. \tag{7.4.19}$$

For example, if $a_\nu = \nu$, $\nu = 1, 2, \ldots$, then $\lambda = 1$. In the same way as (7.4.6), we define the counting functions, $n(r)$ and $N(r)$ of $\{a_\nu\}_{\nu=1}^{\infty}$.

(7.4.20) LEMMA. *Let $\{a_\nu\}_{\nu=1}^{\infty}$, λ, $n(r)$, and $N(r)$ be as above. Then*

$$\lambda = \overline{\lim_{r\to\infty}} \frac{\log n(r)}{\log r} = \overline{\lim_{r\to\infty}} \frac{\log N(r)}{\log r}.$$

PROOF. It follows from Example (7.3.5) and Theorem (7.3.4) that for $\mu > 0$

$$(7.4.21) \qquad \sum_{\nu=1}^{\infty} \frac{1}{|a_\nu|^\mu} = \lim_{r\to\infty} \int_0^r \frac{1}{t^\mu} dn(t) = \lim_{r\to\infty} \left\{ \frac{n(r)}{r^\mu} + \mu \int_0^r \frac{n(t)}{t^{\mu+1}} dt \right\}.$$

If $\sum_{\nu=1}^{\infty} |a_\nu|^{-\mu} < \infty$, it follows from (7.4.21) that for an arbitrary $\epsilon > 0$ there is an $r_0 > 0$, satisfying

$$(7.4.22) \qquad \epsilon > \mu \int_r^\infty \frac{n(t)}{t^{\mu+1}} dt \geqq \mu n(r) \int_r^\infty \frac{dt}{t^{\mu+1}} = \frac{n(r)}{r^\mu}, \qquad r \geqq r_0.$$

Setting $\rho = \overline{\lim}_{r\to\infty} (\log n(r))/\log r$, one obtains $\rho \leqq \mu$. Therefore, $\rho \leqq \lambda$.

Conversely, we arbitrarily take $\mu > \rho$ and $\rho < \mu' < \mu$. Then, there is an $r_1 > 0$ such that $n(t) < t^{\mu'}$ for $t > r_1$. Hence, we have

$$\frac{n(r)}{r^\mu} < \frac{1}{r^{\mu-\mu'}} \to 0, \qquad r \to \infty,$$
$$\int_{r_1}^\infty \frac{n(t)}{t^{\mu+1}} dt \leqq \int_{r_1}^\infty \frac{1}{t^{\mu-\mu'+1}} dt < \infty.$$

By (7.4.21), $\lambda \leqq \mu$, and so $\lambda \leqq \rho$. Thus, the first equality has been proved.

The second equality follows from the following:

$$n(r) \leqq \frac{1}{\log 2} \int_r^{2r} \frac{n(t)}{t} dt \leqq \frac{1}{\log 2} N(2r),$$
$$\log^+ N(r) = \log^+ \int_0^r \frac{n(t)}{t} dt = \log^+ \left(\int_1^r \frac{n(t)}{t} dt + O(1) \right)$$
$$\leqq \log^+ (n(r) \log r) + O(1)$$
$$\leqq \log^+ n(r) + \log^+ \log r + O(1). \qquad \square$$

(7.4.23) THEOREM. *Let f be a meromorphic function of order ρ on* **C**. *Let $a \in \widehat{\mathbf{C}}$ and let λ be the exponent of convergence of the solutions of $f(z) = a$ with counting multiplicities, that is, a-points of f. Then, $\lambda \leqq \rho$.*

PROOF. This follows from Corollary (7.4.10) and Lemma (7.4.20). $\square$

By Picard's theorem of Chapter 6 (more directly, by Corollary (6.8.14)), there are infinitely many a-points of a transcendental meromorphic function f except for at most two values $a \in \widehat{\mathbf{C}}$. If the number of a-points of f is finite, a is called a *Picard exceptional value*. E. Borel proved that there are at most two $a \in \widehat{\mathbf{C}}$ such that the exponent of convergence of a-points of f is less than the order ρ of

f; such an a is called a *Borel exceptional value.* Of course, a Picard exceptional value is Borel. R. Nevanlinna defined the *defect* $\delta(a)$ of a by

$$\delta(a) = 1 - \varlimsup_{r\to\infty} \frac{N(r, 1/(f-a))}{T(r,f)}.$$

Corollary (7.4.10) implies that
i) $0 \leqq \delta(a) \leqq 1$;
ii) if a is a Picard exceptional value, then $\delta(a) = 1$.
In fact, Nevanlinna established an even deeper theorem:

$$\sum_{a\in\widehat{\mathbf{C}}} \delta(a) \leqq 2.$$

This formula means that there are at most countably many $a \in \widehat{\mathbf{C}}$ with $\delta(a) > 0$, and that their sum is not greater than two. The above inequality is called *Nevanlinna's defect relation,* and immediately implies that the number of Picard exceptional values is not greater than two.

EXERCISE 2. Let f_j, $j = 1, 2, \ldots, n$, be meromorphic functions on $\mathbf{C}$. Show the following:
i) $T\left(r, \sum_{j=1}^{n} f_j\right) \leqq \sum_{j=1}^{n} T(r, f_j) + \log n$,
ii) $T\left(r, \prod_{j=1}^{n} f_j\right) \leqq \sum_{j=1}^{n} T(r, f_j)$.

EXERCISE 3. Show the following for $f(z) = e^z$.
i) $N(r, f) = 0$ and $m(r, f) = r/\pi$. (Thus, $T(r, f) = r/\pi$.)
ii) For $a \in \mathbf{C}^*$, we have

$$N\left(r, \frac{1}{f-a}\right) = \frac{r}{\pi} + O(\log r), \qquad m\left(r, \frac{1}{f-a}\right) = O(\log r).$$

EXERCISE 4. What is the exponent of convergence of $a_n = n^\alpha$, $n = 1, 2, \ldots$, with $\alpha > 0$?

EXERCISE 5. i) For $f(z) = \sin z, \cos z$, show that

$$\frac{2r}{\pi} + O(\log r) \leqq T(r, f) \leqq \frac{2r}{\pi} + O(1).$$

ii) For $f(z) = \tan z$, show that $T(r, f) = \dfrac{2r}{\pi} + O(1)$.

7.5. Weierstrass' Product

Let f be a meromorphic function on $\mathbf{C}$. By Theorem (7.4.23) the exponent of convergence of zeros of f does not exceed the order of f. Conversely, for a discrete sequence $\{a_\nu\}$ of $\mathbf{C}$, whose exponent of convergence is ρ, we may ask whether there exists a meromorphic function of order ρ whose zeros are exactly $\{a_\nu\}$. In the present section we describe Weierstrass' product, which answers this question.

Let $p \in \mathbf{Z}^+$ and define *Weierstrass' irreducible factor* $E(z;p)$ by

$$E(z;0) = 1 - z,$$
$$E(z;p) = (1-z)e^{z+\frac{z^2}{2}+\cdots+\frac{z^p}{p}}, \qquad p \geqq 1.$$

(7.5.1) LEMMA. i) *For* $|z| \leqq 1/2$, $|\log E(z;p)| \leqq 2|z|^{p+1}$.

ii) *For an arbitrary* $z \in \mathbf{C}$ *we have*

$$\log|E(z;0)| \leqq \log(1+|z|),$$
$$\log|E(z;p)| \leqq A(p)\min\{|z|^p, |z|^{p+1}\}, \qquad p \geqq 1,$$

where $A(p) = 2(2+\log p)$.

PROOF. i) For $|z| \leqq 1 - 1/2(p+1) (\geqq 1/2)$ we have by (2.5.3)

$$\begin{aligned}
|\log E(z;p)| &= \left|\log(1-z) + z + \frac{z^2}{2} + \cdots + \frac{z^p}{p}\right| \\
&= \left|\frac{z^{p+1}}{p+1} + \frac{z^{p+2}}{p+2} + \cdots\right| \\
&\leqq \frac{|z|^{p+1}}{p+1}(1 + |z| + |z|^2 + \cdots) \\
&= \frac{|z|^{p+1}}{(p+1)(1-|z|)} \leqq 2|z|^{p+1}.
\end{aligned} \tag{7.5.2}$$

ii) The case of $p = 0$ is clear. Assume that $p > 0$. For $|z| \geqq 1$ we have

$$\begin{aligned}
\log|E(z;p)| &\leqq \log(1+|z|) + |z| + \frac{|z|^2}{2} + \cdots + \frac{|z|^p}{p} \\
&\leqq 2|z| + \frac{|z|}{2} + \frac{|z|^2}{2} + \cdots + \frac{|z|^p}{p} \\
&\leqq \left(2 + \frac{1}{2} + \cdots + \frac{1}{p}\right)|z|^p \\
&\leqq (2 + \log p)|z|^p.
\end{aligned} \tag{7.5.3}$$

By (7.5.2) and (7.5.3) it remains to show the given estimate for z in the annulus $\{1 - 1/2(p+1) < |z| < 1\}$. For $|z| = 1$, (7.5.3) implies that $\log|E(z;p)| \leqq 2 + \log p$. Since $\log|E(z;p)|$ is harmonic in $\Delta(1)$, the maximum principle (Theorem (3.6.10), iv)) implies that

$$\log|E(z;p)| \leqq (2 + \log p), \qquad |z| \leqq 1.$$

If $1 - 1/2(p+1) < |z| < 1$, then $|z|^{p+1} \geqq (1 - 1/2(p+1))^{p+1} \geqq 1/2$, and hence

$$\log|E(z;p)| \leqq (2 + \log p) \leqq 2(2 + \log p)|z|^{p+1}. \qquad \square$$

(7.5.4) LEMMA. *Let $\{a_\nu\}_{\nu=1}^\infty$ be a discrete sequence of $\mathbf{C}$ such that $0 < |a_\nu| \leqq |a_{\nu+1}|$. Let $p \in \mathbf{Z}^+$ be the minimum such that $\sum_{\nu=1}^\infty 1/|a_\nu|^{p+1} < \infty$. Then, the infinite product*

$$\Pi(z) = \prod_{\nu=1}^\infty E\left(\frac{z}{a_\nu}; p\right) \tag{7.5.5}$$

converges absolutely and uniformly on compact subsets, and hence defines an entire function satisfying

$$\log|\Pi(z)| = (p+1)A(p)\left\{|z|^p \int_0^{|z|} \frac{n(t)}{t^{p+1}} + |z|^{p+1}\int_{|z|}^\infty \frac{n(t)}{t^{p+2}}dt\right\}. \tag{7.5.6}$$

Here, $n(r)$ is the counting function of $\{a_\nu\}$, $A(0) = 1$, and $A(p) = 2(2 + \log p)$ for $p > 0$.

PROOF. Let $R > 0$ and $|z| \leqq R/2 \leqq |a_{\nu_0}|/2$. It follows from Lemma (7.5.1) that

$$\sum_{\nu=\nu_0}^\infty \left|\log E\left(\frac{z}{a_\nu}; p\right)\right| \leqq \sum_{\nu=\nu_0}^\infty 2\left|\frac{z}{a_\nu}\right|^{p+1} \leqq \sum_{\nu=\nu_0}^\infty \frac{R^{p+1}}{2^p|a_\nu|^{p+1}} < \infty.$$

Thus, the infinite product converges absolutely and uniformly on compact subsets.

Now, we prove the estimate (7.5.6). Assume that $p = 0$. Then,

$$\begin{aligned}
\log|\Pi(z)| &\leqq \sum_{\nu=1}^\infty \log\left(1 + \left|\frac{z}{a_\nu}\right|\right) = \int_0^\infty \log\left(1 + \frac{|z|}{t}\right) dn(t) \\
&= \left[n(t)\log\left(1 + \frac{|z|}{t}\right)\right]_0^\infty + \int_0^\infty \frac{|z|n(t)}{t(t+|z|)}dt \\
&= \lim_{r\to\infty} n(r)\log\left(1 + \frac{|z|}{r}\right) + \int_0^\infty \frac{|z|n(t)}{t(t+|z|)}dt
\end{aligned}$$

It follows from (7.4.22) that

$$0 < n(r)\log\left(1 + \frac{|z|}{r}\right) \leqq n(r)\frac{|z|}{r} \to 0, \qquad r \to \infty.$$

Therefore, one gets

$$\begin{aligned}
\log|\Pi(z)| &\leqq |z|\int_0^\infty \frac{n(t)}{t(t+|z|)}dt \\
&\leqq |z|\int_0^{|z|} \frac{n(t)}{t(t+|z|)}dt + |z|\int_{|z|}^\infty \frac{n(t)}{t(t+|z|)}dt \\
&\leqq \int_0^{|z|} \frac{n(t)}{t}dt + |z|\int_{|z|}^\infty \frac{n(t)}{t^2}dt.
\end{aligned}$$

By (7.4.21) the integral of the last term is finite. Assume that $p > 0$. Then, from Lemma (7.5.1) it follows that

$$\begin{aligned}\log|\Pi(z)| \leqq& A(p)\left\{\sum_{|a_\nu|<|z|}\left|\frac{z}{a_\nu}\right|^p + \sum_{|a_\nu|\geqq|z|}\left|\frac{z}{a_\nu}\right|^{p+1}\right\}\\ =& A(p)\left\{\int_0^{|z|}\frac{|z|^p}{t^p}dn(t) + \int_{|z|}^{\infty}\frac{|z|^{p+1}}{t^{p+1}}dn(t)\right\}\\ =& A(p)\left\{\left[\frac{|z|^p}{t^p}n(t)\right]_0^{|z|} + p|z|^p\int_0^{|z|}\frac{n(t)}{t^{p+1}}dt\right.\\ &\left.+\left[\frac{|z|^{p+1}}{t^{p+1}}n(t)\right]_{|z|}^{\infty} + (p+1)|z|^{p+1}\int_{|z|}^{\infty}\frac{n(t)}{t^{p+2}}dt\right\}.\end{aligned}$$

Note that $\lim_{t\to\infty} n(t)t^{-p-1} = 0$ by (7.4.22). Hence,

$$\log|\Pi(z)| \leqq (p+1)A(p)\left\{|z|^p\int_0^{|z|}\frac{n(t)}{t^{p+1}}dt + |z|^{p+1}\int_{|z|}^{\infty}\frac{n(t)}{t^{p+2}}dt\right\}.$$

By (7.4.21) the integral of the last term is finite. □

The above $\Pi(z)$ is called *Weierstrass' product* or the *canonical product* associated with the sequence $\{a_\nu\}$. Let λ be the exponent of convergence of $\{a_\nu\}$, and let ρ be the order of $\Pi(z)$. Then, $p \leqq \lambda \leqq p+1$. We take λ' so that if $\lambda < p+1$, $\lambda < \lambda' < p+1$, and if $\lambda = p+1$, $\lambda' = p+1$. In the case of $\lambda' < p+1$, it follows from Lemma (7.4.20) and (7.5.6) that

$$\begin{aligned}\log|\Pi(z)| \leqq& O\left(|z|^p\int_1^{|z|}t^{\lambda'-p-1}dt + |z|^{p+1}\int_{|z|}^{\infty}t^{\lambda'-p-2}dt\right)\\ \leqq& O\left(|z|^{\lambda'} + |z|^{\lambda'}\right) \leqq O\left(|z|^{\lambda'}\right).\end{aligned}$$

In the case of $\lambda' = p+1$, it follows from (7.4.22) and (7.5.6) that $\log|\Pi(z)| \leqq o(|z|^{\lambda'})$, where $o(|z|^{\lambda'})$ stands for a term such as $\lim_{|z|\to\infty} o(|z|^{\lambda'})/|z|^{\lambda'} = 0$. Therefore, by (7.4.17) $\rho \leqq \lambda'$. Letting $\lambda' \searrow \lambda$, one gets $\rho \leqq \lambda$. Combining this with Theorem (7.4.23), we see that

$$p \leqq \lambda = \rho \leqq p+1. \tag{7.5.7}$$

Note that $\Pi(0) = 1$, and if $p \geqq 1$, then about 0

$$\log E\left(\frac{z}{a_\nu}; p\right) = \frac{z^{p+1}}{a_\nu^{p+1}(p+1)} + \cdots.$$

From this and the estimate of $\log|\Pi(z)|$ obtained above, it follows that

$$\frac{d^k}{dz^k}\log\Pi(0) = 0, \qquad 0 \leqq k \leqq p, \qquad T(r,\Pi) = o(r^{p+1}). \tag{7.5.8}$$

(7.5.9) THEOREM. *Let $\{a_\nu\}_{\nu=1}^{\infty}$ be a discrete sequence in $\mathbf{C}$ such that $0 < |a_\nu| \leqq |a_{\nu+1}|$, and let λ be its exponent of convergence. Let $p \in \mathbf{Z}^+$ be the minimum such that $\sum_{\nu=1}^{\infty} 1/|a_\nu|^{p+1} < \infty$. Then, the order of Weierstrass' product $\Pi(z)$ associated with $\{a_\nu\}$ is λ. Moreover, if an entire function $f(z)$ of order λ has exactly the zeros $\{a_\nu\}$ and satisfies (7.5.8), then $f(z) = \Pi(z)$.*

PROOF. It remains to prove the uniqueness. Set

$$g(z) = \log \frac{f(z)}{\Pi(z)}, \qquad e^{g(z)} = \frac{f(z)}{\Pi(z)}.$$

It follows from Exercise 2 in the previous section, the first main Theorem (7.4.9), and (7.5.8) that

$$T(r, e^{g(z)}) \leqq T(r, f) + T(r, \Pi) = o(r^{p+1}).$$

Thus, by Theorem (7.4.12)

$$\max\{\operatorname{Re} g(z); |z| \leqq r\} = o(r^{p+1}).$$

In the same way, for $-g(z)$ we have that $\max\{-\operatorname{Re} g(z); |z| \leqq r\} = o(r^{p+1})$. Therefore,

$$\max\{|\operatorname{Re} g(z)|; |z| \leqq r\} = o(r^{p+1}).$$

It follows that

$$\int_0^{2\pi} \left| g(re^{i\theta}) + \overline{g(re^{i\theta})} \right|^2 d\theta = o(r^{2(p+1)}).$$

Setting $g(z) = \sum_{n=0}^{\infty} c_n z^n$, one obtains

$$\sum_{n=0}^{\infty} |z_n|^2 r^{2n} = o(r^{2(p+1)}).$$

Thus, for $n > p$, $c_n = 0$, and so $g(z) = \sum_{n=0}^{p} c_n z^n$. Then, condition (7.5.8) implies that $c_n = 0$ for $0 \leqq n \leqq p$, and that $g \equiv 0$. □

REMARK. The existence of an entire function with given zeros was already established in Theorem (7.2.9). It is an important point of Weierstrass' product $\Pi(z)$ that it is obtained with estimate (7.5.6). W. Stoll generalized this to the case of several complex variables (1953).

EXAMPLE. We prove the infinite product expression:

$$\sin z = z \prod_{-\infty}^{\infty}{}' \left(1 - \frac{z}{n\pi}\right) e^{z/n\pi} = z \prod_{n=1}^{\infty} \left(1 - \frac{z^2}{n^2\pi^2}\right).$$

Here, $\prod'$ stands for the product excluding $n = 0$. The zeros of $(\sin z)/z$ are $n\pi, n \in \mathbf{Z} \setminus \{0\}$. Since $\sum n^{-1} = \infty$ and $\sum n^{-2} < \infty$, we set $p = 1$, and

$$\begin{aligned} \Pi(z) &= \prod_{n=1}^{\infty} \left(1 + \frac{z}{n\pi}\right) e^{-z/n\pi} \cdot \left(1 - \frac{z}{n\pi}\right) e^{z/n\pi} \\ &= \prod_{-\infty}^{\infty}{}' \left(1 - \frac{z}{n\pi}\right) e^{z/n\pi}, \end{aligned}$$

which converges absolutely and uniformly on compact subsets in $\mathbf{C}$. Since the exponent of convergence of the zeros is 1, the order of $\Pi(z)$ is 1. It follows from Exercise 5 in the previous section that the order of $\sin z$ is 1. Since

$$\frac{\sin z}{z} = 1 - \frac{z^2}{3!} + \cdots, \qquad \frac{d}{dz}\Big|_{z=0} \frac{\sin z}{z} = 0,$$

we have

$$\frac{\sin z}{z} = \prod_{-\infty}^{\infty}{}' \left(1 - \frac{z}{n\pi}\right) e^{z/n\pi}.$$

(7.5.10) EXAMPLE. The gamma function $\Gamma(z)$ is defined in $\operatorname{Re} z > 0$ by

$$\Gamma(z) = \int_0^{\infty} e^{-t} t^{z-1} dt. \tag{7.5.11}$$

The integral is uniformly convergent on compact subsets in $\operatorname{Re} z > 0$. Integrating by parts, we have

$$\Gamma(z+1) = z\Gamma(z). \tag{7.5.12}$$

Since $\Gamma(1) = 1$, $\Gamma(n) = n!$, $n \in \mathbf{N}$.

For $x > 0$ we set

$$\phi_n(x) = \int_0^n \left(1 - \frac{t}{n}\right)^n t^{x-1} dt, \qquad n = 1, 2, \dots.$$

We show that

$$\phi_n(x) < \phi_{n+1}(x) \to \Gamma(x), \qquad n \to \infty. \tag{7.5.13}$$

For that purpose we show that the following convergence is monotone and uniform on compact subsets of $[0, \infty)$:

$$\left(1 - \frac{t}{n}\right) \leqq \left(1 - \frac{t}{n+1}\right)^{n+1} \to e^{-t}, \qquad n \to \infty, \quad t \geqq 0. \tag{7.5.14}$$

It follows from (2.5.3) that for $n > t \geqq 0$

$$\log e^{-t} - \log \left(1 - \frac{t}{n}\right)^n = \sum_{\nu=2}^{\infty} \frac{t^{\nu}}{\nu n^{\nu-1}}.$$

The right side is clearly monotone decreasing in n. Let $k > 0$ be an arbitrary number. Take n_0 so that $0 \leqq t \leqq k < n_0/2$.

$$\begin{aligned}\sum_{\nu=2}^{\infty} \frac{t^\nu}{\nu n^{\nu-1}} &\leqq \sum_{\nu=2}^{\infty} \frac{n_0{}^\nu}{\nu n^{\nu-1}} \left(\frac{1}{2}\right)^\nu \\ &\leqq \sum_{\nu=2}^{n_0} \frac{n_0{}^\nu}{\nu n^{\nu-1}} \left(\frac{1}{2}\right)^\nu + \sum_{\nu=n_0+1}^{\infty} \left(\frac{1}{2}\right)^\nu \\ &= \sum_{\nu=2}^{n_0} \frac{n_0{}^\nu}{\nu n^{\nu-1}} \left(\frac{1}{2}\right)^\nu + \left(\frac{1}{2}\right)^{n_0}.\end{aligned}$$

For an arbitrary $\epsilon > 0$, we take $n_0 > 2k$ such that $2^{-n_0} < \epsilon$. Moreover, take $n_1 > n_0$ so that

$$\sum_{\nu=2}^{n_0} \frac{n_0{}^\nu}{\nu n^{\nu-1}} \left(\frac{1}{2}\right)^\nu < \epsilon, \qquad n > n_1.$$

Then, for $n > n_1$ and $t \in [0, k]$

$$0 \leqq \log e^{-t} - \log \left(1 - \frac{t}{n}\right)^n < 2\epsilon.$$

Thus, the uniform convergence on compact subsets is proved.

Now, we prove the convergence of (7.5.13). Let $I \subset (0, \infty)$ be an arbitrary bounded and closed interval. Take an arbitrary $\epsilon > 0$ and $n_0 \in \mathbf{N}$ so that

$$\int_{n_0}^{\infty} e^{-t} t^{x-1} dt < \epsilon, \qquad x \in I.$$

This and (7.5.14) imply that for $n > n_0$ and $x \in I$

$$\begin{aligned}0 < \phi_n(x) - \int_0^{n_0} \left(1 - \frac{t}{n}\right)^n t^{x-1} dt \\ = \int_{n_0}^{n} \left(1 - \frac{t}{n}\right)^n t^{x-1} dt < \int_{n_0}^{\infty} e^{-t} t^{x-1} dt < \epsilon.\end{aligned}$$

There is an $n_1 > n_0$ such that for all $n > n_1$

$$0 < \int_0^{n_0} e^{-t} t^{x-1} dt - \int_0^{n_0} \left(1 - \frac{t}{n}\right)^n t^{x-1} dt < \epsilon.$$

Therefore,

$$0 < \Gamma(x) - \phi_n(z) < 3\epsilon, \qquad n > n_1, \quad x \in I.$$

By repeating the integration by parts for $\phi_n(x)$, we get

$$\phi_n(x) = \frac{n^x n!}{x(x+1) \cdots (x+n)}.$$

By making use of the expression of the right side, we extend the defining domain of $\phi_n(x)$ to $\operatorname{Re} z > 0$, and define

$$\phi_n(z) = \frac{n^z n!}{z(z+1)\cdots(z+n)}, \qquad \operatorname{Re} z > 0.$$

It follows from the above definition of $\phi_n(z)$ that

$$\frac{1}{\phi_n(z)} = ze^{z(1+\frac{1}{2}+\cdots+\frac{1}{n}-\log n)} \prod_{\nu=1}^{n} \left(1+\frac{z}{\nu}\right) e^{-\frac{z}{\nu}}.$$

It is well known that $\sum_{\nu=1}^{n} \frac{1}{\nu} - \log n$ converges to a positive constant C, called Euler's number. It follows from Lemma (7.5.4) that the convergence

$$\frac{1}{\phi_n(z)} \to ze^{Cz} \prod_{\nu=1}^{\infty} \left(1+\frac{z}{\nu}\right) e^{-\frac{z}{\nu}} \tag{7.5.15}$$

is absolute and uniform on compact subsets of $\mathbf{C}$; thus the limit is an entire function. It follows from (7.5.13) and the identity theorem (sharing the same values on the real positive line) that

$$\frac{1}{\Gamma(z)} = ze^{Cz} \prod_{\nu=1}^{\infty} \left(1+\frac{z}{\nu}\right) e^{-\frac{z}{\nu}}. \tag{7.5.16}$$

Therefore, we deduce that $\Gamma(z)$ is analytically continued to a meromorphic function on $\mathbf{C}$ which has poles $n \in \mathbf{Z}$, $n \leqq 0$, of order one, and has no pole nor zeros elsewhere. It also follows from (7.5.15) that

$$\Gamma(z) = \lim_{n\to\infty} \frac{n^z n!}{z(z+1)\cdots(z+n)}. \tag{7.5.17}$$

Equation (7.5.16) is called Weierstrass' infinite product expression, and (7.5.17) is called Gauss' infinite product expression.

The gamma function $\Gamma(z)$ plays an important role in studying a number of other special and important functions, e.g., *Riemann's zeta function* $\zeta(z)$. Here, $\zeta(z)$ is defined by

$$\zeta(z) = \sum_{n=1}^{\infty} \frac{1}{n^z}, \qquad \operatorname{Re} z > 1.$$

It is known that $\zeta(z)$ can be analytically continued to a meromorphic function on $\mathbf{C}$ which has an pole of order one only at $z = 1$, and has zeros of order one at the negative even integers (called trivial zeros). It is the famous olden Riemann hypothesis that all the other zeros of $\zeta(z)$ lie on the line $\operatorname{Re} z = \frac{1}{2}$; this has been an open problem since 1859.

EXERCISE 1. Obtain the infinite product expression of $\cos z$.

EXERCISE 2. What are the exponents of convergence of the following sequences, and their Weierstrass' products?

i) $a_n = n(\log n)^2$, $n = 2, 3, \ldots$. ii) $a_n = n!$, $n = 1, 2, \ldots$.

EXERCISE 3. Show that for an arbitrary $\rho \geqq 0$ there is an entire function of order ρ.

EXERCISE 4 (H. Bohr and J. Mollerup). Show that if a function $\phi(x)$ in $x > 0$ satisfies the following three conditions, then $\phi(x) = \Gamma(x)$:

i) $\phi(1) = 1$,
ii) $\phi(x+1) = x\phi(x)$,
iii) $\log \phi(x)$ is a convex function.

7.6. Elliptic Functions

Let f be a meromorphic function on $\mathbf{C}$. If a number $\omega \in \mathbf{C}$ satisfies

$$f(z+\omega) = f(z), \qquad z \in \mathbf{C},$$

ω is called a *period* of f. For example, $f(z) = e^z$ has a period $2\pi i$, and $f(z) = \sin z, \cos z$ have a period 2π. Let Ω denote the set of all periods of f. Then Ω forms an additive group; that is, $0 \in \Omega$, and if $\omega_1, \omega_2 \in \Omega$, then $\omega_1 - \omega_2 \in \Omega$. We call Ω the *period group* of f.

(7.6.1) LEMMA. *Let f and Ω be as above, and assume that f is not constant. Then Ω is one of the following three cases:*

i) $\Omega = \{0\}$.
ii) *There is an $\omega_1 \in \mathbf{C} \setminus \{0\}$ such that $\Omega = \{n\omega_1; n \in \mathbf{Z}\}$.*
iii) *There are $\mathbf{R}$-linearly independent $\omega_1, \omega_2 \in \Omega$ such that*

$$\Omega = \{n_1\omega_1 + n_2\omega_2; n_1, n_2 \in \mathbf{Z}\}.$$

PROOF. Assume that $\Omega \neq \{0\}$. Since f is not constant, by the identity Theorem (2.4.14) Ω is discrete. Take $\omega_1 \in \Omega \setminus \{0\}$ such that

$$|\omega_1| = \min\{|\omega|; \omega \in \Omega\}. \tag{7.6.2}$$

Take an arbitrary element $\omega \in \Omega \setminus \{0\}$, and set $\tau = \omega/\omega_1$. Assume that $\tau \in \mathbf{R}$. By making use of Gauss' symbol $[\tau]$, we have that $\omega = [\tau]\omega_1 + (\tau - [\tau])\omega_1$. Hence, $(\tau - [\tau])\omega_1 \in \Omega$, and $|(\tau - [\tau])\omega_1| < |\omega_1|$. It follows from (7.6.2) that $\tau = [\tau] \in \mathbf{Z}$. Therefore, if there is no element of Ω which is linearly independent of ω_1 over $\mathbf{R}$, Ω is of case ii).

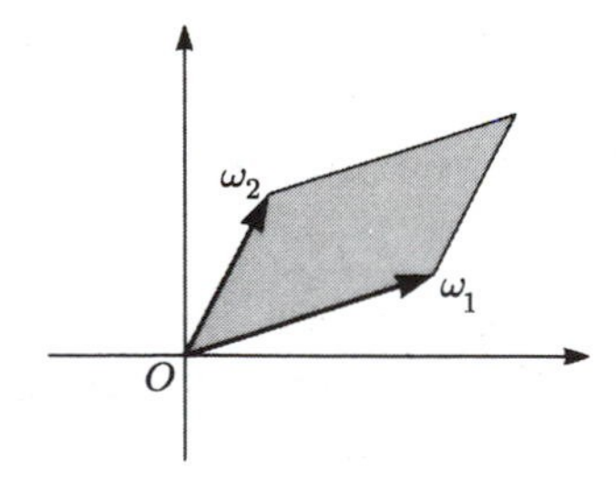

Figure 89

Assume that $\Omega' = \{\omega \in \Omega; \omega/\omega_1 \notin \mathbf{R}\} \neq \emptyset$. Let $\omega \in \Omega'$, and denote by $Q[\omega_1, \omega]$ the parallelogram with vertices, $0, \omega_1, \omega, \omega_1 + \omega$. Let $|Q[\omega_1, \omega]|$ be the area of $Q[\omega_1, \omega]$. Take $\omega_2 \in \Omega'$ such that

$$|Q[\omega_1, \omega_2]| = \min_{\omega \in \Omega'} |Q[\omega_1, \omega]|.$$

Let $\omega \in \Omega$ be an arbitrary element. Then there are unique $\tau_1, \tau_2 \in \mathbf{R}$ such that

$$\begin{aligned}\omega &= \tau_1\omega_1 + \tau_2\omega_2 \\ &= [\tau_1]\omega_1 + [\tau_2]\omega_2 + (\tau_1 - [\tau_1])\omega_1 + (\tau_2 - [\tau_2])\omega_2.\end{aligned}$$

We want to show that $\tau_1, \tau_2 \in \mathbf{Z}$. By the above equation, it suffices to show that if $0 \leqq \tau_j < 1$, $j = 1, 2$, then $\tau_j = 0$. Since $|Q[\omega_1, \omega]| = \tau_2 |Q[\omega_1, \omega_2]|$, the definition of ω_2 implies that $\tau_2 = 0$. Thus, $\omega = \tau_1\omega_1$. The choice of ω_1 implies $\tau_1 = 0$. In this case, iii) holds. □

The above Ω is a discrete subgroup of $\mathbf{C}$, and acts on $\mathbf{C}$ by

$$\mathbf{C} \ni z \to z + \omega \in \mathbf{C}, \qquad \omega \in \Omega.$$

As described in Chapter 5, §3, the set $\mathbf{C}/\Omega$ of all Ω-orbits $[z] = \{z + \omega; \omega \in \Omega\}$ gives rise to a Riemann surface. If $\Omega = \{0\}$, then $\mathbf{C}/\Omega = \mathbf{C}$, and this is the most likely case for f. For instance, if $f(z)$ is a non-constant polynomial, or rational function, then $\Omega = \{0\}$. For, if there exists an $\omega \in \Omega \setminus \{0\}$, then for $z_0 \in \mathbf{C}$ $f(z_0 + n\omega) = f(z_0)$, $n \in \mathbf{Z}$, and hence $f(z) \equiv f(z_0)$.

Next we consider the case of ii). Take the universal covering (cf. Figure 90)

$$\pi : \mathbf{C} \ni z \to e^{2\pi i z/\omega_1} \in \mathbf{C}^*.$$

Thus, $\mathbf{C}/\Omega = \mathbf{C}^*$. Then, $f(z)$ is expressed by a meromorphic function $g(w)$ in $w = e^{2\pi i z/\omega_1}$. We take a ring domain $R(r_1, r_2) = \{r_1 < |z| < r_2\}$ so that it does not contain any pole of $g(w)$. Then, $g(w)$ is expanded there to a Laurent series

$$g(w) = \sum_{n=-\infty}^{\infty} c_n w^n.$$

Therefore, we have the expression,

$$f(z) = \sum_{n=-\infty}^{\infty} c_n e^{2\pi i z/\omega_1}.$$

If $f(z)$ is entire, the above expansion converges uniform on compact subsets in $\mathbf{C}$.

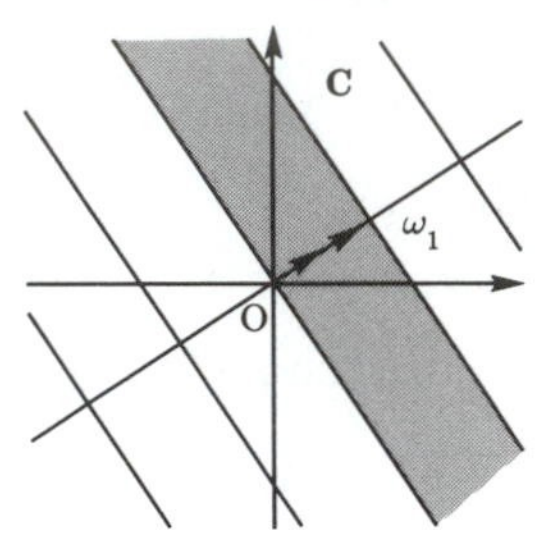

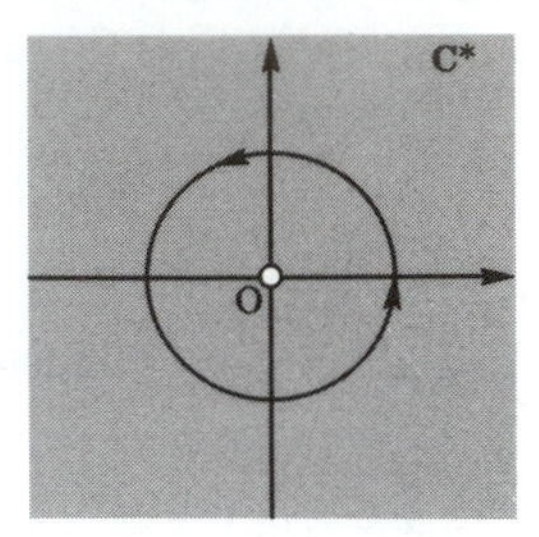

FIGURE 90

We next deal with the remaining case of iii), which is the the main theme of the present section. In what follows, we assume that Ω is of case iii). Then, f is called a *doubly periodic function* or an *elliptic function*; ω_1 and ω_2 are called the *fundamental periods*. (The origin of the name, "elliptic function" will be explained at the end of the section.) The set $Q[\omega_1, \omega_2]$ is called the *period parallelogram* of f, or of Ω. In the sequel, we assume that $Q[\omega_1, \omega_2]$ contains the edges $t\omega_1$ and $t\omega_2$, $0 \leqq t < 1$.

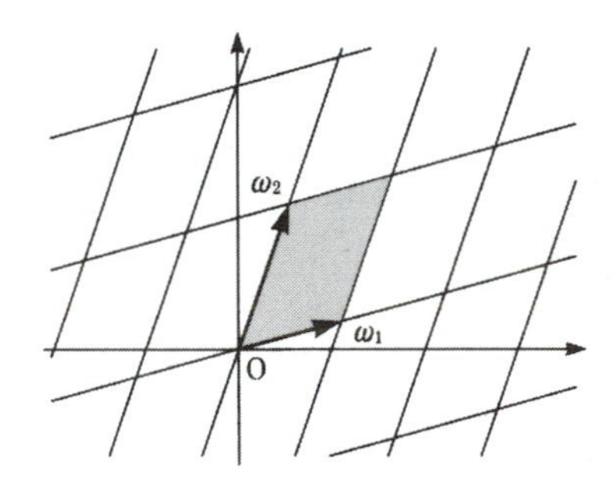

FIGURE 91

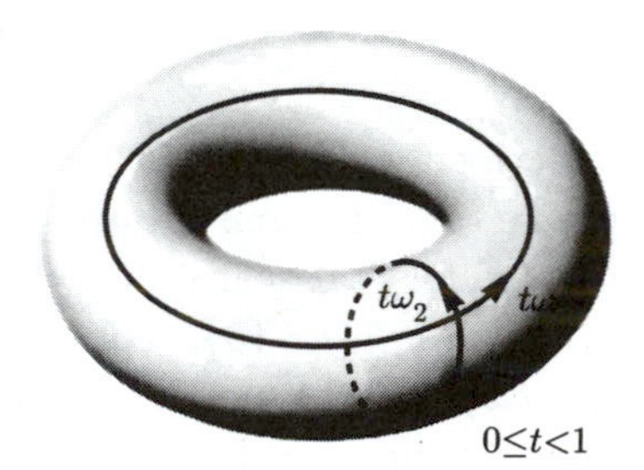

FIGURE 92

Let

$$\pi : \mathbf{C} \to \mathbf{C}/\Omega$$

be the universal covering. Then, $\pi(Q[\omega_1, \omega_2]) = \pi(\overline{Q[\omega_1, \omega_2]}) = \mathbf{C}/\Omega$, and so $\mathbf{C}/\Omega$ is compact. Topologically, $\mathbf{C}/\Omega$ is homeomorphic to the surface of a doughnut as in Figure 92. We call $\mathbf{C}/\Omega$ a 1-dimensional *complex torus*, or an *elliptic curve*. The periods ω_1, ω_2 are called *generators* of Ω, and we write

$$\Omega = \Omega[\omega_1, \omega_2].$$

The subgroup Ω is also called a *lattice* of $\mathbf{C}$, and every element $\omega \in \Omega$ is called a *lattice point*. Using $-\omega_2$ if necessary, we may assume that

$$\operatorname{Im} \frac{\omega_1}{\omega_2} > 0.$$

For two elements $[a]$ and $[b]$ of $\mathbf{C}/\Omega$, we define the addition by

$$[a]+[b]=[a+b].$$

For an arbitrary $[a]\in\mathbf{C}/\Omega$, the transition

$$\mathbf{C}/\Omega\ni[z]\to[z]+[a]\in\mathbf{C}/\Omega \tag{7.6.3}$$

is biholomorphic.

Let $\omega_1',\omega_2'\in\Omega$ be other generators. Then,

$$\begin{aligned}\omega_1' &= a\omega_1+b\omega_2, & \omega_1 &= a'\omega_1'+b'\omega_2',\\ \omega_2' &= c\omega_1+d\omega_2, & \omega_2 &= c'\omega_1'+d'\omega_2',\end{aligned}$$

where $a,b,c,d,a',b',c',d'\in\mathbf{Z}$. Set $A=\left(\begin{smallmatrix}a&b\\c&d\end{smallmatrix}\right)$ and $A'=\left(\begin{smallmatrix}a'&b'\\c'&d'\end{smallmatrix}\right)$. Then, $AA'=\left(\begin{smallmatrix}1&0\\0&1\end{smallmatrix}\right)$. Let $GL(2,\mathbf{Z})$ denote the group of all 2×2 invertible matrices with elements in $\mathbf{Z}$. Then we have

$$A=\begin{pmatrix}a&b\\c&d\end{pmatrix}\in GL(2,\mathbf{Z}),\qquad \begin{pmatrix}\omega_1'\\ \omega_2'\end{pmatrix}=\begin{pmatrix}a&b\\c&d\end{pmatrix}\begin{pmatrix}\omega_1\\ \omega_2\end{pmatrix}. \tag{7.6.4}$$

Conversely, if ω_1' and ω_2' are given by (7.6.4), then ω_1' and ω_2' are the generators of Ω. If $\operatorname{Im}\omega_1/\omega_2>0$ and $\operatorname{Im}\omega_1'/\omega_2'>0$, then $A\in SL(2,\mathbf{Z})$. Summarizing the above, we have

(7.6.5) LEMMA. *Let $\Omega[\omega_1,\omega_2]$ and $\Omega[\omega_1',\omega_2']$ be lattices of $\mathbf{C}$. Then, $\Omega[\omega_1,\omega_2]=\Omega[\omega_1',\omega_2']$ if and only if there is a matrix $\left(\begin{smallmatrix}a&b\\c&d\end{smallmatrix}\right)\in GL(2,\mathbf{Z})$ such that*

$$\begin{pmatrix}\omega_1'\\ \omega_2'\end{pmatrix}=\begin{pmatrix}a&b\\c&d\end{pmatrix}\begin{pmatrix}\omega_1\\ \omega_2\end{pmatrix}.$$

In particular, if $\operatorname{Im}\omega_1/\omega_2>0$ and $\operatorname{Im}\omega_1'/\omega_2'>0$, then $\left(\begin{smallmatrix}a&b\\c&d\end{smallmatrix}\right)\in SL(2,\mathbf{Z})$.

Setting $\tau=\omega_1/\omega_2$, one gets a holomorphic automorphism $F:\mathbf{C}\ni z\to z\omega\in\mathbf{C}$, and so a biholomorphic mapping

$$f:\mathbf{C}/\Omega[\omega_1,\omega_2]\to\mathbf{C}/\Omega[1,\tau]. \tag{7.6.6}$$

Let $\pi:\mathbf{C}\to\mathbf{C}/\Omega[\omega_1,\omega_2]$ and $\pi_1:\mathbf{C}\to\mathbf{C}/\Omega[1,\tau]$ be the universal covering mappings. Then,

$$f\circ\pi=\pi_1\circ F,\qquad \begin{array}{ccc}\mathbf{C}&\xrightarrow[F]{}&\mathbf{C}\\ \downarrow\pi&&\downarrow\pi_1\\ \mathbf{C}/\Omega[\omega_1,\omega_2]&\xrightarrow{f}&\mathbf{C}/\Omega[1,\tau].\end{array} \tag{7.6.7}$$

(7.6.8) THEOREM. *Two complex tori, $\Omega[1,\tau_j]$, $j=1,2$, where $\operatorname{Im}\tau_j>0$, are biholomorphic to each other if and only if*

$$\tau_1=\frac{a\tau_2+b}{c\tau_2+d},\qquad \begin{pmatrix}a&b\\c&d\end{pmatrix}\in SL(2,\mathbf{Z}).$$

PROOF. The "if" part follows from Lemma (7.6.5). Conversely, suppose that there is a biholomorphic mapping

$$f : \mathbf{C}/\Omega[1,\tau_1] \to \mathbf{C}/\Omega[1,\tau_2].$$

By (7.6.3) we may assume without loss of generality that $f([0]) = [0]$. Let $\pi_j : \mathbf{C} \to \mathbf{C}/\Omega[1,\tau_j]$ be the universal covering mappings. As in (7.6.7) there is a lifting $F : \mathbf{C} \to \mathbf{C}$ of $f \circ \pi_1$ such that F is biholomorphic, $F(0) = 0$, and $f \circ \pi_1 = \pi_2 \circ F$. It follows from Theorem (6.1.1) that $F(z) = \alpha z$ with $\alpha \in \mathbf{C}^*$. Then, $F(\Omega[1,\tau_1]) = \Omega[1,\tau_2]$; in particular, $F(1) = \alpha \in \Omega[1,\tau_2]$, and $F(\tau_1) = \alpha\tau_1 \in \Omega[1,\tau_2]$. Thus, α and $\alpha\tau_1$ generate $\Omega[1,\tau_2]$. By Lemma (7.6.5) there is a matrix $\left(\begin{smallmatrix} a & b \\ c & d \end{smallmatrix}\right) \in SL(2,\mathbf{Z})$ such that

$$\begin{pmatrix} \alpha\tau_1 \\ \alpha \end{pmatrix} = \begin{pmatrix} a & b \\ c & d \end{pmatrix} \begin{pmatrix} \tau_2 \\ 1 \end{pmatrix}.$$

Therefore, $\tau_1 = (a\tau_2 + b)/(c\tau_2 + d)$. □

We investigate holomorphic and meromorphic functions on 1-dimensional complex tori.

(7.6.9) THEOREM. *A doubly periodic holomorphic function f is necessarily constant.*

PROOF. Let $\Omega[\omega_1,\omega_2]$ be the period group of f. Then,

$$|f(z)| \leqq \max\left\{|f(w)|; w \in \overline{Q[\omega_1,\omega_2]}\right\}, \qquad z \in \mathbf{C}.$$

Therefore, f is bounded on $\mathbf{C}$, and by Liouville's Theorem (3.5.22) f is a constant. □

By this theorem we see that for a given lattice $\Omega[\omega_1,\omega_2]$ there is no non-constant holomorphic function with period group $\Omega[\omega_1,\omega_2]$. One may next naturally ask about the existence of non-constant doubly periodic meromorphic functions. Later, we will construct such functions by means of Weierstrass, but for now we deal with the general properties of doubly periodic meromorphic functions.

Let $\Omega = \Omega[\omega_1,\omega_2]$ be a lattice, and let f be a meromorphic function on $\mathbf{C}$ such that every $\omega \in \Omega$ is a period of f. In this case, f is said to be *Ω-invariant.* We define the *degree* $\deg f$ of f as the number of poles of f in the period parallelogram $Q[\omega_1,\omega_2]$, counting multiplicities.

(7.6.10) THEOREM. *Let f and $\Omega = \Omega[\omega_1,\omega_2]$ be as above. Then we have the following.*

i) $\sum_{a\in Q[\omega_1,\omega_2]} \operatorname{Res}(a; f) = 0$.
ii) *If f is not constant, then* $\deg f \geqq 2$.
iii) *For an arbitrary $\alpha \in \widehat{\mathbf{C}}$, the number of α-points of f, counting multiplicities, is* $\deg f$.

PROOF. i) Let a_i, $i = 1, 2. \ldots n$, be poles of f contained in $Q[\omega_1, \omega_2]$ ($n \leqq \deg f$).

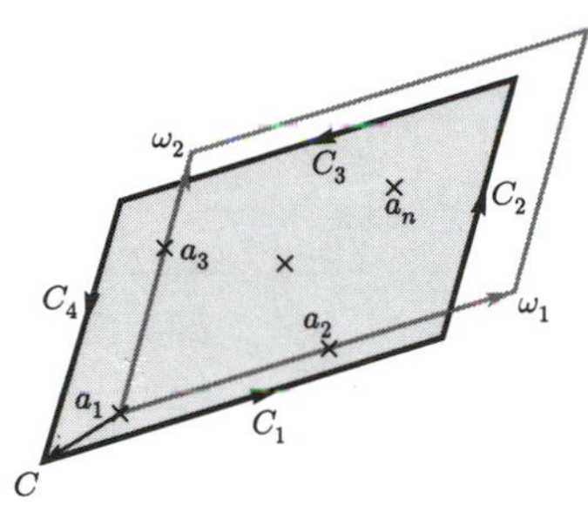

FIGURE 93

Take a small $c \in \mathbf{C}$ so that the interior of the shifted parallelogram $c + Q[\omega_1, \omega_2]$ contains all a_i. Let $C_j, 1 \leqq j \leqq 4$, be its perimeter, so that $C_2 = \omega_1 + C_4$ and $C_3 = \omega_2 + C_1$, and the orientation of $\sum_{j=1}^{4} C_j$ is anti-clockwise. Then it follows from the residue Theorem (4.2.7) (cf. (4.2.14)) that

$$\begin{aligned}
\sum_{i=1}^{n} \operatorname{Res}(a_i; f) &= \frac{1}{2\pi i} \int_{\sum C_j} f(z) dz = \frac{1}{2\pi i} \sum_{j=1}^{4} \int_{C_j} f(z) dz \\
&= \frac{1}{2\pi i} \sum_{j=1}^{2} \left\{ \int_{\sum C_j} f(z) dz + \int_{\sum C_{j+2}} f(z) dz \right\} \\
&= \frac{1}{2\pi i} \left\{ \int_{C_1} (f(z) - f(z + \omega_2)) dz + \int_{C_4} (f(z) - f(z + \omega_1)) dz \right\} \\
&= 0.
\end{aligned}$$

In the last equality, the periodicity of f was used.

ii) By Theorem (7.6.9) it suffices to show that there is no f with $\deg f = 1$. If so, f has only one pole $a_1 \in Q[\omega_1, \omega_2]$ with order 1. Then,

$$\sum_{a \in Q[\omega_1, \omega_2]} \operatorname{Res}(a; f) = \operatorname{Res}(a_1; f) \neq 0.$$

This contradicts i).

iii) Let b_i, $1 \leqq i \leqq m$, be points of $f^{-1}(\alpha)$, counting multiplicities. As in the proof of i), we may suppose that not only the poles of f, but also all b_j are contained in the interior of $Q[\omega_1, \omega_2]$. It follows from the argument principle, Theorem (4.3.4) and the periodicity of f and f' that

$$m - \deg f = \frac{1}{2\pi i} \int_{\sum_{j=1}^{4} C_j} \frac{f'(z)}{f(z) - \alpha} dz = 0. \qquad \square$$

EXERCISE 1. Let $f(z)$ be an Ω-invariant meromorphic function on $\mathbf{C}$. If $f(z)$ is an even function, then $f'(z)$ has a pole or zero at ω such that $2\omega \in \Omega$.

In what follows, the notation $\sum'_{\omega\in\Omega}$ stands for summation over $\omega \in \Omega \setminus \{0\}$.

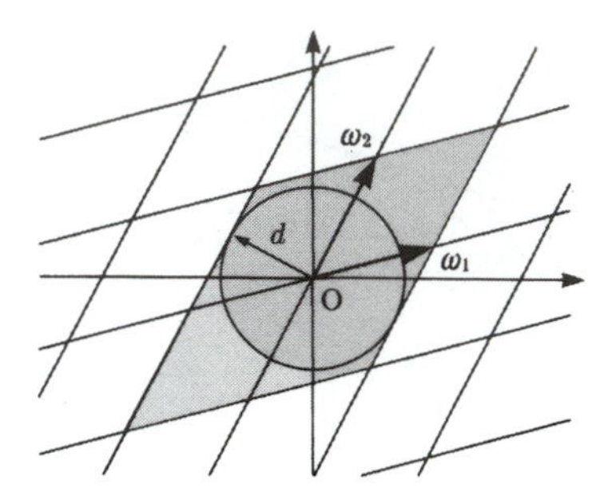

FIGURE 94

(7.6.11) LEMMA. *For $\alpha > 2$, the sum $\sum'_{\omega\in\Omega} |\omega|^{-\alpha}$ is convergent.*

PROOF. Set $\Omega_k = \{n\omega_1 + m\omega_2; |n| \leqq k, |m| \leqq k\}$. Then the number of lattice points of Ω_k is $(2k+1)^2$, and that of $\Omega_k \setminus \Omega_{k-1}$ is $8k$. Let d denote the shortest distance from the origin 0 to the perimeter of the parallelogram with vertices $\pm\omega_1$ and $\pm\omega_2$. Then one has

$$\begin{aligned}\sum_{\omega\in\Omega}{}' \frac{1}{|\omega|^\alpha} &= \sum_{k=1}^{\infty} \sum_{\omega\in\Omega_k\setminus\Omega_{k-1}} \frac{1}{|\omega|^\alpha} \leqq \sum_{k=1}^{\infty} \sum_{\omega\in\Omega_k\setminus\Omega_{k-1}} \frac{1}{(kd)^\alpha} \\ &= \sum_{k=1}^{\infty} \frac{8}{d^\alpha} \frac{1}{k^{\alpha-1}} < \infty. \qquad \square\end{aligned}$$

(7.6.12) THEOREM. *The infinite sum*

$$\wp(z) = \frac{1}{z^2} + \sum_{\omega\in\Omega}{}' \left\{ \frac{1}{(z-\omega)^2} - \frac{1}{\omega^2} \right\}, \qquad z \in \mathbf{C}, \tag{7.6.13}$$

converges absolutely and uniformly on every compact subset except for a finite number of terms, and defines a doubly periodic meromorphic function whose period group is Ω.

PROOF. Take an arbitrary $R > 0$. For $|z| \leqq R$ and $|\omega| \geqq 2R$ we have

$$\begin{aligned}\left| \frac{1}{(z-\omega)^2} - \frac{1}{\omega^2} \right| &= \left| \frac{2z\omega - z^2}{(z-\omega)^2\omega^2} \right| \\ &\leqq R \frac{2 + \frac{|z|}{|\omega|}}{\left(1 - \frac{|z|}{|\omega|}\right)^2} \frac{1}{|\omega|^3} \leqq 10R \frac{1}{|\omega|^3}.\end{aligned}$$

Therefore, by Lemma (7.6.11) we get the convergence stated above.

We show the periodicity. Termwise differentiation implies

$$\wp'(z) = -2 \sum_{\omega\in\Omega} \frac{1}{(z-\omega)^3}. \tag{7.6.14}$$

Thus, the period group of $\wp'(z)$ is clearly Ω. For an arbitrary $\omega_0 \in \Omega$

$$\wp(z+\omega_0) - \wp(z) = C, \tag{7.6.15}$$

where C is a constant. By definition (7.6.13) $\wp(z)$ is an even function; i.e.,

$$\wp(-z) = \wp(z), \qquad z \in \mathbf{C}. \tag{7.6.16}$$

Setting $z = -\omega_0/2$ in (7.6.15), one obtains

$$\wp\left(\frac{\omega_0}{2}\right) - \wp\left(-\frac{\omega_0}{2}\right) = \wp\left(\frac{\omega_0}{2}\right) - \wp\left(\frac{\omega_0}{2}\right) = 0 = C.$$

Hence, the periodicity, $\wp(z+\omega_0) = \wp(z)$ follows. Since the poles of $\wp(z)$ are exactly at the lattice points of Ω, Ω is the period group of $\wp(z)$. $\square$

It follows from (7.6.13) and (7.6.14) that $\wp(z)$ is even, $\wp'(z)$ is odd, and

$$\deg \wp = 2, \qquad \deg \wp' = 3. \tag{7.6.17}$$

The above obtained doubly periodic meromorphic function $\wp(z)$ associated with Ω is called *Weierstrass' pe-function*, and it plays a fundamental role in the study of the 1-dimensional complex torus (elliptic curve) $\mathbf{C}/\Omega$. In the rest of this section we describe a very introductory part of the theory.

(7.6.18) THEOREM. i) *For two points $z, z' \in \mathbf{C}$, $\wp(z) = \wp(z')$ if and only if $[z] = [\pm z']$.*

ii) *The poles of $\wp'(z)$ are the points of Ω, and the zeros of $\wp'(z)$ are the Ω-orbits of $\omega_1/2$, $\omega_2/2$, and $(\omega_1+\omega_2)/2$.*

iii) *Set $e_1 = \wp(\omega_1/2)$, $e_2 = \wp(\omega_2/2)$, and $e_3 = \wp((\omega_1+\omega_2)/2)$. Then these are distinct, and*

$$\wp(z)^2 = 4(\wp(z) - e_1)(\wp(z) - e_2)(\wp(z) - e_3).$$

iv) *Set $g_2 = 60\sum'_{\omega\in\Omega} \omega^{-4}$ and $g_3 = 140\sum'_{\omega\in\Omega}\omega^{-6}$. Then we have*

$$\wp'(z)^2 = 4\wp(z)^3 - g_2\wp(z) - g_3.$$

PROOF. i) By (7.6.13) $\wp(z)$ is even, and its period group is Ω. Therefore, if $[z] = [\pm z']$, then $\wp(z) = \wp(z')$. On the other hand, assume that $\wp(z) = \wp(z')$. Since $\deg \wp = 2$ and $\wp(z) = \wp(-z)$, it follows from Theorem (7.6.10) that $[z'] = [\pm z]$.

ii) Let ω_0 be $\omega_1/2$, $\omega_2/2$, or $(\omega_1+\omega_2)/2$. By Exercise 1 of the present section, $\wp'(\omega_0) = 0$. Since $\deg \wp = 2$, it follows from Theorem (7.6.10), iii) that $\wp(z)$ has $\wp(\omega_0)$ as double point, and so $\wp(z)$ does not take the value $\wp(\omega_0)$ on $Q[\omega_1, \omega_2]$ except for ω_0. Hence, e_1, e_2, and e_3 are distinct.

iii) Set

$$f(z) = (\wp(z) - e_1)(\wp(z) - e_2)(\wp(z) - e_3).$$

The set of poles of $\wp'(z)^2$ and $f(z)$ is Ω, and the orders of those poles are 6. Hence, $\wp'(z)^2/f(z)$ is entire, and by Theorem (7.6.9) there is a constant $c \in \mathbf{C}$ such that

$$\wp'(z)^2 = cf(z).$$

We look at the Laurent series of both sides about 0:

$$\wp'(z)^2 = \frac{4}{z^6} + \cdots,$$
$$cf(z) = \frac{c}{z^6} + \cdots.$$

Therefore, $c = 4$.

iv) By making use of (2.4.9) we have the Laurent series of $\wp(z)$ about $z = 0$:

$$\text{(7.6.19)} \qquad \wp(z) = \frac{1}{z^2} + \sum_{\omega\in\Omega}' \left\{ \frac{1}{\omega^2} \sum_{n=0}^{\infty} (n+1) \left(\frac{z}{\omega}\right)^2 - \frac{1}{\omega^2} \right\}$$
$$= \frac{1}{z^2} + \sum_{n=1}^{\infty} (2n+1) G_{2n+2} z^{2n}.$$

Here, $G_k = \sum'_{\omega\in\Omega} \omega^{-k}$, $k \geqq 3$. If k is odd, $G_k = 0$. It follows from (7.6.19) that

$$\text{(7.6.20)} \qquad \wp'(z) = \frac{-2}{z^3} + 6G_4 z + 20G_6 z^3 + \cdots,$$
$$\wp'(z)^2 = \frac{4}{z^6} - \frac{24G_4}{z^2} - 80G_6 + \cdots,$$
$$4\wp(z)^3 = \frac{4}{z^6} + \frac{36G_4}{z^2} + 60G_6 + \cdots.$$

Noting that $g_2 = 60G_4$ and $g_3 = 140G_6$, one has

$$\wp'(z)^2 - 4\wp(z)^3 + g_2\wp(z) = -g_3 + \cdots.$$

Thus, the above right side is a doubly periodic entire function, which must be the constant $-g_3$. □

EXERCISE 2. Show that $\wp(z)$ satisfies the differential equation

$$\wp''(z) = 6\wp(z)^2 - \frac{g_2}{2}.$$

We are going to show that an Ω-invariant meromorphic function can be expressed in terms of $\wp(z)$ and $\wp'(z)$.

(7.6.21) LEMMA. *Let $f(z)$ be an Ω-invariant meromorphic function on* $\mathbf{C}$. *If $f(z)$ is even, then* $\deg f$ *is even.*

PROOF. We may assume that $f(z)$ is not constant. There are only finitely many zeros of $f'(z)$ contained in the period parallelogram $Q[\omega_1, \omega_2]$ of Ω. Take a value $\alpha \in \mathbf{C}^*$ which $f(z)$ does not take at zeros of f' and on the boundary of $Q[\omega_1, \omega_2]$. Set $f^{-1}(\alpha) \cap Q[\omega_1, \omega_2] = \{a_i\}_{i=1}^d$ with $d = \deg f$. Since f is even, $f(-a_i) = f(z_i) = \alpha$. If $[-a_i] = [a_i]$, then $2a_i \in \Omega$. By Exercise 1, $f'(a_i) = 0$, which contradicts the choice of α. Therefore, the Ω-orbits of a_i and $-a_i$ are distinct, and hence, there exist $j \neq i$ and $\omega \in \Omega$ such that

$$a_j = -a_i + \omega.$$

It follows that d is even. $\square$

(7.6.22) THEOREM. *For an Ω-invariant meromorphic function f on $\mathbf{C}$, there are polynomials $R_i(T)$, $i = 1, 2$, in T such that*

$$f(z) = R_1(\wp(z)) + \wp'(z)R_2(\wp(z)).$$

PROOF. Note that $f(z)$ can be written as the sum of even and odd functions: $f(z) = (f(z) + f(-z))/2 + (f(z) - f(-z))/2$. If $f(z)$ is odd, $f(z)/\wp'(z)$ is even. Therefore, it suffices to prove that if $f(z)$ is a non-constant even function, $f(z)$ can be written as a rational function of $\wp(z)$. In that case, $\deg f$ is even. We set $\deg f = 2k$ with $k \in \mathbf{N}$, and take α as in the proof of Lemma (7.6.21); in the same, we take $\beta \in \mathbf{C}, \beta \neq \alpha$. Let $a_1, \ldots, a_k \in Q[\omega_1, \omega_2] \cap f^{-1}(\alpha)$ be such that

$$f^{-1}(\alpha) = \bigcup_{j=1}^{k} (a_j + \Omega) \cup (-a_j + \Omega).$$

In the same way as above, let $b_1, \ldots, b_k \in Q[\omega_1, \omega_2] \cap f^{-1}(\beta)$ be such that

$$f^{-1}(\beta) = \bigcup_{j=1}^{k} (b_j + \Omega) \cup (-b_j + \Omega).$$

The zeros of $\prod_{j=1}^{k}(\wp(z) - \wp(a_j))$ are of order 1, and coincide with those of $f(z) - \alpha$. Set

$$\frac{f(z) - \alpha}{f(z) - \beta} = c\frac{\prod_{j=1}^{k}(\wp(z) - \wp(a_j))}{\prod_{j=1}^{k}(\wp(z) - \wp(b_j))}.$$

Then, c is a non-zero constant. Thus, $f(z)$ is expressed by a rational function of $\wp(z)$. $\square$

REMARK. The set $\mathrm{Mer}(\Omega)$ of all Ω-invariant meromorphic functions forms a field, and coincides with the meromorphic function field of the Riemann surface $\mathbf{C}/\Omega$. One sees by Theorems (7.6.22) and (7.6.18), iv) that $\mathrm{Mer}(\Omega)$ is a field of an algebraic extension of $\mathbf{C}(\wp)$ by adding $\wp'$ which is of order 2, where $\mathbf{C}(\wp)$ denote the transcendental extension of $\mathbf{C}$ by adding $\wp$.

The functions $X = \wp(z)$ and $Y = \wp'(z)$ give a parameter expression of the solutions of the following algebraic equation:

$$Y^2 = 4X^3 - g_2X - g_3 \left(= 4\prod_{j=1}^{3}(X - e_j) \right). \tag{7.6.23}$$

EXERCISE 3. Show that an arbitrary polynomial $P(X)$ of degree 3 can be reduced, by a change of variable $X = Y + a$ with $a \in \mathbf{C}$, to the form

$$P(Y + a) = a_0 Y^3 + a_2 Y + a_3.$$

A polynomial equation of degree 3 of the type (7.6.23) is called *Weierstrass' canonical form.* Now, we consider the Riemann surface S determined by the multi-valued function $Y = Y(X)$ defined by (7.6.23). The function $Y(X)$ has branch points e_1, e_2 and e_3 whose order of the branch is 1.

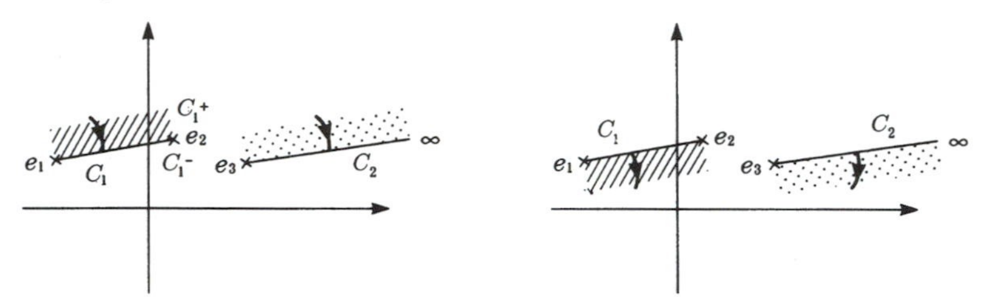

FIGURE 95

Let C_1 be the line segment from e_1 to e_2, and let C_2 be a line from e_3 to ∞ of $\widehat{\mathbf{C}}$, disjoint from C_1. Prepare two copies of such a Riemann sphere, and consider C_i as double, $C_i^{\pm}$. Then, identify one C_i^+ with another C_i^-. The resulting space S is topologically homeomorphic to a doughnut (precisely, its surface), which is the same as a 1-dimensional complex torus.

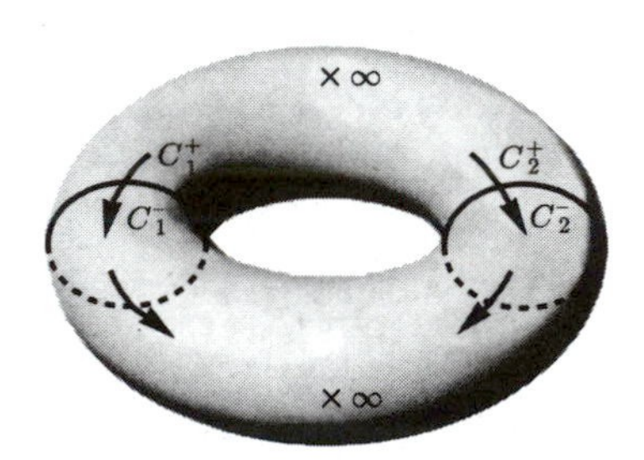

FIGURE 96

In fact, let $\pi(z)$ denote the point expressed by $X = \wp(z)$ and $Y = \wp'(z)$. Then the mapping

$$\pi : z \in \mathbf{C} \to \pi(z) \in S \tag{7.6.24}$$

is holomorphic. Since $\wp'(z)$ is odd, it follows from Theorem (7.6.18), i) that the restriction $\pi|Q[\omega_1, \omega_2]$ of π to $Q[\omega_1, \omega_2]$ is injective. Therefore, one sees that (7.6.24) gives a universal covering, and

$$\mathbf{C}/\Omega[\omega_1, \omega_2] = S. \tag{7.6.25}$$

That is, (7.6.23) gives an algebraic representation of the 1-dimensional complex torus $\mathbf{C}/\Omega[\omega_1, \omega_2]$. The study of Weierstrass' pe-function $\wp(z)$ associated with

Ω, which is an analytic object, is equivalent to the study of the 1-dimensional complex torus $\mathbf{C}/\Omega$, which is a geometric object, and also to that of the algebraic equation given by (7.6.23), which is an algebraic object. They are the same study from three different viewpoints.

(7.6.26) LEMMA. *Let g_2, g_3, e_j, $1 \leqq j \leqq 3$, be as in Theorem* (7.6.18). *Then*

$$g_2^3 - 27g_3^2 = 16\prod_{i<j}(e_i - e_j)^2 \neq 0.$$

PROOF. This is a special case of the discriminant of a general algebraic equation, and follows from the relations between the roots and the coefficients. In fact, we have

$$e_1 + e_2 + e_3 = 0, \qquad e_1e_2 + e_2e_3 + e_3e_1 = -\frac{g_2}{4}, \qquad e_1e_2e_3 = \frac{g_3}{4}.$$

Thus, it follows from $(e_1 - e_2)^2 = (e_1 + e_2)^2 - 4e_1e_2 = e_3^2 - 4e_1e_2$, etc. that

$$\begin{aligned}(e_1 - e_2)^2(e_2 - e_3)^2(e_3 - e_1)^2 &= (e_3^2 - 4e_1e_2)(e_1^2 - 4e_2e_3)(e_2^2 - 4e_1e_3)\\ &= -63e_1^2e_2^2e_3^2 - 4(e_1^3e_2^3 + e_2^3e_3^3 + e_3^3e_1^3)\\ &\quad + 16e_1e_2e_3(e_1^3 + e_2^3 + e_3^3),\\ e_1^3 + e_2^3 + e_3^3 &= (e_1 + e_2 + e_3)(e_1^2 + e_2^2 + e_3^2\\ &\quad - e_1e_2 - e_2e_3 - e_3e_1) + 3e_1e_2e_3\\ &= \frac{3}{4}g_3,\\ e_1^3e_2^3 + e_2^3e_3^3 + e_3^3e_1^3 &= (e_1e_2 + e_2e_3 + e_3e_1)\{(e_1^2e_2^2 + e_2^2e_3^2 + e_3^2e_1^2\\ &\quad - e_1e_2e_3(e_1 + e_2 + e_3)\} + 3e_1^2e_2^2e_3^2\\ &= -\frac{g_2}{4}\{(e_1e_2 + e_2e_3 + e_3e_1)^2\\ &\quad - 2e_1e_2e_3(e_1 + e_2 + e_3)\} + \frac{3}{16}g_3^2\\ &= -\frac{g_2^3}{64} + \frac{3}{16}g_3^2.\end{aligned}$$

Getting together the above equations, we have the required identity. □

Assume that $\tau = \omega_2/\omega_1$, and $\operatorname{Im}\tau > 0$. Then we set

$$g_2 = g_2(\omega_1, \omega_2) = 60\sum_{m,n}{}' \frac{1}{(m\omega_1 + n\omega_2)^4} = \frac{1}{\omega_1^4}g_2(1, \tau),$$

$$g_3 = g_3(\omega_1, \omega_2) = 140\sum_{m,n}{}' \frac{1}{(m\omega_1 + n\omega_2)^6} = \frac{1}{\omega_1^6}g_3(1, \tau).$$

By Lemma (7.6.26), one can define

$$J(\tau) = \frac{g_2(1,\tau)^3}{g_2(1,\tau)^3 - 27g_3(1,\tau)^2} = \frac{g_2(\omega_1,\omega_2)^3}{g_2(\omega_1,\omega_2)^3 - 27g_3(\omega_1,\omega_2)^2} \tag{7.6.27}$$

The function $J(\tau)$ is holomorphic in $\mathbf{H}$. By (7.6.6) the 1-dimensional complex torus $\mathbf{C}/\Omega[\omega_1,\omega_2]$ is biholomorphic to $\mathbf{C}/\Omega[1,\tau]$. By Theorem (7.6.8) $\mathbf{C}/\Omega[1,\tau]$ is biholomorphic to $\mathbf{C}/\Omega[1,\tau']$ if and only if

$$\tau' = \frac{a\tau+b}{c\tau+d}, \qquad \begin{pmatrix} a & b \\ c & d \end{pmatrix} \in SL(2,\mathbf{Z}).$$

In this case, we compare $J(\tau)$ with $J(\tau')$:

$$\begin{aligned} g_2(1,\tau') &= 60{\sum_{m,n}}' \frac{1}{(m+n\tau')^4} = 60{\sum_{m,n}}' \left(m + n\frac{a\tau+b}{c\tau+d}\right)^{-4} \\ &= (c\tau+d)^4 60{\sum_{m,n}}' \{(na+mc)\tau + nb + md\}^{-4}. \end{aligned}$$

Since $(na+mc, nb+md) = (n,m)\left(\begin{smallmatrix} a & b \\ c & d \end{smallmatrix}\right), \left(\begin{smallmatrix} a & b \\ c & d \end{smallmatrix}\right) \in SL(2,\mathbf{Z})$,

$$g_2\left(1, \frac{a\tau+b}{c\tau+d}\right) = (c\tau+d)^4 g_2(1,\tau). \tag{7.6.28}$$

Similarly, we have

$$g_3\left(1, \frac{a\tau+b}{c\tau+d}\right) = (c\tau+d)^6 g_3(1,\tau). \tag{7.6.29}$$

In general, if a holomorphic function $g(\tau)$ in $\mathbf{H}$ satisfies

$$g\left(\frac{a\tau+b}{c\tau+d}\right) = (c\tau+d)^k g(\tau), \qquad \begin{pmatrix} a & b \\ c & d \end{pmatrix} \in SL(2,\mathbf{Z}),$$

where $k \in \mathbf{Z}$, then g is called an *automorphic form* of weight k (with respect to $SL(2,\mathbf{Z})$). The ratio $G(\tau)$ of two automorphic forms of the same weight is an $SL(2,\mathbf{Z})$-invariant meromorphic function in $\mathbf{H}$; i.e.,

$$G\left(\frac{a\tau+b}{c\tau+d}\right) = G(\tau), \qquad \begin{pmatrix} a & b \\ c & d \end{pmatrix} \in SL(2,\mathbf{Z}).$$

It follows from (7.6.28) and (7.6.29) that $g_2(\tau)$ is an automorphic form of weight 4, and $g_3(\tau)$ is of weight 6. Hence, in (7.6.27) the denominator and numerator of $J(\tau)$ are of weight 12, and so $J(\tau)$ is $SL(2,\mathbf{Z})$-invariant. In fact, we have the following:

(7.6.30) THEOREM. *The function $J(\tau)$ maps $\mathbf{H}/SL(2,\mathbf{Z})$ bijectively onto $\mathbf{C}$. Moreover, two 1-dimensional complex tori, $\mathbf{C}/\Omega[\omega_1,\omega_2]$ and $\mathbf{C}/\Omega[\omega_1',\omega_2']$ with $\tau=\omega_2/\omega_1\in\mathbf{H}$ and $\tau'=\omega_2'/\omega_1'\in\mathbf{H}$, are biholomorphic to each other if and only if $J(\tau)=J(\tau')$.*

We have given the proof only of a part of the statement. For the complete proof, cf. [1] and [4]. We collect 1-dimensional complex tori which are biholomorphic to each other. We consider them as one point, and form a set $\mathcal{M}$. The above Theorem (7.6.30) shows that the mapping

$$J:\mathcal{M}=\mathbf{H}/SL(2,\mathbf{Z})\to\mathbf{C}$$

gives rise to a biholomorphic mapping; the fundamental domain of the action of $SL(2,\mathbf{Z})$ on $\mathbf{H}$ is described in Figure 97.

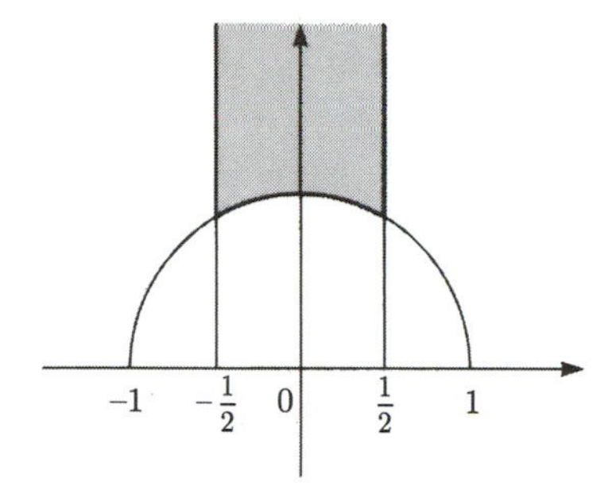

FIGURE 97

EXERCISE 4. Show that $SL(2,\mathbf{Z})$ is generated by $A=\begin{pmatrix}0&1\\-1&0\end{pmatrix}$ and $B=\begin{pmatrix}1&-1\\0&1\end{pmatrix}$, and then determine the above fundamental domain.

Finally, we explain why the algebraic curve S defined by (7.6.23) is called an "elliptic" curve, and doubly periodic meromorphic functions such as $\wp(z)$ are called "elliptic" functions. The arc length θ of the circle of radius 1 is given by

$$\theta=\int_x^1\frac{dx}{\sqrt{1-x^2}}=\int_0^y\frac{dy}{\sqrt{1-y^2}}.$$

As the inverse functions of $\theta=\theta(x)$ and $\theta=\theta(y)$, we have

$$x=\cos\theta,\qquad y=\sin\theta.$$

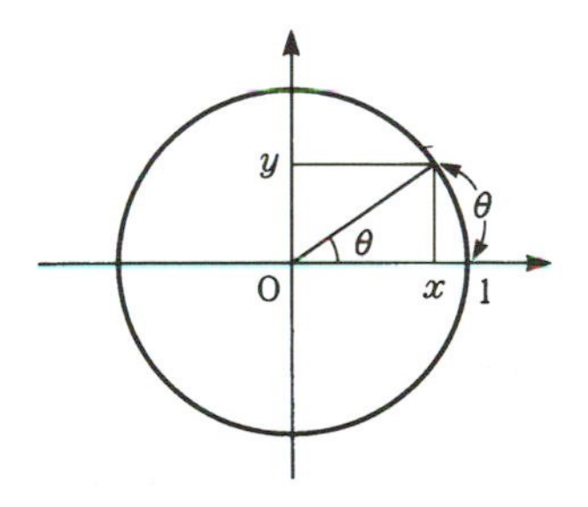

FIGURE 98

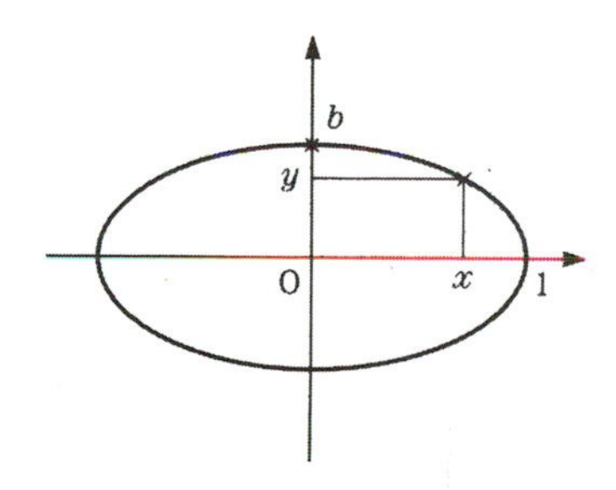

FIGURE 99

Instead of a circle, we consider the case of an ellipse

$$x^2 + \frac{y^2}{b^2} = 1, \qquad b^2 \neq 1.$$

Let $F(x)$ be the length of the curve from $(1,0)$ to (x,y). Then

$$\begin{aligned} F(x) &= \int_x^1 \sqrt{\frac{1+(b^2-1)x^2}{1-x^2}}dx \\ &= \int_x^1 \frac{1+(b^2-1)x^2}{\sqrt{(1-x^2)(1+(b^2-1)x^2)}}dx. \end{aligned} \tag{7.6.31}$$

Thus, the integrand contains the square root of a polynomial of degree 4. In general, let $P(X)$ be a polynomial of degree 3 or 4, and let $R(X,Y)$ be a rational function of two variables. Then, the integral

$$G(X) = \int^X R(X, \sqrt{P(X)})dX$$

is called the *elliptic integral.* The inverse function $X = G^{-1}(z)$ of $X = G(z)$ is called an *elliptic function.* As in the case of trigonometric functions, it is easier to deal with $G^{-1}(z)$ than with $G(X)$. The idea of extending the defining domain of elliptic integrals to complex domains, and of considering their inverse functions is due to Abel and Jacobi, and was a landmark of mathematics. It not only gave a transparent viewpoint to the complicated theory dealt with only over real numbers, but also made it possible to grow to a great theory of mathematics which has had a primary influence on the advancement of mathematics since then.

For example, we consider the following integral:

$$G(x) = \int^x \frac{dx}{\sqrt{4x^3 - g_2 x - g_3}}. \tag{7.6.32}$$

This is called an elliptic integral of the first kind, whereas (7.6.31) is called an elliptic integral of the second kind; it has rather more complicated properties. Substituting $x = \wp(z)$ in (7.6.32), we have by Theorem (7.6.18), iv)

$$\begin{aligned} G(x) &= \int^{\wp^{-1}(z)} \frac{d\wp(z)}{\sqrt{\wp'(z)^2}} = \int^{\wp^{-1}(z)} \frac{\wp'(z)}{\wp'(z)}dz \\ &= \wp^{-1}(z) + C \qquad (C \in \mathbf{C}). \end{aligned}$$

That is, G is essentially the inverse function of $\wp(z)$.

We know the addition theorems of trigonometric functions; for example,

$$\sin(u+v) = \sin u \cos v + \cos u \sin v.$$

The elliptic function $\wp(z)$ satisfies a sort of an addition theorem, too, but it is not so simple. Take $v \in \mathbf{C}$ such that $2v \notin \Omega$. By Theorem (7.6.18), ii), $\wp'(v) \neq 0$. Set

$$f(u) = \frac{1}{2}\frac{\wp'(u) + \wp'(v)}{\wp(u) - \wp(v)}.$$

The poles of f are points of Ω and the Ω-orbit $[v]$ of v, and their order is 1. It follows from (7.6.19) and (7.6.20) that about $u = 0$

$$\begin{aligned} f(u) &= \frac{1}{2}\frac{-\frac{2}{u^3} + 6G_4 u + 20G_6 u^3 + \cdots + \wp'(v)}{\frac{1}{u^2} + \sum_{n=1}^{\infty}(2n+1)G_{2n+2}u^{2n} - \wp(v)} \\ &= -\frac{1}{u} - \wp(v)u + \frac{1}{2}\wp'(v)u^2 + \cdots . \end{aligned}$$

Hence, one obtains

$$f(u)^2 = \frac{1}{u^2} + 2\wp(v) - \wp'(v)u + \cdots . \tag{7.6.33}$$

Next, we find the Laurent series of $f(u)$ about v. The point v is a pole of f of order 1, and $\operatorname{Res}(v; f) = 1$. We have

$$f(u) - \frac{1}{u - v} = \frac{(u - v)(\wp'(u) + \wp'(v)) - 2(\wp(u) - \wp(v))}{2(\wp(u) - \wp(v))(u - v)}.$$

Let $g(u)$ be the numerator of the above right side. Then, $g(v) = 0$, and

$$g'(u) = -\wp'(u) + \wp'(v) + (u - v)\wp''(u).$$

Thus, $g'(v) = 0$. Since $g''(u) = (u - v)\wp'''(u)$, $g''(v) = 0$. Therefore, we get

$$f(u) = \frac{1}{u - v} + c_1(u - v) + \cdots ,$$

$$f(u)^2 = \frac{1}{(u - v)^2} + 2c_1 + \cdots . \tag{7.6.34}$$

One infers from (7.6.13), (7.6.33) and (7.6.34) that $\wp(u - v) + \wp(u) - f(u)^2$ is a doubly periodic holomorphic function on $\mathbf{C}$, and so a constant. By making use of (7.6.33) and (7.6.19), we calculate the constant term of its expansion about $u = 0$:

$$\begin{aligned} \wp(u - v) + \wp(u) - f(u)^2 &= \wp(-v) - 2\wp(v) + \cdots \\ &= -\wp(v) + \cdots . \end{aligned}$$

Hence, $\wp(u - v) + \wp(u) - f(u)^2 \equiv -\wp(v)$; i.e.,

$$\wp(u - v) = \frac{1}{4}\left(\frac{\wp'(u) + \wp'(v)}{\wp(u) - \wp(v)}\right)^2 - \wp(u) - \wp(v).$$

Replacing v by $-v$, we have

$$\wp(u + v) = \frac{1}{4}\left(\frac{\wp'(u) - \wp'(v)}{\wp(u) - \wp(v)}\right)^2 - \wp(u) - \wp(v). \tag{7.6.35}$$

This is called the *addition theorem for the pe-function.*

EXERCISE 5. Write $\wp(2z)$ and $\wp'(2z)$ as rational functions of $\wp(z)$ and $\wp'(z)$.

The elliptic integral described in the present section is extended to the so-called Abelian integral on a general Riemann surface (cf. the reference books [10]~[14]).

Problems

1. Prove the following.

i) $\displaystyle \frac{1}{\cos z} = 1 + \sum_{\nu=-\infty}^{\infty} (-1)^{\nu} \left(\frac{1}{z - \frac{2\nu-1}{2}\pi} + \frac{2}{(2\nu-1)\pi} \right).$

ii) $\displaystyle \tan z = -\sum_{\nu=-\infty}^{\infty} \left(\frac{1}{z - \frac{2\nu-1}{2}\pi} + \frac{2}{(2\nu-1)\pi} \right).$

2. Let $D \subset \mathbf{C}$ be a bounded domain whose boundary is a closed Jordan curve. By making use of Jordan's Theorem (3.3.8), show that a holomorphic function defined in a neighborhood of $\overline{D}$ is uniformly approximated by polynomials on $\overline{D}$.

3. Prove that the exponent of convergence of a discrete sequence $\{a_n\}_{n=0}^{\infty}$ ($a_n \neq 0$) is given by

$$\overline{\lim_{n\to\infty}} \frac{\log n}{\log |a_n|}.$$

4. (Blaschke's theorem) (a) Let $f(z) \not\equiv 0$ be a bounded holomorphic function on $\Delta(1)$, and let $\{z_j\}_{j=1}^{\infty}$ be the zeros of f, counting multiplicities. Then, prove that

$$\sum_{j=1}^{\infty} (1 - |z_j|) < \infty.$$

(b) Let $\{a_j\}_{j=1}^{\infty}$ be a sequence in $\Delta(1)$ such that $\sum_{j=1}^{\infty}(1 - |a_j|) < \infty$. Set

$$g(z) = \prod_{j=1}^{\infty} \eta_j \frac{z - a_j}{1 - \bar{a}_j z},$$

$$\eta_j = -\frac{|a_j|}{a_j} \qquad (\text{if } a_j = 0, \eta_j = 1).$$

Then, prove that $g(z)$ converges absolutely and uniformly on compact subsets of $\Delta(1)$, and defines a holomorphic function with $|g(z)| < 1$, whose zeros are $\{a_j\}$.

5. Let $f(z)$ be an entire function without zeros, whose order is ρ. Show that $\rho \in \mathbf{N}$ and there is a polynomial of order ρ such that $f(z) = e^{P(z)}$.

6. Let $\lambda \in \mathbf{C}^*$ and let $P(z) \not\equiv 0$ be a polynomial. Show that $e^{\lambda z} + P(z)$ has infinitely many zeros.

7. Let $p \geqq 2$ run over all prime numbers, and set

$$\zeta(z) = \prod \left(1 - \frac{1}{p^z}\right)^{-1}, \qquad \operatorname{Re} z > 1.$$

Prove that this infinite product converges absolutely and uniformly on compact subsets, and moreover, show that it coincides with Riemann's zeta function.

8. Let $f(z)$ be a meromorphic function, and let $\left(\begin{smallmatrix} a & b \\ c & d \end{smallmatrix}\right) \in SL(2, \mathbf{Z})$. Show that

$$T\left(r, \frac{af+b}{cf+d}\right) = T(r, f) + O(1).$$

9. Let $\Omega = \Omega[\omega_1, \omega_2]$ be a lattice of $\mathbf{C}$, and let f be an Ω-invariant nonconstant meromorphic function. Let $a_1, \ldots, a_d$ (resp., $b_1, \ldots, b_d$) be zeros (resp., poles) of f in $Q[\omega_1, \omega_2]$, counting multiplicities ($d = \deg f$). Prove that

$$a_1 + \cdots + a_d - (b_1 + \cdots + b_d) \in \Omega.$$

10. Let $e_i \in \mathbf{C}, 1 \leqq i \leqq 4$, be distinct, and set $P(w) = \prod_{i=1}^{4}(w - e_i)$. Show that by a linear transformation of the variable, the elliptic integral

$$\int \frac{1}{\sqrt{P(w)}} dw$$

is reduced to an elliptic integral

$$\int \frac{1}{\sqrt{Q(z)}} dz,$$

where $Q(z)$ is a polynomial of degree 3.

11. Let $\wp(z)$ be Weierstrass' pe-function associated with a lattice Ω of $\mathbf{C}$. Show that if $u + v + w \in \Omega$,

$$\begin{vmatrix} \wp(u) & \wp'(u) & 1 \\ \wp(v) & \wp'(v) & 1 \\ \wp(w) & \wp'(w) & 1 \end{vmatrix} = 0.$$

12. Observe that the solution of the dynamics of the pendulum shown below is described by an elliptic integral.

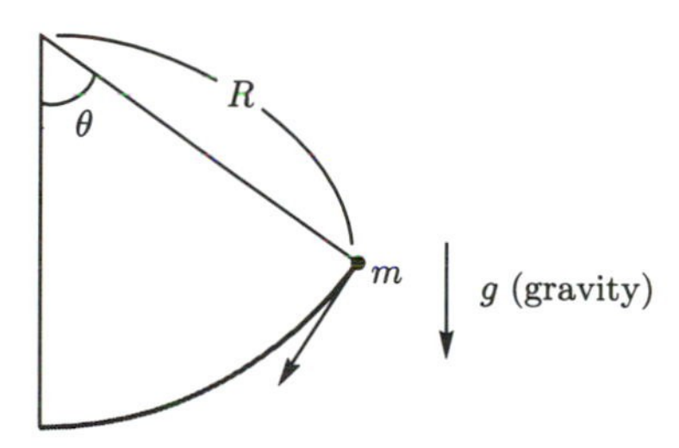

FIGURE 100

Hints and Answers

Here we give hints and answers to some of the exercises and problems presented in the text. The reader is urged to think over the exercises and the problems for at least a week before looking at these hints and answers.

Chapter 1

§2. Ex. 2. Assume that $A = [T_0, T_1]$ is covered by two open subsets, U_0 and U_1, such that $U_0 \cap U_1 \cap A = \emptyset$. We may assume that $T_0 \in U_0$. It suffices to show that $A \subset U_0$. Set $\alpha = \sup\{t \in [T_0, T_1]; [T_0, t] \subset U_0\}$. If $\alpha \notin U_0$, $\alpha \in U_1$, which leads to a contradiction (why?). Therefore, $\alpha \in U_0$. In the same way, if $\alpha < T_1$, a contradiction follows; hence, $\alpha = T_1$, and $[T_0, T_1] \subset U_0$.

§3. Ex. 2. For $z \in U_{\alpha_i}$, take $\Delta(z; r(z)) \subset U_{\alpha_i}$, $r(z) > 0$. Moving z and α_i, we obtain an open covering $\{\Delta(z; r(z)/2)\}$ of A. There are finitely many $\{\Delta(z_\lambda; r(z_\lambda)/2)\}$ which cover A. For U_{α_i}, let V_i be the union of all $\{\Delta(z_\lambda; r(z_\lambda)/2)\}$ such that $\Delta(z_\lambda; r(z_\lambda)) \subset U_{\alpha_i}$.

Problems.

1. E.g., set $\sqrt{i} = x + iy$. Then, $x^2 - y^2 = 0$, and $2xy = 1$. The first implies $x = \pm y$, but by the second, $x = y$. Hence, $x = y = \pm 1/\sqrt{2}$.

2. $u = \pm\sqrt{\sqrt{x^2+y^2}+x}\Big/\sqrt{2}$, $v = \pm\sqrt{2}y\Big/\sqrt{\sqrt{x^2+y^2}+x}$.

4. Use the identity $(z-1)\prod_{j=1}^{n-1}(z - \mathrm{P}_j) = z^n - 1 = (z-1)(z^{n-1} + \cdots + 1)$.

5. Use the absolute convergence.

6. $\lim z_n = (z_1 - az_0)/(1-a)$.

7. Take an arbitrary $\alpha_0 \in \Gamma$. If $\bigcap E_\alpha = \emptyset$, $E_{\alpha_0} \subset \bigcup_{\alpha\in\Gamma}(\mathbf{C} \setminus E_\alpha)$. Thus, there are finitely many E_{α_i}, $1 \leqq i \leqq l$, such that $E_{\alpha_0} \subset \bigcup_{i=1}^{l}(\mathbf{C} \setminus E_{\alpha_i})$, and so $\bigcap_{i=i}^{l} E_{\alpha_i} = \emptyset$.

Chapter 2

§1. Ex. 3. $\delta = \sqrt{2 - \sqrt{4 - \epsilon^2}}$.

§2. Ex. 1. We consider them at $z = 1$. For any $\delta > 0$, there is an irrational number $\theta > 0$ such that $|e^{2\pi i\theta} - 1| < \delta$. Since $\{e^{2n\pi i\theta}\}_{n=0}^{\infty}$ is dense in $C(0; 1)$, there is a subsequence $e^{2n_\nu\pi i\theta} \to -1$, $\nu \to \infty$. Thus, $|(e^{2\pi i\theta})^{n_\nu} - 1| \to 2$, and so

$\{f_n\}$ is not equicontinuous.

§3. Ex. 1. $|\sum_{k=m}^n f_k(z)| \leqq \sum_{k=m}^n |f_k(z)| \leqq \sum_{k=m}^n M_k$. Then, use Theorem (2.3.2).

§4. Ex. 1. Set $r_n = (n^{\log n})^{1/n}$. $\log r_n = \frac{1}{n}(\log n)^2 \to 0$. Hence, $r_n \to 1$, and the radius of convergence is 1.

Ex. 2. Use Theorem (2.4.4). $n!n^{-n}/(n+1)!(n+1)^{-(n+1)} = (1+1/n)^n \to e$. The radius of convergence is e.

§5. Ex. 2. $z' = 1/\overline{z}$.

§6. Ex. 1. It follows from (2.6.3) and Theorem (2.6.4).

Ex. 2. $\sum_{n=1}^\infty \left|-\frac{z}{n^2}\right| = |z| \sum_{n=1}^\infty \frac{1}{n^2} < \infty$. Then, use Theorem (2.6.6).

§7. Ex. 2. Using (2.7.3), we have $d(\mathrm{P}, \mathrm{P}') = 2|z - z'|/\sqrt{(1+|z|^2)(1+|z'|^2)}$. Letting $|z'| \to \infty$, we get $d(\mathrm{P}, \mathrm{N}) = 2/\sqrt{1+|z|^2}$.

Ex. 3. Setting $z = x + iy$, by (2.7.3) we have $T_1 = 2x/(1+|z|^2)$, $T_2 = 2y/(1+|z|^2)$, $1 + T_3 = 2|z|^2/(1+|z|^2)$, and $1 - T_3 = 2/(1+|z|^2)$. A circle in the complex plane is written as $a|z|^2 + bx + cy + d = 0$. Multiplying this by $2/(1+|z|^2)$, one gets $bT_1 + cT_2 + (a-d)T_3 + a + d = 0$. This is the equation of a hyperplane in $\mathbf{R}^3$, and its intersection with a sphere is a circle.

§8. Ex. 1. Let e' be a unit element, too. Then, $e' = e \cdot e' = e$. Let b' be the inverse of a. Then, $bab' = eb' = b'$, and $ab' = e$, so that $b = b'$.

Problems.

1. Use Taylor series expansions about $z = 0$. All are 1.

2. $\sup |f| = n$, and $\inf |f| = 0$.

3. As far as the author knows, the proof requires Lebesgue integration. If the reader has not learned it, he may just read this proof as a story, keeping this fact in mind. Assume that there is a subsequence $\{f_{n_\nu}\}_{\nu=1}^\infty$ which converges at every point of $[0, 2\pi]$. Set $f(x) = \lim f_{n_\nu}(x)$. Lebesgue's bounded convergence theorem implies that $\int_0^x f(t)dt = \lim \int_0^x f_{n_\nu}(t)dt$. A simple computation yields $\int_0^x f_{n_\nu}(t)dt \to 0$. Thus, $\int_0^x f(t)dt \equiv 0$. By Lebesgue's theorem on differentiation and integration, $\int_0^x f(t)dt$ is differentiable almost everywhere, and equals $f(x)$ almost everywhere. Therefore, $f(x) = 0$ almost everywhere, and so $\int_0^{2\pi} |f(t)|dt = 0$. It follows again from Lebesgue's bounded convergence theorem that $\lim \int_0^{2\pi} |f_{n_\nu}(t)|dt = \int_0^{2\pi} |f(t)|dt = 0$. On the other hand, $\int_0^{2\pi} |\sin nt|dt = \frac{1}{n}\int_0^{2n\pi} |\sin t|dt = \int_0^{2\pi} |\sin t|dt = 2\int_0^\pi \sin t dt = 4$. This is a contradiction.

4. i) $n^p/(n+1)^p \to 1$. ii) $1/\sqrt[n]{|q|^{n^2}} = 1/|q|^n \to \infty$. iii) $\log(n!)^{1/n^2} = \frac{1}{n^2} = \frac{1}{n^2}\sum_{j=1}^n \log j < \frac{1}{n^2}\int_1^{n+1} \log x dx = \frac{1}{n^2}(n+1)(\log(n+1) - 1) \to 0$. The radius of convergence is 1. iv) $\frac{\gamma}{\alpha\beta}$.

5. By the change of variable $z \to \zeta z$, we may assume that $\zeta = 1$. Set $s_n = \sum_{\nu=0}^n a_\nu$ and $s = \lim_{n\to\infty} s_n = \sum_{n=0}^\infty a_n$. For $z \in \Delta(1)$, $f(z)/(1-z) = \sum_{n=0}^\infty z^n \sum_{n=0}^\infty a_n z^n = \sum_{n=0}^\infty s_n z^n$, and then $f(z) - s = (1-z)\sum_{n=0}^\infty (s_n - s) z^n$. For $z \in D(1; \cos\tau)$ there is a positive constant K $(= 2\sec\tau = 2/\cos\tau)$ such that $|1 - z| < K(1 - |z|)$. For an arbitrary $\epsilon > 0$, there is an $n_0 \in \mathbf{N}$ such that $|s_n - s| <$

ϵ, $n \geqq n_0$. Then, $|f(z)-s| \leqq |1-z| \left(\sum_{n=0}^{n_0} |s_n - s| \cdot |z|^n + \sum_{n=n_0+1}^{\infty} \epsilon |z|^n\right) < |1-z| \left(\sum_{n=0}^{n_0} |s_n - s| + \epsilon \sum_{n=0}^{\infty} |z|^n\right) = |1-z| \sum_{n=0}^{n_0} |s_n - s| + \epsilon|1-z|/(1-|z|) < |1-z| \sum_{n=0}^{n_0} |s_n - s| + \epsilon K$. Taking z so that $|1-z| \sum_{n=0}^{n_0} |s_n - s| < \epsilon$, we have $|f(z)-s| < (K+1)\epsilon$.

6. Let $f(z) = e^{i\theta}\frac{z-a}{-\bar{a}z+1}$, $|a| < 1$, $|z| \leqq r < 1$, and $|z'| \leqq r$. $|f(z) - f(z')| \leqq (1-|a|^2)|z-z'|/|1-r|^2 \leqq |z-z'|/|1-r|^2$.

7. Since $\sin(z+\pi) = -\sin z$ and $\cos(z+\pi) = -\cos z$, $\tan(z+\pi) = \tan z$. Suppose that there is $0 < \pi' < \pi$ such that $\tan(z+\pi') = \tan z$. Then, $\sin \pi' = 0$. It follows that $\sin(z+2\pi') = \sin z$. This contradicts the fact that 2π is the fundamental period of $\sin z$.

10. By induction, $\prod_{\nu=0}^{n}(1+z^{2\nu}) = \sum_{\nu=0}^{2^{n+1}-1} z^{\nu}$. Let $n \to \infty$.

11. Set $f(z) = (z-z_3)(z_2-z_4)/(z-z_4)(z_2-z_3)$. This is a linear transformation, and $f(z_2) = 1, f(z_3) = 0, f(z_4) = \infty$. By Theorem (2.8.4) such f is unique, and so $f(z) = (z, z_2, z_3, z_4)$.

12. Set $g(z) = (f(z), f(z_2), f(z_3), f(z_4))$. Then, $g(z)$ is a linear transformation, and $g(z_2) = 1$, $g(z_3) = 0$, $g(z_4) = \infty$. By the above problem, $g(z_1) = (z_1, z_2, z_3, z_4)$.

17. $g(z) = R^2 e^{i\theta}(z-a)/(-\bar{a}z + R^2), a \in \Delta(R), \theta \in \mathbf{R}$.

18. i) $f(z) = ((2+i)z+i)/(z+1)$. ii) $f(z) = (6iz+2)/(z+3i)$.

Chapter 3

§1. Ex. 1. For $n \geqq 0$, $((z+h)^n - z^n)/h = \sum_{j=1}^{n} \binom{n}{j} z^{n-j}h^{j-1} \to nz$ $(h \to 0)$. For $n < 0$, use $(1/f)' = -f'/f^2$.

Ex. 2. Set $h = h_1 + ih_2$. By the Cauchy-Riemann equations, $f(z+h) - f(z) = u(x+h_1, y+h_2) - u(x,y) + i\{v(x+h_1, y+h_2) - v(x,y)\} = \frac{\partial u}{\partial x}h_1 + \frac{\partial u}{\partial y}h_2 + i\frac{\partial v}{\partial x}h_1 + i\frac{\partial v}{\partial y}h_2 + o(|h|) = \left(\frac{\partial u}{\partial x} + i\frac{\partial v}{\partial x}\right)h + o(|h|)$.

Ex. 3. Let A be the Jacobian matrix of f at (x,y). For two arbitrary unit vectors X, Y, $\|AX\| = \|AY\|$. Because the angle formed by X and $Z = (X+Y)/2$ is equal to that formed by Z and Y, by the conformality, the angle formed by AX and AZ is equal to that by AZ and AY. Therefore, $\|AX\| = \|AY\|$. There are $r > 0$ and $\theta \in \mathbf{R}$ such that $\begin{pmatrix} u_x & u_y \\ v_x & v_y \end{pmatrix} = r\begin{pmatrix} \cos\theta & -\sin\theta \\ \sin\theta & \cos\theta \end{pmatrix}$. Thus, $u_x = v_y$, and $u_y = -v_x$.

Ex. 4. Use the Taylor expansions.

Ex. 5. Use the mean value theorem in two variables.

§2. Ex. 3. Assume that $|\phi(t) - \phi(t')| \leqq K|t-t'|$. For $(d): T_0 = t_0 \leqq t_1 \leqq \cdots \leqq t_l = T_1$, $L(C;(d)) = \sum_{j=1}^{l} |\phi(t_j) - \phi(t_{j-1})| \leqq K\sum_{j=1}^{l} t_j - t_{j-1} = K(T_1 - T_0)$.

Ex. 7. $\int_C x dz = 4i$.

§4. Ex. 2. Use (2.4.9). The primitive function is $\sum_{n=1}^{\infty} \frac{n}{2n-1} z^{2n-1}$.

§5. Ex. 2. Use the maximum principle for $|f_n(z) - f_m(z)|$.

Ex. 4. Set $f(z) = \sum a_m z^m$. By Theorem (3.5.20), ii), $\sum_{m=0}^{\infty} |a_m|^2 r^{2m} \leqq \frac{1}{2\pi}\int_0^{2\pi} |f(re^{i\theta})|^2 d\theta \leqq M^2 r^{2n}$. Let $r \to \infty$. $a_m = 0, m > n$.

Ex. 5. We may assume that $f(z) \equiv 0$ on γ. Take finitely many suitable $0 < \alpha_i < 2\pi$, $1 \leqq i \leqq l$, so that $C(0;1) = \bigcup e^{i\alpha_i}\gamma$. Set $g(z) = \prod_i f(e^{i\alpha_i}z)$.

Then, g is holomorphic in $\Delta(1)$ and continuous on$\overline{\Delta(1)}$. Since $|g| \equiv 0$ on $\partial\Delta(1)$, $g(z) \equiv 0$. By Theorem (2.4.16) some $f(e^{i\alpha_i}z) \equiv 0$, and so $f(z) \equiv 0$.

Ex. 6. Apply the identity theorem.

§6. Ex. 1. $\Delta(x^2 - y^2) = 0$. $x^2 - y^2 = \frac{1}{2}(z^2 + \bar{z}^2)$. Thus, $2\frac{\partial}{\partial z}(x^2 - y^2) = 2z$. The primitive function is $z^2 = x^2 - y^2 + i(2xy)$. Thus, the harmonic adjoint function is $2xy$.

Ex. 2. Use directly the rule of partial differentiations of a composite function, and the Cauchy-Riemann equations. Or, use locally the facts that a harmonic function is a real part of a holomorphic function, and that a composite of holomorphic functions is holomorphic.

Ex. 3. Let $|z| < r < R$. Applying the Poisson integral, and using $u \geqq 0$, we have $((r - |z|)/(r + |z|))u(0) \leqq u(z) \leqq ((r + |z|)/(r - |z|))u(0)$. Now, let $r \nearrow R$.

Ex. 4. In Ex. 3, let $R \to \infty$. Then, $u(z) = u(0)$.

Ex. 5. $u(z) = \frac{r^2 - |z|^2}{2\pi} \int_0^{\pi} \left\{ \frac{1}{|re^{i\theta} - z|^2} - \frac{1}{|re^{i\theta} + z|^2} \right\} d\theta$.

Problems.

1. Check the Cauchy-Riemann equations.
2. Check by definition: $\frac{\partial}{\partial z}\overline{f(\bar{z})} = \overline{\frac{\partial}{\partial z} f(\bar{z})} = 0$.
4. For an arbitrary $\epsilon > 0$, there is a partition $(d_0) : a = t_0 < t_1 < \cdots < t_k = b$ such that $L(C) - \epsilon < L(C; (d_0)) \leqq L(C)$. Take a partition $(d) : a = s_0 < s_1 < \cdots < s_l = b$ with $\delta = |(d)| < \min\{t_j - t_{j-1}\} = \delta_1$. For each t_j there is only one $[s_{\nu_j - 1}, s_{\nu_j})$ containing t_j. Let $(\tilde{d})$ be the partition formed by all t_i and s_j. Then, $0 \leqq L(C; (\tilde{d})) - L(C; (d)) = \sum_{j=1}^{k} \{|\phi(s_{\nu_j}) - \phi(t_j)| + |\phi(t_j) - \phi(s_{\mu_j - 1})| - |\phi(s_{\nu_j}) - \phi(s_{\nu_j - 1})|\}$. By the uniform continuity of ϕ, there is a $\delta_2 \in (0, \delta_1)$ such that for $|t - t'| < \delta_2$, $|\phi(t) - \phi(t')| < \epsilon$. For $0 < \delta < \delta_2$, $0 \leqq L(C; (\tilde{d})) - L(C; (d)) \leqq 2k\epsilon$. Thus, $L(C) - (2k + 1)\epsilon \leqq L(C; (d)) \leqq L(C)$, $0 < \delta < \delta_2$.
5. $9(1 - i)r^3$.
6. i) $f \circ \tilde{z} = (1 - \alpha\tilde{z})/(1 - \beta\tilde{z})$, which is holomorphic about $\tilde{z} = 0$.
7. Use (3.5.4).
8. Consider f and $1/f$. The assertion does not hold if f takes the value zero: for example, let $D = \Delta(1)$ and $f(z) = z$.
9. By Ex. 4 in §6, $\operatorname{Re} f$ is constant, and hence Theorem (3.1.8) implies that f is constant.
10. Let $p, q \in \mathbf{N}$, and set $g(z) = z^p f(z)^q$. For $r_1 < |z| = r_2 < r_3$, the maximum principle implies that $|g(z)| \leqq \max\{r_1^p M(r_1)^q, r_3^p M(r_3)^q\}$. $r_2^{p/q} M(r_2) \leqq \max\{r_1^{p/q} M(r_1), r_3^{p/q} M(r_3)\}$. Take $\alpha > 0$ so that $r_1^\alpha M(r_1) = r_3^\alpha M(r_3)$. Then, let $p/q \to \alpha$. $r_2^\alpha M(r_2) \leqq r_1^\alpha M(r_1)$. Take the logarithms of both sides, and substitute $\alpha = (\log M(r_1) - \log M(r_3))/(\log r_3 - \log r_1)$.
11. We have $\frac{1}{2\pi i} \int_{C(0;1)} \frac{\overline{f(z)}}{z - a} dz = \frac{1}{2\pi i} \int_0^{2\pi} \frac{\overline{f(e^{i\theta})}}{e^{i\theta} - a} ie^{i\theta} d\theta = \frac{1}{2\pi i} \int_0^{2\pi} \frac{\overline{f(\overline{e^{-i\theta}})}}{1 - ae^{-i\theta}} i \cdot d\theta = \frac{1}{2\pi i} \int_{C(0;1)} \frac{\overline{f(\bar{\zeta})}}{1 - a\zeta} d\zeta$. If $|a| < 1$, $\overline{f(\bar{\zeta})}/(1 - a\zeta)$ is holomorphic in a neighborhood of $\overline{\Delta(1)}$, and so the integral equals $\overline{f(0)}$. If $|a| > 1$, use $1/(1 - a\zeta) = 1/\zeta - 1/(\zeta - 1/a)$.
12. Use the mean value theorem.

13. $y - x + 2xy$.
14. $f(z) = ze^z$.
15. Use the Poisson integral.
16. Use $\psi(z) = (z-i)/(z+i)$, and (3.6.7).
17. Note that $\mathrm{Re}\,\frac{tz+1}{i(t-z)}\frac{1}{t^2+1} = y/|t-z|^2$.

Chapter 4

§2. Ex. 4. Let C be the union of circles around every point of E. Then, $\int_C \omega = 0$.

Ex. 5. Let a_ν (resp., b_μ) be the zeros (resp., poles) of f of order m_ν (resp., n_μ). Then, $g(z) = f(z)\prod(z-a_\nu)^{-m_\nu}\prod(z-b_\mu)^{n_\nu}$. Here, set $(z-a_\nu)^{-m_\nu} = z^{m_\nu}$ if $a_\nu = \infty$, and $(z-b_\mu)^{n_\mu} = z^{-n_\mu}$ if $b_\mu = \infty$. Then, $g(z)$ is a non-zero constant.

Ex. 6. Let $g(z)$ be the right side of the last equation. Then, $h(z) = f(z) - g(z)$ is holomorphic on $\widehat{\mathbf{C}}$. Thus, $h(z) \equiv c \in \mathbf{C}$. Comparing the Laurent series expansion of $f \circ \tilde{z}$ about $\tilde{z} = 0$ with that of $g \circ \tilde{z}$, one sees that $h(\infty) = 0$.

Ex. 7. $\mathrm{Res}(\pm i; f) = 1/2$.

Ex. 8. $\mathrm{Res}(\pm ai; f) = \mp i/4a^3$.

Ex. 9. $\mathrm{Res}(0; f) = \sum_{n=0}^\infty 1/n!(n+1)!$, $\mathrm{Res}(\infty; f) = -\sum_{n=0}^\infty 1/n!(n+1)!$.

§3. Ex. 2. Set $\omega = (f'(z)/f(z))dz$. Let M (resp., P) be the number of zeros (resp., poles) of f, counting multiplicities. Then, $M - P = \sum_{a\in\widehat{\mathbf{C}}} \mathrm{Res}(a;\omega) = 0$.

Ex. 3. $z = 2\sqrt{w} + w + \sum_{n=1}^\infty (-1)^{n-1}(2n-1)!!2^{-2n+1}(\sqrt{w})^{2n+1}/n!$.

§4. Ex. 1. i) $2\pi/(1-a^2)$. ii) $\pi/2\sqrt{a^2+a}$. iii) $2\pi\{(2a^2-1)/\sqrt{a^2-1} - 2a\}$.

Ex. 2. i) 0. ii) $\pi/2\sqrt{2}$. iii) $\pi/3\sqrt{2}$. iv) $\pi(2n-2)!/2^{2n-1}((n-1)!)^2$.

Ex. 3. i) $2\pi i e^{-\sqrt{3}/2+i/2}/3 + \pi i e^{-i}/3$. ii) $(\pi\cos a)/2$. iii) $\pi/2e$. iv) $\pi/4$.

Ex. 4. i) $\pi/2\sin(\alpha\pi/2)$. ii) $\alpha(\alpha+1)\cdots(\alpha+n-2)\pi/(n-1)!\sin\alpha\pi$. iii) $\pi/2a^{1+\alpha}\cos(\alpha\pi/2)$.

Ex. 5. i) $(1+\log a)/2a$. ii) $(\log^2 a - \log^2 b)/2(a-b)$. iii) $-\pi/4$.

Problems

1. i) $\sum_{n=-\infty}^\infty \left(\sum_{k=\max\{0,n\}}^\infty \frac{1}{(k-n)!}\right) z^n$. ii) $\sin 1 + \cos 1\frac{1}{z-1} - \frac{\sin 1}{2}\frac{1}{(z-1)^2} - \cdots + \frac{(-1)^n \sin 1}{(2n)!}\frac{1}{(z-1)^{2n}} + \frac{(-1)^n\cos 1}{(2n+1)!}\frac{1}{(z-1)^{2n+1}} + \cdots$. iii) $\sum_{n=0}^\infty (-1)^n \tilde{z}^{2n-1}$.

2. i) When $n > 0$, the pole is at ∞, and the residue is 0; when $n = -1$, the pole is at 0, and the residues are 1 at 0 and -1 at ∞; when $n < -1$, the pole is at 0, and the residues are 0 at 0 and at ∞. ii) The poles are $\pi/2 + n\pi, n \in \mathbf{Z}$, and the residues are all 1. iii) No pole. The residues are 1 at 0, and -1 at ∞. iv) The poles are -2 with residue -2, and -3 with residue 3. The residue at ∞ is 1. v) The poles are -1 with residue 6, and ∞ with residue -6.

3. i) π. ii) $2\pi i/(n+1)!$. iii) When $n \geqq 0$ and $m \geqq 0$, 0; when $n < 0$ and $m \geqq 0$, 0 for $m \leqq -n-2$, and $2\pi i\binom{m}{-n-1}(-1)^{m+n+1}$ for $m \geqq -n-1$; when $n \geqq 0$ $m < 0$, 0 for $n \leqq -m-2$, and $2\pi i\binom{n}{-m-1}$ for $n \geqq -m-1$; when $n < 0$ and $m < 0$, 0. iv) $2\pi i(-1)^{n-1}(m+n-2)!/(n-1)!(m-1)!(a-b)^{m+n-1}$. v) 0.

4. i) On $C(0;1)$, $|7z^4| - |z^6 - 2z^5 + z^3 - z + 1| \geqq 7 - 6 = 1$. Apply Rouché's theorem. ii) If $|z| \geqq 2$, $|z^4| - 6|z| - 3 \geqq 1$, and so there is no solution there. Thus, there are four solutions in $|z| < 2$. On $|z| = 1$, $|6z| - |z^4 + 3| \geqq 2$, so that

there is one solution there. Thus, there are three solutions in $R(1,2)$.

5. On $|z| = R'$, $|z| - |f(z)| \geqq |z| - |f(z)| \geqq R' - \theta R$. Hence, for $R' < R$ sufficiently close to R, the number of zeros of z and $z - f(z)$ in $\Delta(R')$ are the same. Thus, there is one $z \in \Delta(R)$ such that $z - f(z) = 0$.

6. Use Rouché's theorem on $|z| = 1$.

7. Let $f(z_0) = 0$. Apply Rouché's theorem for $f(z)$ and for $f(z) + (f_n(z) - f(z)) = f_n(z)$ on a small circle $|z - z_0| = \delta$.

8. i) $(\sqrt{(a+1)/a} - 1)\pi/2$. ii) π/ae^a. iii) $2\pi i e^{\sqrt{3}\pi}/3 - \pi i/3$. iv) $2\pi(1 - a/\sqrt{a^2-1})$. v) $\pi/2e^a$. vi) $\pi\left(\cos\frac{a}{2}\pi + b\sin\frac{a}{2}\pi\right)/\sin a\pi$. vii) Use the curve in Figure 101. $-\pi\tan\frac{\alpha}{2}\pi$. viii) Use the equality $\log^3 z - (\log z + 2\pi i)^3 = -12\pi\log z + 6\pi i\log^2 z - 8\pi i$. $\frac{2}{3}\pi + \frac{13}{24}\pi^3$. ix) Use the curve in Figure 101. $-\pi^2/4$.

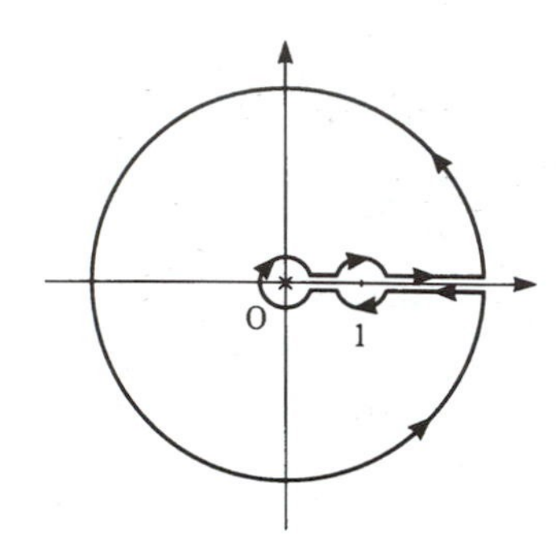

FIGURE 101

Chapter 5

§1. Ex. 1. Use Theorem (2.8.9).

Ex. 2. Apply the reflection principle to extend f meromorphically to all $\widehat{\mathbf{C}}$. Hence, f is rational and holomorphic in $\mathbf{C}$, so that f is polynomial.

Ex. 3. Let $f(z) = \sum a_n z^n$. The first condition implies $a_n \in \mathbf{R}$. The second condition implies that $a_n = 0$ for even n. Thus, $f(z)$ is odd.

§3. Ex. 3. Use Theorem (3.5.14).

Problems

1. Glue two $\widehat{\mathbf{C}}$ along $[0,1]$.

5. Let $|f(\mathrm{P})| = \max\{|f|\} = M$. Show that $\{\mathrm{Q} \in X; |f(\mathrm{Q})| = M\}$ is open and closed.

7. The length is $\log\frac{R+r}{R-r}$. The area is $4\pi r^2/(R^2 - r^2)$.

Chapter 6

§2. Ex. 1. Use $\psi(z) = (z-i)/(z+i)$, which maps $\mathbf{H}$ biholomorphically onto $\Delta(1)$, and maps geodesics to geodesics.

Ex. 3. $d_{\mathbf{H}}(z_1,z_2) = \log\{(1 + |z_1 - z_2|/|\overline{z}_1 - z_2|)/(1 - |z_1 - z_2|/|\overline{z}_1 - z_2|)\}$.

§3. Ex. 2. For arbitrary $z_j \in \Delta(1)$, $j = 1,2$, such that $\pi(z_j) = w_j$, $d_{\Delta(1)}(z_1,z_2) \geqq d_D(w_1,w_2)$ by Theorem (6.3.10). For an arbitrary $\epsilon > 0$, there is a piecewise C^1 curve $C \subset D$ from w_1 to w_2 such that $L_D(C) < d_D(w_1,w_2) + \epsilon$. Fix z_1 as above. Let $\hat{C}$ be the lifting of C with the initial point z_1. Let z_2 be the terminal point of $\hat{C}$. Then, $\pi(z_2) = w_2$, and $L_D(C) = L_{\Delta(1)}(\hat{C})$ (cf. (6.3.2)). Therefore, $D_{\Delta(1)}(z_1,z_2) \leqq L_{\Delta(1)}(\hat{C}) = L_D(C) < d_D(w_1,w_2)+\epsilon$. Hence,

$d_{\Delta(1)}(z_1, z_2) \leqq d_D(w_1, w_2)$.

§4. Ex. 1. Using Lemma (2.2.7) or (7.1.12), find countable open coverings, $U_\nu \Subset V_\nu$ of D such that $\{f_n | V_\nu\}_{n=1}^{\infty}$ is normal. Then, apply the diagonal argument used in the proof of Theorem (2.2.6).

Ex. 2. Apply Schwarz' lemma to $g \circ f^{-1} : \Delta(1) \to \Delta(1)$.

Ex. 3. Use the uniqueness of Theorem (6.4.4).

Ex. 4. Parametrize the parabola $y^2 = 4c^2(x + c^2)$ by $x = t^2 - c^2, y = 2ct$. Set $z = x + iy = (t + ic)^2$. Replace t by a complex variable w, and take $\sqrt{z}$ such that $\sqrt{-1} = i$. Set $w = \sqrt{z} - ic$. As $w = t \in \mathbf{R}$ runs from $-\infty$ to ∞, $y(t)$ with $z = z(t) = x(t) + iy(t)$ runs from $-\infty$ to ∞. Thus, $w = \sqrt{z} - ic$ is the required mapping.

§5. Ex. 1. $f(z) = (z - 1)/(z + 1)$.

Ex. 2. $f(z) = (iz + 1)/(-iz + 1)$.

Ex. 3. It follows from Theorem (6.5.7) and Theorem (3.6.11).

§6. Ex. 1. If $D \neq \widehat{\mathbf{C}}, \mathbf{C}, \mathbf{C}^*$, then $\widehat{\mathbf{C}} \setminus D$ contains at least three points, which may be assumed to be $0, 1, \infty$. Set $D_0 = \widehat{\mathbf{C}} \setminus \{0, 1, \infty\}$. The inclusion mapping $\iota : D \ni z \to z \in D_0$ is holomorphic, and has a lifting $\iota \tilde{\circ} \pi : \tilde{D} \to \Delta(1)$, where $\Delta(1)$ is the universal covering of D_0 (Theorem (6.6.7)). Thus, there is a bounded non-constant holomorphic function on $\tilde{D}$. It follows from Theorem (6.4.6) that $\tilde{D}$ is biholomorphic to $\Delta(1)$.

§7. Ex. 1. Considering $f - i\alpha$, we may assume $\alpha = 0$. Using (2.8.12), we have $|\psi \circ f(z)| < 1$ for $f \in \mathcal{F}, z \in D$. Thus, $\{\psi \circ f; f \in \mathcal{F}\}$ is normal, and so $\mathcal{F}$ is normal.

Ex. 2. Let $f(z)$ be a non-constant entire function. Assume that $f(z)$ does not take three values $0, 1, \infty$. Set $f_\nu(z) = f(\nu z)$, $\nu = 1, 2, \ldots$. Then, f_ν does not take $0, 1, \infty$, and hence $\{f_\nu\}$ is a normal family. Since $f_\nu(0) = f(0)$, $\{f_\nu\}$ contains a subsequence, converging uniformly on compact subsets. On the other hand, $\frac{1}{2\pi} \int_0^{2\pi} |f_\nu(e^{i\theta})|^2 d\theta = \sum |a_n|^2 |\nu|^{2n} \to \infty$ $(\nu \to \infty)$, where $f(z) = \sum a_n z^n$. This implies that any subsequence of $\{f_\nu\}$ does not converges uniformly on $C(0; 1)$, and so a contradiction.

Ex. 3. Set $f_\nu(z) = e^{\nu z}$, $\nu = 1, 2, \ldots$, which do not take the values $0, \infty$. Since $f_\nu(0) = 1$ and $f_\nu(0) = \nu$, any subsequence of $\{f_\nu\}$ does not converge uniformly on compact subsets.

Problems

1. $2 \log 3$. $\log \frac{25 + 4\sqrt{34}}{9}$.

2. Let $\pi_j : \Delta(1) \to D_j$ be the universal coverings, and let $F : \Delta(1) \to \Delta(1)$ be a lifting of $f \circ \pi_1$. Take $z_0 \in \Delta(1)$ such that $\pi_1(z_0) = a$. The assumption implies $F^* g_1(z_0) = g_1(z_0)$, and by Theorem (6.2.3) $F^* g_1 = g_1$ at all points. Hence, $f^* h_{D_2} = h_{D_1}$.

3. Use Exercise 2 of §3, and take a geodesic in the universal covering $\Delta(1)$.

4. Take $w \in \mathbf{C}$ so that the cardinality of $f^{-1}(w)$ is the maximum. Set $f^{-1}(w) = \{z_j\}_{j=1}^{l}$. Then, $f'(z_j) \neq 0$. There are neighborhoods $\Delta(w; \epsilon_0)$ of w and U_j of z_j such that $f | U_j : U_j \to \Delta(w; \epsilon_0)$ are biholomorphic. We may assume

that U_j are bounded. Take $R > 0$ such that $\Delta(R) \supset \cup_j U_j$. By the choice of w, $|f(z) - w| \geqq \epsilon_0$ on $\mathbf{C} \setminus \Delta(R)$. Hence, in a neighborhood of ∞, $1/(f(z) - w)$ is bounded and holomorphic. By Riemann's extension Theorem (5.1.1), $f(z)$ has at most a pole at ∞. Thus, f is rational.

5. Define a biholomorphic mapping $f : \Delta(1) \to \{z \in \mathbf{C}; \operatorname{Re} z > 0\}$ by $f(z) = (1+z)/(1-z)$. The real axis in $\Delta(1)$ is mapped by f onto the real positive axis. Thus, $f(D) = \{z; \operatorname{Re} z > 0, \operatorname{Im} z > 0\}$, and $w = f(z)^2$ has the required property.

6. Compose $g : D \ni z \to z^{1/\alpha} = e^{(\log z)/\alpha} \in \{w \in \Delta(1); \operatorname{Im} w > 0\}$ $(\log 1 = 0)$ with the one obtained in 5, and then with $\psi(w) = (w - i)/(w + i)$.

7. Set $z = e^{i\theta}$. Then $w = 2\cos\theta$. We see that $C(0;1)$ is mapped onto $[-2, 2]$. Using the branch $\sqrt{1} = 1$, we have $z = (w - \sqrt{w^2 - 4})/2$, which expands to $z = \frac{1}{w} + \frac{2}{w^3} + \cdots$. Thus, $z = 0$ is corresponding to ∞. Thus, $\Delta(1)$ is mapped univalently onto D. One sees also that $\widehat{\mathbf{C}} \setminus \Delta(1)$ is mapped to D.

10. Suppose that $\{f_n\}$ does not converge uniformly on a compact subset $K \subset D$. Then, there is an $\epsilon_0 > 0$ such that for every $N \in \mathbf{N}$ there are $\nu_N, \mu_N > N$ and $z_N \in K$ with $|f_{\nu_N}(z_N) - f_{\mu_N}(z_N)| > \epsilon_0$. Run $N = 1, 2, \ldots$, and choose subsequences to get $f_{\nu_N} \to g$, $f_{\mu_N} \to h$, and $z_N \to z_0 \in K$. Thus, we have $|g(z_0) - h(z_0)| \geqq \epsilon_0$. On the other hand, $g = h$ on E, and hence $g \equiv h$. This is a contradiction.

11. Apply (3.6.7).

12. Let $\pi : \Delta(1) \to D$ be the universal covering, and let $\hat{f} : \Delta(1) \to \Delta(1)$ be a lifting of f. We may assume that $\pi(0) = z_0$ and $\hat{f}(0) = 0$. By the assumption and Lemma (6.3.11), $\hat{f} \in \operatorname{Aut}(\Delta(1))$, and $\hat{f}(z) = e^{2\pi i\theta}z, 0 \leqq \theta < 1$. If $\theta \in \mathbf{Q}$, then for some $m \in \mathbf{N}$ $\hat{f}^m = \hat{f} \circ \cdots \circ \hat{f}(m \text{ times}) = \mathrm{id}_{\Delta(1)}$. Thus, $f^m = \mathrm{id}_D$, and $f^{-1} = f^{m-1}$. Let $\theta \notin \mathbf{Q}$. Take an increasing sequence $m_\nu < m_{\nu+1} \in \mathbf{N}, \nu = 1, 2, \ldots$, so that $e^{2m_\nu \pi i\theta} \to 1$. Applying Theorem (6.7.6) to choose a subsequence, we may assume that $f^{m_\nu} \to \mathrm{id}_D$ and $f^{m_\nu - 1} \to g$. Then, $f \circ g = \mathrm{id}_D$.

13. The first half follows immediately from Theorem (6.5.7). Then, using Schwarz' reflection principle, extend f to $\tilde{f} \in \operatorname{Aut}(\mathbf{C})$. One sees that $f(z) = az + b$ $(a \neq 0)$. Thus, D_1 is similar to D_2.

17. Use the little and big Picard theorems.

19. We may assume that $U = \Delta(1)$ and $\mathrm{P} = 0$. We also may assume that f omits $0, 1, \infty$. Let $z_\nu \in \Delta^*(1)$ be such that $|z_\nu| < 1/2$, $\lim z_\nu = 0$, and $\lim f(z_\nu) = \alpha$. If $\alpha = \infty$, we consider $1/f(z)$, so we may assume that $\alpha \in \mathbf{C}$. Set $f_\nu(\zeta) = f(z_\nu \zeta), \nu = 1, 2, \ldots$. Then, $f_\nu : \Delta^*(2) \to \mathbf{C} \setminus \{0, 1\}$, $\nu = 1, 2, \ldots$, form a normal family by Theorem (6.7.6). Hence, we may assume that $\{f_\nu\}$ converges uniformly on $C(0;1)$. Set $f(z) = \sum_{-\infty}^{\infty} a_n z^n$ (Laurent series). Then $\sum_{-\infty}^{\infty} a_n z_\nu^n z^n$, $\nu = 1, 2, \ldots$, converge uniformly on $C(0;1)$. There is an $M > 0$ such that $\frac{1}{2\pi}\int_0^{2\pi} |f_\nu(e^{i\theta})|^2 d\theta = \sum_{-\infty}^{\infty} |a_n|^2 |z_\nu|^{2n} < M$. Thus, for $n < 0$, $a_n = 0$.

Chapter 7

§2. Ex. 1. Let $g(z)$ denote the right side. It follows that $g(z+1) = g(z)$. Set $f(z) = \pi \cot \pi z - g(z)$. Looking at the principal part at $z = n \in \mathbf{Z}$, one sees that $f(z)$ is entire. We want to show the boundedness of $f(z)$. By making use of the periodicity, $f(-z) = -f(z)$, and of $f(\overline{z}) = \overline{f(z)}$, it suffices to show that $f(z)$ is bounded on the set of $z = x + iy, 0 \leqq x \leqq 1/2, y \geqq 1$. Since $|\cot \pi z| \leqq (e^{\pi y} + e^{-\pi y})/(e^{\pi y} - e^{-\pi y})$, $\cot \pi z$ is bounded for $y \geqq 1$. Note that $|g(z)| \leqq 1 + \sum_{n=1}^{\infty} \frac{2|z|}{|z-n|^2} \leqq 1 + 2\sqrt{2}\sum_{n=1}^{\infty} \left(\frac{1}{2} + y\right) / \left\{\left(n - \frac{1}{2}\right)^2 + y^2\right\} \leqq 1 + \sqrt{2}(1 + 2y)/\left(\frac{1}{4} + y^2\right) + \sqrt{2}(1 + 2y) \int_1^{\infty} \frac{1}{(t-1/2)^2+y^2} dt \leqq C_1 + \sqrt{2}(1 + 2y) \int_0^{\infty} \frac{1}{t^2+y^2} dt = C_1 + \sqrt{2}(1+2y)2\pi/y < C_2 < \infty$. Therefore, $f(z)$ is constant. The constant term of the expansion of $f(z)$ about $z = 0$ is 0, and hence $f(z) \equiv 0$.

Ex. 2. Differentiate both sides of the equation obtained in Ex. 1. Check that termwise differentiation is possible.

Ex. 3. Let $z = e^{2\pi i\theta}$ with $\theta \in \mathbf{Q}$. For $0 < r < 1$, $f(re^{2\pi i\theta})$ is, up to finite terms, the sum of $r^{n!} + r^{(n+1)!} + \cdots$. Thus, $\lim_{r\to 1} f(re^{2\pi i\theta}) = +\infty$. Since such points are dense in the boundary of $\Delta(1)$, $f(z)$ cannot be analytically continued over any boundary point of $\Delta(1)$.

Ex. 4. There is a sequence of points $z_n \in \mathbf{H}$ such that it accumulates to an arbitrary point of $\mathbf{R}$, and $\lambda(z_n) = \lambda(z_1)$.

§4. Ex. 1. We have $\log^+ \sum_{i=1}^n x_i \leqq \log^+ n \max\{x_i\} \leqq \max\{\log^+ x_i\} + \log n \leqq \sum_{i=1}^n \log^+ x_i + \log n$.

Ex. 3 i) $N(r, f) = 0$. $m(r, f) = \frac{1}{\pi} \int_0^{\pi/2} r \cos\theta d\theta = \frac{r}{\pi}$. ii) $n(r, 1/(f - a)) = \frac{r}{\pi} + O(1)$, and hence $N(r, 1/(f - a)) = \frac{r}{\pi} + O(\log r)$. It follows from i) that $m(r, 1/(f - a)) = O(\log r)$.

Ex. 4. $1/\alpha$.

Ex. 5. Use Ex. 3.

§5. Ex. 1. $\cos z = \prod_{n=-\infty}^{\infty} \left(1 - \frac{z}{n\pi+\pi/2}\right) e^{z/(n\pi+\pi/2)}$.

Ex. 2. i) $\sum 1/|a_n|^{\mu}$ is convergent if and only if $\int^{\infty} 1/x^{\mu}(\log x)^{2\mu} dx$ converges. Thus, it converges for $0 \leqq \mu < 1$, and diverges for $\mu > 1$. Weierstrass' product is $\prod_{n=2}^{\infty} \left(1 - \frac{z}{n(\log n)^2}\right) e^{z/n(\log n)^2}$. ii) $\lambda = 0$. $\prod_{n=1}^{\infty}(1 - z/n!)$.

Ex. 3. Take Weierstrass' product associated with $a_n = n^{1/\rho}, n = 1, 2, \ldots$.

Ex. 4. It follows from i) and ii) that $\phi(n) = (n-1)!$ for $n \in \mathbf{N}$. Condition iii) implies that for $0 < x_1 < x_2 < x_3 < x_4$, $(\log \phi(x_2) - \log \phi(x_1))/(x_2 - x_1) \leqq (\log \phi(x_3) - \log \phi(x_2))/(x_3 - x_2) \leqq (\log \phi(x_4) - \log \phi(x_3))/(x_4 - x_3)$. For $0 < x < 1$, set $x_1 = n - 1 < x_2 = n < x_3 = n + x < x_4 = n + 1$, and substitute them in the above expression: $\log(n-1) \leqq (\log \phi(n+x) - \log(n-1)!)/x \log n$. Combine this with $\phi(n+x) = (n+x-1)\cdots x\phi(x)$ to get $(n-1)^x(n-1)!/x(x+1)\cdots(x+n-1) \leqq \phi(x) \leqq n^x(n-1)!/x(x+1)\cdots(x+n-1)$. Therefore, $(n/(n+x))\phi(x) \leqq n^x n!/x(x+1)\cdots(x+n) \leqq \phi(x)$. Let $n \to \infty$. Then, by (7.5.17) $\phi(x) = \Gamma(x)$.

§6. Ex. 1. Since $f(z)$ is even, $f'(z)$ is odd. Suppose that ω is not a pole of f', nor f. Since $f'(\omega) = f'(\omega - 2\omega) = f'(-\omega) = -f'(\omega)$, $f'(\omega) = 0$.

Ex. 2. Differentiate the equation of Theorem (7.6.18), iv).

Ex. 4. $A^4 = \left(\begin{smallmatrix} 1 & 0 \\ 0 & 1 \end{smallmatrix}\right), A^3 = A^{-1} = \left(\begin{smallmatrix} 0 & -1 \\ 1 & 0 \end{smallmatrix}\right)$. $B^n = \left(\begin{smallmatrix} 1 & -n \\ 0 & 1 \end{smallmatrix}\right), n \in \mathbf{N}$. Take $S = \left(\begin{smallmatrix} a & b \\ c & d \end{smallmatrix}\right) \in SL(2, \mathbf{Z})$. $SB^n = \left(\begin{smallmatrix} a & b-na \\ c & d-nc \end{smallmatrix}\right) = \left(\begin{smallmatrix} a & b' \\ c & d' \end{smallmatrix}\right)$. a) The case of $a = 0$. Then $c = \pm 1$. Take n such that $d - nc = 0$. Then, $SB^n = \left(\begin{smallmatrix} 0 & b \\ c & 0 \end{smallmatrix}\right) = A$ or A^{-1}. b) The case of $a \neq 0$. If b is a multiple of a, take n such that $b' = 0$. Then $SB^n = \left(\begin{smallmatrix} a & 0 \\ c & d' \end{smallmatrix}\right)$. Then, $\left(\begin{smallmatrix} a & 0 \\ -d' & c \end{smallmatrix}\right) A = \left(\begin{smallmatrix} 0 & a \\ -d' & c \end{smallmatrix}\right)$. This is reduced to case a). If b is not a multiple of a, we may take n so that $|b'| \leqq |a|/2$. Set $\left(\begin{smallmatrix} a & b' \\ c & ' \end{smallmatrix}\right) A = \left(\begin{smallmatrix} a'' & b'' \\ c'' & d'' \end{smallmatrix}\right)$. Then, $|a''| \leqq |a|/2$. Repeat this. It is reduced to case a).

Ex. 5. Use (7.6.35), Ex. 2, Theorem (7.6.18), iv).

Problems

1. i) Let $f(z)$ be the right side. For ν such that $\left|\left(\nu - \frac{1}{2}\right)\pi\right| > 2|z|$, $\left|\frac{1}{z - \frac{2\nu - 1}{2}\pi} + \frac{2}{(2\nu-1)\pi}\right| \leqq \sum_{k=1}^{\infty} |z|^k \left|\left(\nu - \frac{1}{2}\right)\pi\right|^{-k-1} \leqq \frac{2|z|}{\left|\left(\nu - \frac{1}{2}\right)\pi\right|^2}$. Hence, $f(z)$ converges absolutely and uniformly on compact subsets except for finitely many terms. It follows that $f(-z) = f(z)$, $f(z + 2\pi) = f(z)$. Set $F(z) = \frac{1}{\cos z} - f(z)$. As in the computation of $1/\sin z$, one sees that $F(z)$ is constant. Since $F(0) = 0$, $F(z) \equiv 0$. ii) is similar to i).

2. Take $K = \overline{D}$ in Corollary (7.1.10).

3. Let ρ_0 be the exponent of convergence of $\{a_n\}$. Set $\rho_1 = \overline{\lim}(\log n)/\log|a_n|$. Let $n(t)$ be the counting function of $\{a_n\}$. If $\sum |a_n|^{-\alpha} < \infty$, (7.4.21) implies that $\int^{\infty} n(t) t^{-\alpha - 1} dt < \infty$. For all large $r > 0$, we have $1 > \int_r^{2r} n(t) t^{-\alpha-1} dt \geqq (1 - 2^{-\alpha}) n(r)/\alpha r^{\alpha}$. Setting $r = |a_n|$, one obtains $n/|a_n|^{\alpha} \leqq C$ (a constant). Hence, $\rho_1 \leqq \rho_0$. If $\rho_1 = \infty$, then $\rho_0 = \infty$. Suppose that $\rho_1 < \infty$. For an arbitrary $\beta > \rho_1$, take $\rho_1 < \beta' < \beta$. For all large n, $\log n/\log|a_n| < \beta' < \beta$. Take $\epsilon > 0$ small enough so that $(1+\epsilon)(\log n)/\log|a_n| < \beta$. Thus, $|a_n|^{-\beta} < n^{-1-\epsilon}$, and $\sum |a_n|^{\beta} < \infty$. We have $\rho_1 \geqq \rho_0$.

4. i) Use (7.4.2). ii) Use (7.4.2), $\sum \log|z_j| = \sum \log\{1 - (1 - |z_j|)\}$, and Theorem (2.6.4). ii) Use $\eta_j(z - a_j)/a_j(\overline{a}_j z - 1) = |a_j|\{1 + z(1 - |a_j|^2)/a_j(\overline{a}_j z - 1)\}$.

5. Let $g(z)$ be a branch of $\log f(z)$ on $\mathbf{C}$. Then, $f(z) = e^{g(z)}$. By the assumption, $|\operatorname{Re} g(z)| = \log|f(z)| \leqq O(|z|^{\rho+1})$. Using (3.6.7), one sees that $|g(z)| = O(|z|^{\rho+1})$. Thus, $T(r, g) = O(\log r)$. Lemma (7.4.13) implies that $g(z)$ is a polynomial. Let p be the degree of $g(z)$. Then, $\log M(r, f) = c_0 r^p (1 + o(1))$ with $c_0 \neq 0$, and so $\rho = p$.

6. The order of $e^{\lambda z} + P(z)$ is 1. Suppose it has only finitely many zeros. Then, $e^{\lambda z} + P(z) = Q(z) e^{\mu z}$, where $Q(z)$ is a polynomial and $\mu \in \mathbf{C}^*$. Differentiate both sides $\deg P + 1$ times. Then, $e^{\lambda z} = R(z) e^{\mu z}$, where $R(z)$ is a polynomial. Hence, $\lambda = \mu$, and $P(z) = (Q(z) - 1) e^{\lambda z}$, and so $Q \equiv 1$. Thus, $P \equiv 0$, which is a contradiction.

7. Since $\operatorname{Re} z > 0$, $\sum p^{-z}$ converges absolutely, and so is $\prod(1 + p^{-z})$. The uniform convergence on compact subsets is proved in the same way. Let $2 = p_1 < p_2 < \cdots$ be primes. Then, $(1 - p_1^{-z})^{-1}(1 - p_2^{-z})^{-1} \cdots (1 - p_n^{-z})^{-1} =$

$(1+p_1^{-z}+p_1^{-2z}+\cdots)(1+p_2^{-z}+p_2^{-2z}+\cdots)\cdots(1+p_n^{-z}+p_n^{-2z}+\cdots)=1+p_1^{-z}+(p_1p_2)^{-z}+p_2^{-z}+p_1^{-2z}+(p_1^2p_2)^{-z}+\cdots$. Letting $n\to\infty$, we get $\prod(1-p_n^{-z})=\sum_{n=1}^{\infty}n^{-z}=\zeta(z)$.

8. This follows from Theorem (7.4.9), (7.4.8), and (2.8.5).

9. Let $Q[\omega_1,\omega_2]$ be the period parallelogram of Ω. Take $a\in\mathbf{C}$ so that the boundary of $E=a+Q[\omega_1,\omega_2]$ contains no point of $a_j+\Omega$ nor $b_j+\Omega$. Thus, we may assume that a_j and b_j are contained in the interior of E. $\sum a_j-\sum b_j=\frac{1}{2\pi i}\int_{\partial E}z\frac{f'(z)}{f(z)}dz$. Using the periodicity, we have $\frac{1}{2\pi i}\int_{\partial E}z\frac{f'(z)}{f(z)}dz=m\omega_1+n\omega_2\in\Omega$.

10. Set $v=1/(w-e_4)$ and $e_j'=e_j-e_4, j=1,2,3$. Then, $\int dw/\sqrt{P(w)}=\int\frac{-1}{\sqrt{(1-e_1'v)\cdots(1-e_3'v)}}dv$.

11. The meromorphic function $f(z)=\wp'(z)-\alpha\wp(z)-\beta$ has the period group Ω, and has degree 3. Let $a_j, j=1,2,3$, be zeros of f. Since the poles of f are on Ω, $a_1+a_2+a_3\in\Omega$ by problem 9. For arbitrary u,v, there are suitable α,β such that $f(u)=f(v)=0$. Then, necessarily $f(-u-v)=0$. Therefore, if $u+v+w\in\Omega$, there are α,β such that $f(u)=f(v)=f(w)=0$. The determinant of a system of linear equations with non-trivial solutions must be 0.

12. The dynamical equation is $d^2\theta/dt^2=-(g/R)\sin\theta$. Multiplying both sides by $d\theta/dt$ and integrating, we have $(d\theta/dt)^2=(4g/R)((v_0/\sqrt{4gR})^2-\sin^2\theta/2)$, where $v_0=R\theta'(0),\theta(0)=0$ (the initial condition). Set $0<k=v_0/\sqrt{4gR}<1$ (to avoid the rotation). Letting $x=k^{-1}\sin\theta/2$, we get $(dx/dt)^2=(g/R)(1-x^2)(1-k^2x^2)$. Thus, $t=\sqrt{g/R}\int_0^x dx/\sqrt{(1-x^2)(1-k^2x^2)}$, which is an elliptic integral.

References

In writing this book the author has referred to a number of books already published. Some of them are listed below:

[1] A. Hurwitz and R. Courant, Funktionentheorie, Springer Verlag, Berlin, 1929.

[2] L. Ahlfors, Complex Analysis, McGraw-Hill, Auckland et al., 1979; Japanese Translation by K. Kasahara, Gendaisugakusha, Tokyo, 1982.

[3] H. Cartan, Théorie Élémentaire de Fonctions Analytiques d'Une ou Plusieures Variables Complexes, Hermann, Paris, 1961; Japanese Translation by R. Takahashi, Iwanami Shoten, Tokyo, 1965.

[4] K. Kasahara, Complex Analysis–Analytic Functions of One Variable (in Japanese), Jikkyoshuppansha, Tokyo, 1978.

[5] Y. Komatsu, Introduction to Analysis [I] (in Japanese), Hirokawa Shoten, Tokyo, 1962.

[6] Y. Komatsu, Function Theory (in Japanese), Asakura Shoten, Tokyo, 1960.

[7] Y. Komatsu, Exercises in Function Theory (in Japanese), Asakura Shoten, Tokyo, 1960.

[8] R. Takahashi, Complex Analysis (in Japanese), University of Tokyo Press, Tokyo, 1990.

[9] T. Ochiai and J. Noguchi, Geometric Function Theory in Several Complex Variables (in Japanese), Iwanami Shoten, Tokyo, 1984; English Translation by Noguchi and Ochiai, Amer. Math. Soc., Providence, Rhode Island, 1990.

For the readers who want to advance to a further study on complex analysis, the author would like to give a list of books, which is not intended to be complete.

For Riemann surfaces, I recommend the following.

[10] H. Weyl, Die Idee der Riemannsche Fläche, B.G. Teubner, Stuttgart, 1913; Japanese Translation by J. Tamura, Iwanami Shoten, 1974. (This is a famous classical book, settling the concept of Riemann surfaces.)

[11] K. Iwasawa, Algebraic Functions (in Japanese), Iwanami Shoten, Tokyo, 1952; 2nd ed., 1972; English Translation by G. Kato, Amer. Math. Soc., Providence, Rhode Island, 1993. (This is a famous introductory book from algebra.)

[12] R.C. Gunning, Lectures on Riemann Surfaces, Princeton University Press, Princeton, 1966. (Comprehensive lecture notes, using sheaf theory.)

[13] Y. Kusunoki, Function Theory (in Japanese), Asakura Shoten, Tokyo, 1973. (This is a nice introductory book of closed and open Riemann surfaces to the research level.)

[14] Y. Imayoshi and M. Taniguchi, An Introduction to Teichmüller Spaces (in Japanese), Nipponhyoronsha, Tokyo, 1989; English Translation by Y. Imayoshi and M. Taniguchi, Springer-Verlag, Tokyo et al., 1992. (This deals with the deformation theory of the complex structure of Riemann surfaces, which is called Teichmüller theory.)

For value distribution theory, I recommend

[15] R. Nevanlinna, Le Théorème de Picard-Borel et la Théorie de Fonctions Méromorphes, Gauthier-Villars, Paris, 1939. (This is a monograph by the creator of Nevanlinna theory, and it clearly unrolls the development of the mathematical theory in front of your eyes.)

[16] W.K. Hayman, Meromorphic Functions, Oxford University Press, Oxford, 1964. (This is an introduction to Nevanlinna theory in one variable, up to the research level.)

[17] M. Ozawa, Modern Function Theory I – Theory of Value Distribution (in Japanese), Morikitashuppansha, Tokyo, 1976. (This is a monograph at the research level.)

[18] S. Kobayashi, Hyperbolic Manifolds and Holomorphic Mappings, Marcel Dekker, New York, 1970. (This is an introduction to the theory of Kobayashi hyperbolic manifolds by Kobayashi himself, and has had a deep influence on complex analysis since then.)

[19] W. Stoll, Value Distribution Theory for Meromorphic Maps, Aspects of Math. **E7**, Vieweg, Braunschweig, 1985.

[20] P.A. Griffiths, Entire Holomorphic Mappings in One and Several Complex Variables, Ann. Math. Studies **85**, Princeton University Press, Princeton, 1976.

For complex analysis in several variables and the theory of complex manifolds, I recommend

[21] L. Hörmander, Introduction to Complex Analysis in Several Variables, van Nostrand, New York, 1966. (This is an introduction to the theory from the viewpoint of elliptic differential equations and functional analysis. This book established one research direction of the theory.)

[22] A. Weil, Introduction à l'Étude des Variétés kähleriennes, Hermann, Paris, 1958. (This is a comprehensive course in the theory of Kähler manifolds by one of the greatest mathematicians of this century. Mathematically it is very nourishing.

[23] S. Nakano, Complex Function Theory in Several Variables – Differential Geometric Approach (in Japanese), Asakura Shoten, Tokyo, 1982. (This is a good book on function theory on Kähler manifolds by the author, who is famous for the Kodaira-Nakano vanishing theorem.)

[24] T. Nishino, Function Theory of Several Complex Variables (in Japanese), Tokyo University Press, Tokyo, 1997. (The author, who was a student of K. Oka, the founder of the theory, gives a treatise based on Oka's idea and work. An English translation is in preparation.)

Index

Symbols

Selected Titles in This Series

(Continued from the front of this publication)

(See the AMS catalog for earlier titles)